INTRODUCTION TO SUSTAINABILITY

# Introduction to Sustainability
Road to a Better Future

*by*

Nolberto Munier
*Ottawa, Ontario, Canada*

 Springer

A C.I.P. Catalogue record for this book is available from the Library of Congress.

ISBN-10 1-4020-3557-8 (PB) Springer, Dordrecht, Berlin, Heidelberg, New York
ISBN-13 978-1-4020-3557-9 (PB) Springer, Dordrecht, Berlin, Heidelberg, New York
ISBN-10 1-4020-3556-X (HB) Springer, Dordrecht, Berlin, Heidelberg, New York
ISBN-10 1-4020-3558-6 (e-book) Springer, Dordrecht, Berlin, Heidelberg, New York
ISBN-13 978-1-4020-3556-2 (HB) Springer, Dordrecht, Berlin, Heidelberg, New York
ISBN-13 978-1-4020-3558-6 (e-book) Springer, Dordrecht, Berlin, Heidelberg, New York

Published by Springer,
P.O. Box 17, 3300 AA Dordrecht, The Netherlands.

*Printed on acid-free paper*

All Rights Reserved
© 2005 Springer
No part of this work may be reproduced, stored in a retrieval system, or transmitted
in any form or by any means, electronic, mechanical, photocopying, microfilming, recording
or otherwise, without written permission from the Publisher, with the exception
of any material supplied specifically for the purpose of being entered
and executed on a computer system, for exclusive use by the purchaser of the work.

Printed in the Netherlands.

# TABLE OF CONTENTS

CHAPTER 1 - BASIC INFORMATION ON SUSTAINABLE ISSUES........ 1
1.1   The purpose of this book .................................................................. 1
1.2   Defining sustainability.................................................................... 10
1.3   Weak and strong sustainability...................................................... 15
1.4   Sustainable development vs. economic growth............................. 16
1.5   People's participation .................................................................... 21
   1.5.1   Case study: Community participation in Albertslund - Denmark... 22
   1.5.2   Case study - The will of a town - People defending their environment and health...................................................................... 23
1.6   The ecological footprint ................................................................ 24
1.7   The ecological rucksack ................................................................ 28
1.8   Emergy accounting ....................................................................... 30
1.9   Resilience (social, economical, and political) .............................. 32
1.10  Environmental resilience .............................................................. 33
1.11  Externalities................................................................................... 34
1.12  Capital ........................................................................................... 35
1.13  Local Agenda 21............................................................................ 36
1.14  The Bellagio principles.................................................................. 37
Internet references for Chapter 1 ............................................................ 37

CHAPTER 2 – THE CULTURE OF WASTE............................................ 43
2.1   Introduction ................................................................................... 43
2.2   First part: Current generation and treatment of waste ................... 43
   2.2.1   What is waste?...................................................................... 43
   2.2.2   Which are the components of waste?................................... 46
   2.2.3   Where is waste generated?................................................... 49
   2.2.4   How is waste disposed of or treated?................................... 52
   2.2.5   Waste and its effect on the environment.............................. 57
2.3   Hazardous waste ............................................................................ 59
   2.3.1   Vitrification.......................................................................... 59
2.4   Recycling....................................................................................... 60
2.5   Incinerators.................................................................................... 60
   2.5.1   Case study: Heat from incinerator and wind energy for Göteborg, Sweden.............................................................................. 63
2.6   Second part: Decreasing waste generation ................................... 64
2.7   Nature's closed waste cycle........................................................... 64
2.8   Society's open path for wastes ...................................................... 65
2.9   Ecosystem metabolism and metabolism in society ...................... 68
2.10  Why is waste produced?................................................................ 68
   2.10.1   Case study: Generating light out of garbage, Groton, the USA..... 69
2.11  What can be done to correct this situation? ................................. 72
2.12  Conservation of resources ............................................................ 73

2.12.1   Energy reduction .................................................................... 75
  2.12.2   Controlling water usage ........................................................ 77
  2.12.3   The use of water for industry and the reuse of wastewater ........... 78
      2.12.3.1   Examples to follow in water reuse .................................. 79
      2.12.3.2   Case study: Wastewater contributes to maintain a renewable resource: The ingenuity of a town, Clearlake, the USA .. 80
  2.12.4   Keeping rivers clean .............................................................. 82
  2.12.5   Recovering energy from tires .................................................. 85
  2.12.6   Savings in the supply chain .................................................... 86
2.13   Actions to reduce consumption ........................................................ 90
2.14   Working together ............................................................................. 94
Internet references for Chapter 2 ................................................................. 98

CHAPTER 3 – SUSTAINABILITY IN THE BUILT ENVIRONMENT ... 105
3.1   Sustainability at the individual level ................................................ 105
3.2   Sustainability in the household ....................................................... 106
  3.2.1   Water use in the household ..................................................... 107
  3.2.2   Solid waste in the household ................................................... 107
      3.2.2.1   An example to follow: Recovery of carpets material ....... 110
  3.2.3   Energy uses in the household .................................................. 111
  3.2.4   Land use for the household ..................................................... 112
3.3   Urban transportation ....................................................................... 113
  3.3.1   Case study: The role of transportation in sustainable Curitiba, Brazil ............................................................................................... 114
3.4   Upgrading slums in cities ................................................................ 117
3.5   Environmental sustainability ........................................................... 120
  3.5.1   Air in a sustainable environment ............................................ 120
  3.5.2   Water in a sustainable environment ....................................... 124
  3.5.3   Soil in a sustainable environment .......................................... 126
3.6   Team efforts toward sustainable environment ................................. 128
3.7   Sustainability in public administration and in urban life ................ 128
3.8   Sustainability in public health ......................................................... 131
3.9   Sustainability in education .............................................................. 132
3.10   Sustainability in commerce ........................................................... 134
3.11   Reducing energy consumption ...................................................... 137
  3.11.1   Reducing energy consumption in the urban space ............... 140
  3.11.2   Reducing land use ................................................................ 142
Internet references for Chapter 3 ............................................................... 142

CHAPTER 4 - INDUSTRIAL APPROACH TO SUSTAINABILITY ...... 149
4.1   Sustainability in industry ................................................................. 149
      4.1.0.1   Reductions in contamination ........................................... 149
      4.1.0.2   Cleaner Production and other approaches ...................... 151
      4.1.0.3   Waste reduction ............................................................... 151

|       | 4.1.0.4  | Reducing rejection ................................................... 154 |
|       | 4.1.0.5  | Reengineering .......................................................... 154 |
|       | 4.1.0.6  | Life Cycle Assessment (LCA) ............................... 158 |
|       | 4.1.0.7  | Input-Output model ............................................... 159 |
|       | 4.1.0.8  | Environmental Input/Output model ....................... 161 |

4.1.1 Industrial ecology (IE) .............................................................. 164
    4.1.1.1 Eco-efficiency .......................................................... 166
4.1.2 Industrial metabolism .............................................................. 166
4.1.3 Materials flow analysis ............................................................ 168
4.1.4 Industrial integration ............................................................... 171
    4.1.4.1 Case study: Industrial integration, Kalundborg, Denmark ................................................................................ 172
    4.1.4.2 Case study: A metallurgical process using the thermal content of flue gas for heating and to extract commercial products. 173
    4.1.4.3 Case study: Generation of electricity by using residues from other industries. The case of Belize ...................................... 174
4.1.5 Dematerialization .................................................................... 175
4.1.6 Design for the environment (DfE) ........................................... 177
4.1.7 Indicators ................................................................................ 177
4.1.8 Waste exchanges ..................................................................... 178
4.1.9 Comparison of methodologies ................................................. 179
4.1.10 Recycling in industry ............................................................. 180
4.1.11 Conclusions on industry ........................................................ 184

4.2 Sustainability in transportation ......................................................... 187
  4.2.1 Case study: The Transmilenio bus system in Bogotá, Colombia ... 198
4.3 Sustainability in agriculture ............................................................... 200
4.4 Forestry sustainability ....................................................................... 202
4.5 Sustainability in the construction industry ........................................ 203
  4.5.1 Comparison between singles dwellings and multi-family buildings ................................................................................................. 205
  4.5.2 Case study: Sustainability in paradise – The Maho Bay resort complex, Virgin Islands, USA .................................................................... 209

Internet references for Chapter 4 ............................................................. 212

CHAPTER 5 - ENERGY SUSTAINABILITY ........................................ 219
5.1 Introduction ....................................................................................... 219
5.2 Brief technical information on energy conversion equipment      223
  5.2.1 Coal-fired, gas-fired or oil-fired power plants ........................ 223
  5.2.2 Nuclear power plants ............................................................. 224
  5.2.3 Gas turbines ............................................................................ 224
  5.2.4 Wind turbines ......................................................................... 224
  5.2.5 Diesel engines ........................................................................ 225
  5.2.6 Hydropower plants ................................................................. 225
  5.2.7 Biomass .................................................................................. 225

5.2.8 Geothermal.................................................................................. 227
    5.2.8.1 Heat pumps ................................................................. 227
    5.2.8.2 Geothermal energy ..................................................... 231
5.3 Non-conventional sources for energy generation ............................... 233
    5.3.1 Wind energy............................................................................ 234
    5.3.2 Photovoltaics (PV).................................................................. 238
        5.3.2.1 Case study: 1 MW decentralized and building integrated PV system in a new housing area, Amersfoort, the Netherlands...... 241
        5.3.2.2 Case study: Solar modules made integral to hypermarket roof - Tampere, Finland................................................................. 241
    5.3.3 Solar collectors ....................................................................... 242
    5.3.4 Biomass................................................................................... 243
        5.3.4.1 Methanol....................................................................... 244
        5.3.4.2 Ethanol.......................................................................... 245
        5.3.4.3 Biodiesel ....................................................................... 245
        5.3.4.4 Methane ........................................................................ 245
        5.3.4.5 Pyrolisis ........................................................................ 247
    5.3.5 Fuel cells................................................................................. 247
        5.3.5.1 The fuel cell in automobiles ........................................ 248
        5.3.5.2 PAFC – Phosphoric acid fuel cell............................... 252
        5.3.5.3 PEM – Proton exchange membrane ........................... 253
        5.3.5.4 MCFC – Molten carbonate ......................................... 253
        5.3.5.5 SOFC – Solid oxide ..................................................... 253
        5.3.5.6 AFC – Alkaline............................................................ 254
        5.3.5.7 DMFC – Direct methanol fuel cells ........................... 254
        5.3.5.8 Regenerative ................................................................ 255
    5.3.6 The sea as a source of energy ................................................ 255
Internet references for Chapter 5 .............................................................. 258

**CHAPTER 6 –MEASURING SUSTAINABILITY** ...................................... 265
6.1 Types of indicators ............................................................................... 265
6.2 Approach for choosing indicators........................................................ 267
6.3 Sustainable vs. common indicators ..................................................... 268
6.4 Indicator uses........................................................................................ 269
6.5 Indicator linkages ................................................................................. 274
6.6 Integration of sustainable indicators.................................................... 275
6.7 Weight of indicators ............................................................................. 276
6.8 The choice of indicators ....................................................................... 277
6.9 Multipliers ............................................................................................. 281
6.10 Framework for indicators ................................................................... 284
6.11 Thresholds ........................................................................................... 286
6.12 Carrying capacity................................................................................ 287
    6.12.1 Carrying capacity in the environment................................. 288
    6.12.2 Carrying capacity in the social fabric ................................. 290

6.12.3 Carrying capacity in the economy .................................................. 291
6.13 Selection of a set of final indicators ................................................. 293
6.14 Monitoring progress ........................................................................ 294
6.15 Indicators for the city ...................................................................... 295
    6.15.1 Case study: Selection of indicators for the city of Guadalajara, Mexico .................................................................................................. 303
Internet references for Chapter 6 ............................................................... 309

CHAPTER 7 – SUSTAINABLE IMPACT ASSESSMENT (SuIA) ............ 315
7.1 Urban and regional sustainability ..................................................... 315
    7.1.1 Assets inventory ....................................................................... 316
    7.1.2 The baseline concept ................................................................ 316
7.2 Agreeing on the goal ........................................................................ 320
7.3 Understanding the problem .............................................................. 322
7.4 Resources inventory ......................................................................... 323
7.5 Plan to accomplish the objective ...................................................... 326
7.6 People's opinion ............................................................................... 326
7.7 Criteria and indicators to gauge projects .......................................... 328
7.8 Application example: A community looks for a sustainable energy option ................................................................................................ 331
    7.8.1 Goals ......................................................................................... 332
    7.8.2 Resources inventory ................................................................. 332
    7.8.3 Criteria ...................................................................................... 333
    7.8.4 Criteria weights ........................................................................ 337
    7.8.5 Threshold selection .................................................................. 337
    7.8.6 Gathering the information ........................................................ 339
    7.8.7 Coefficients .............................................................................. 340
    7.8.8 Alternatives selection ............................................................... 341
    7.8.9 Solving the problem ................................................................. 341
    7.8.10 Objective of this exercise ....................................................... 349
    7.8.11 The database ........................................................................... 350
    7.8.12 Conclusion .............................................................................. 355
Internet references for Chapter 7 ............................................................... 356

CHAPTER 8: CASE EXAMPLE - A COMMUNITY IN SEARCH OF ITS FUTURE .............................................................................................. 359
8.1 Background information for a process ............................................. 359
8.2 Introduction to the sustainable initiative for a community .............. 363
    8.2.1 The system and the process ...................................................... 365
8.3 The process ....................................................................................... 366
    A. Create an agency to be in charge of this whole project .............. 366
    B. Make an inventory of assets and problems, and determine general orientation and sources of information ..................................... 366
    C. Determine a general goal and establish a time limit ................... 368

D. Establish definite objectives...........................................................369
E. Create work commissions and establish responsibilities....................370
F. Set up measures, actions, plans and projects to be executed to accomplish the objectives .......................................................................370
G. Determine type of indicators needed ...............................................375
H. Choose indicators to measure progress of actions in (F) and of targets and goals established..................................................................375
I. Develop a schedule detailing on a bar chart each action, from start and finish, listed in (F), including their interrelationships and sequence...379
J. Determine the economic impacts that tourism and the other undertakings will have on the economy, the environment and society......380
8.4 Impacts created by tourism.................................................................380
8.5 How to measure impacts .....................................................................385
8.6 Conclusions from studies ....................................................................387
K. Establish a reporting mechanism to communicate results to people and for feedback.....................................................................................389
Internet references for Chapter 8 ................................................................390

APPENDIX ...............................................................................................393
A.1 The Zeleny method for determining weights......................................395
A.2 Determination of Return on Investment and Net Present Value ........399
A.3 A guide to strategic planning...............................................................402
A.4 Visualizing progress towards sustainability goals..............................406
A.5 Life Cycle Assessment (LCA)..............................................................410
 A.5.1 Example of application in industrial complex ............................411
A.6 Regression analysis for weights determination .................................415
A.7 Discharges and their effect on the environment .................................421
Internet references for Appendix ................................................................425

GLOSSARY ..............................................................................................427

BIBLIOGRAPHY .....................................................................................435

INDEX........................................................................................................439

# ACKNOWLEDGMENTS

The author gratefully acknowledges the contributions of the following colleagues, who reviewed sections of the manuscript, making valuable technical suggestions and comments, and proposing changes that have improved its readability.

Dr. Dana Vanier, National Research Council of Canada, Ottawa, for his revisions of Chapters 1, 2, 3 and 4.

Dr. Khaled Nigim, Electrical and Computer Engineering, University of Waterloo, Waterloo, Ontario, Canada, who kindly reviewed Chapter 5 'Energy Sustainability'.

I would also like to express my appreciation to Bert Bailey, my friend and language editor, who spent several weeks carefully improving the diction and overall presentation of the final draft, making it far more readable.

To all of them my hearty thanks and gratitude for their advice, time and opinions.

Naturally, I am solely responsible for any errors the reader may find.

Nolberto Munier
Ottawa, November 2004

# CHAPTER 1 - BASIC INFORMATION ON SUSTAINABLE ISSUES

## 1.1 The purpose of this book

Sustainability is a difficult and complex issue, and an elusive one. It is enormously important since it has to do with nothing less than the chances of humankind surviving on this planet. At the rate that the human race is using scarce and limited resources it appears that, unless measures are taken now — and if there is still time — the future of civilization, at least as we understand it now, is uncertain, to say the least. It follows that such a complex subject has no simple and straightforward treatment, especially considering that sustainability is not a **goal** but a **process**. It leads to a better life for the present generation and survival for generations to come, enhancing their ability to cope with the world that they will inherit. As Chief Seattle put it so very well, *"We do not inherit the earth from our parents; we borrow it from our children"*.

In the last 100 years, humankind has very effectively managed to squander the earth's resources, clear-cutting a large amount of existing forests, contaminating the atmosphere, polluting rivers, and even altering our climate. We have also destroyed or caused the disappearance of thousands of species, to say nothing of increasing the risk of exposing the world's population to lethal ultra-violet rays by destroying the ozone layer, courtesy of organochloride chemicals. If historians are to exist in a distant future, they will probably consider the $19^{th}$ and $20^{th}$ centuries as a period as catastrophic as that which precipitated the extinction of the dinosaurs. George Schaller, cited en Raven (see Internet references for Chapter 1) accurately sums up this catastrophic mismanagement by saying, *"We cannot afford another century like this one* [i.e., the 20th century]*"*.

K. Bidwell and P. A. Quinby (see Internet references for Chapter 1) cite the startling observation of *The Western Center for Environmental Decision-making* (Boulder-Colorado), that:

*When the* [twentieth] *century began, neither human numbers nor technology had the power to radically alter planetary systems. As the century closes, not only do vastly increased human numbers and their activities have that power, but major, unintended changes are occurring in the atmosphere, in soils, in waters, among plants and animals, and in the relationships among all of these.* (WCED 1987:343)

How is it possible that the human race has done so much damage to our home planet? Humankind is guilty of mismanagement, greed, and ignorance. Economic forces have stimulated mass consumption of renewable and non-renewable resources, producing, and continuing to create, millions of tons of waste. The 'wars to end all wars' have taken the lives of millions while managing to poison the planet with radioactive rubbish, destroying forests with defoliants, and contaminating the air with lethal gases. If at least that enormous waste had led to an equitable society where everybody had shelter, equal rights, enough food and good education, where diseases had been eradicated and everybody was living a healthy life — well, one might think that it had not been such a waste, but a form of investment of the planet's resources to secure a better living.

In actual fact, a very different scenario is the case. Millions of people currently live in shanty towns, their children share their confined spaces with rodents and pests, some diseases are spreading faster than man is able to treat them, crime due to poverty is soaring, a huge gap exists between developed and developing countries, and humankind continues to use its resources as if these were inexhaustible and free. Just recently, since the 1992 Earth Summit in Rio de Janeiro (see Glossary), society has begun taking notice of this grim scenario. Nevertheless, running quite contrary to what was expected and intended, the countries producing most pollution refuse to adhere to the Kyoto Protocol (see Glossary), and to promote acid rain reductions, for instance, as these are perceived to run contrary to their economic interests.

Many countries carry on with their current lifestyles despite losses of 'natural capital' (section 1.12) that takes place in other regions of the planet. Countries exporting goods and services can suffer social and environmental costs while they persist in measuring their 'progress' by increases in added value and in their exports earnings. This notion of 'progress' still understands it in terms of economic Gross Domestic Product, ignoring the social and environmental damage spent to secure the presumed 'advancement'.

However, nature, a very patient and bountiful mother, is telling us loud and clearly that enough is enough. How? A few well-documented and uncontroversial examples will illustrate the point[1]:

---

[1] If the reader is interested in expanding his/her background information on each example, a footnote directs she/he to a selected address of a file or Web page accessible on the Internet (URL).

- The meteorological problems caused by El Niño.
  *While no proven relationship has been established between this event and human activities, scientists are looking for a correlation between global warming and an increase in the frequency of the El Niño phenomenon*[2]

- The melting of the glaciers.
  *"The shrinking [of glaciers] is reflective of rising global temperatures and is happening to glaciers around the world… As the melting causes sea levels to rise and freshwater supplies to disappear, scientists warn of potential worldwide economic and environmental disaster if the process isn't reversed"*[3]

- Climate change.
  *Produced by the rise of greenhouse gases*[4].

- The disappearance of many animal and plant species.
  *Mainly produced by pollution, population growth, and exploitation.*[5]

- The hole in the ozone layer.
  *Produced by chlorofluorocarbons from industrialized countries.*[6]

- The warming of the planet.
  *Produced by greenhouse gases such as carbon dioxide ($CO_2$), methane ($NH_4$), nitrous oxide ($N_2O$), and man-made gases such as chlorofluorocarbons (CFCs).*[7]

- Contamination of the food chain with mercury and other metals.
  *Mercury contamination comes from pesticides, electric batteries, paints, thermometers, etc. When mercury reaches the sea, via groundwater and rivers, bacteria convert mercury in methyl mercury,*

---

[2] See http://www.davidsuzuki.org/Climate_Change/Impacts/Extreme_Weather/El_Nino.asp

[3] (Stanard - Internet references for Chapter 1)

[4] A clear and succinct explanation can be found in this Internet publication:
http://edugreen.teri.res.in/explore/climate/causes.htm

[5] See http://encarta.msn.com/text_761557586___1/Endangered_Species.html

[6] For concise and well-illustrated information —where one can also observe there the hole in the ozone layer over the Antarctica in September of 2000.
http://www.70south.com/resources/atmosphere/ozone

[7] For information see:
http://www.geocities.com/csango80/gwweb02.htm

which fish ingest. *This affects the human brain and nervous system; evidence also suggests that mercury causes genetic damage.*[8]

- The increasing salinity of soil in many agricultural areas.
  *Increasing soil salinity is a consequence of poor irrigation practices and of clearing trees.*[9] [10]

- Escalating rates of soil erosion.
  *This is a natural process that can become a serious problem when human activity accelerates it.*
  *Its main causes are the removal of vegetation, such as forests, that not only protects soil from wind and rain, but also helps with the retention of humidity.*[11]

- The loss of permeability of soil.
  *This prevents the replenishment of aquifers by rainwater. One of its causes is when human action makes soil impervious with asphalt, concrete, and buildings — thereby essentially sealing land surfaces).*[12]

- The shrinking of large bodies of water such as the Aral Sea.
  *In the case of the Aral Sea, this was the result of diverting the Amu Darya River's waters to irrigate cotton crops.*[13]

It is true that, since the conferences in Rio in 1992, something is being done: most countries are taking measures to reduce these dangers; one can only hope that these remedies are not too late. The problem is that, while in many cases a reduction of noxious contaminant factors has taken place, because our birth rates are still too high, more people are producing more pollution. By 2025, the planet will have about 9,000 million people. Will the next generation be able to cope with their needs for shelter, food, jobs, health care, and education? Most probably not. Well, if not, then what will happen? What is the solution?

---

[8] See publication Environmental Health Perspectives in:
http://ehp.niehs.nih.gov/docs/1996/104-8/focus.html

[9] See Soil salinity in:
http://interactive.usask.ca/ski/agriculture/soils/soilman/soilman_sal.html

[10] See Salinity in:
http://www.landcareaustralia.com.au/admin/upload/B0B1738C-501F-11D6-8813-0002A574AC50.pdf

[11] See Soil erosion in
http://www.botany.uwc.ac.za/Envfacts/facts/erosion.htm

[12] See Sealing of soil surface in
http://www.stadtentwicklung.berlin.de/umwelt/umweltatlas/ed102_01.htm

[13] See "Aral Sea and cotton" (ARAL case) in:http://www.american.edu/ted/aral.htm

One does not need to be a futurologist, or to consult the Oracle at Delphi, for an answer. Humankind needs to **reduce its consumption** of everything: water, minerals, meat, paper, computers, cars, chemicals and land. Society **has to increase the efficiency** of its production, to generate less garbage, to create more natural agriculture, to reduce atmospheric pollution. In all probability, the key idea is that we have to **erase from our vocabulary the word 'waste'**.

Another way of putting this is that the human race has to emulate and try to achieve the same efficiency as nature, where every organism's waste provides the food of another, in a perpetual **repetitive scheme**. Moreover, it may be wise to recall French scientist Antoine Lavoisier (1743-1794), who said that, in nature: *"Nothing is lost, nothing is created, everything is transformed."* It is unfortunate that too few pay any heed to this advice in our daily activities.

The reader is asked to pardon this lengthy introduction, which merely strives to serve as a gentle reminder of the state of our environment, and to pave the way to presenting this book's objective — which will unfortunately also consume precious resources. No single formula can magically be applied to reach sustainability immediately. Sustainability rests on three main pillars: economics, society, and the environment. In fact, there is no direct link between these three pillars, although there is a lot of criss-crossing and looping between them. In other words, sustainability is a linked and looped feedback process which, to complicate things further, is linked and looped at different levels in each of these three pillars.

Sustainability requires a joint proactive effort: without people's participation, it is impossible for the built environment to improve. Yet such participation further complicates the sustainability issue, given the need to assuage partisan positions, which are sometimes apparently irreconcilable. Sustainability is not simply a matter of people complying with regulations and bylaws, as it also has to do with community participation in the efficient management of resources with a view of social equity, including sustainable funding levels, shelter for everybody, etc. It has to do with citizens joining in efforts with local authorities to improve public health, working conditions, and the care of the environment, to mention just a few. But it is not easy to put this combined challenge into practice, since there are numerous and complex relationships involving society, the economy and the environment, involving direct and indirect dependencies, a variety of different interests, and various sometimes diverse goals.

The objective of this book is to analyse these complex relationships by offering ideas and methods that everybody can understand and use. It encourages participation from decision-makers, policy-makers, technical people and grassroots organizations, providing examples of actual applications, and describing the necessary tools to monitor and control adopted compliance measures. The book will thereby try to offer a useful condensation of various techniques, procedures and experience that are available today on the complex issue of sustainability.

**Chapter 1** defines the main concepts listed immediately below. It provides the reader with a ground for discussion by highlighting the main points in the sustainability issue (which discusses in detail terms or concepts that will require attention in the forthcoming chapters).

Some terms that will require attention from the start are:

- Defining sustainability, and all it encompasses.
- Comparing economic growth and sustainable development, vital concepts that many people regard as contradictory or conflicting.
- The perception of natural resources as the natural capital of a country or region.
- The social, economic and ecological environment's carrying capacity: a fundamental idea very broadly defined as the environment's capacity to support life.
- The concept of indicators, which is closely related to the carrying capacity, will also prove to be fundamental for monitoring purposes.
- Multipliers, which provide an idea of the economic impact of certain undertakings or measures.
- The footprint theory, which produced what is considered to be one of the best indicators about sustainability, as it relates needs in an area with the physical surface required to satisfy them.
- Industrial ecology calls for an ecosystem that maximizes the efficiency in the use of industry-based resources while minimizing any contamination and waste production.
- Industrial metabolism tries to mirror nature's metabolism, in that no remains or excess should end up as a waste, but should rather become an input for some other entity.
- Materials flow analyses the flow of material in an economy, considering local production, imports, and unused extraction.

Most indicators do not take into account what are called the 'hidden costs', that is, those social and environmental costs that are not part of the typical cost/benefit equation and are not incorporated in the economics of a project. These costs end up being borne by society at large, or specific organizations. Such costs are, for instance, increases in soil salinity due to extensive crop irrigation; the depletion of fossil fuels due to inefficient heating; the social and health problems deriving from extensive single-passenger automobile use; and air pollution and its effects on people, works of art, etc., resulting from industrial emissions, etc.

Fortunately, something is being done at present to correct this omission regarding such 'hidden costs'. For instance, the European Union (EU) has developed the EXTERNE project to evaluate the damage to both the natural and the built environment, stating: *"EXTERNE allows the comparison of technologies based on their socio-environmental costs"*. See 'Social and environmental impacts for energy use, in Internet references for Chapter 1.[14]

Chapter 1 emphasizes the necessity of people participation with two case studies:

1. A community that has adopted a participatory approach to establish its priorities — Denmark.
2. A community mobilizing to decide what industries they want in their area — Argentina.

The last part of this chapter is devoted to Local Agenda 21, a body that encourages municipalities to be sustainable, an initiative that is a direct spin-off of the Rio Summit and the Bellagio Principles — a set of concepts that are used to measure progress.

**Chapter 2** is mainly about the rational use of resources and 'the culture of waste', or the propensity to produce enormous quantities of waste, and just dump it in landfills, without giving much thought about its potential reutilization.

This chapter will deals with the precept that saving resources requires increasing their efficient use, mainly by producing less waste and by recycling. As different kinds of waste merit different types of treatment, wastes are divided into various categories and this chapter will suggest how to process each kind.

This will be illustrated with three actual cases where waste is put to use:

1. Waste utilized to produce heat by incineration — Sweden.
2. The generation of energy from garbage — the USA.
3. The use of wastewater to generate electricity — the USA.

---

[14] At the end of this chapter — as in the others — a comprehensive bibliography with Internet references on selected publications will be included to enable readers to gain access to the relevant materials. These links were open and active at the time of reviewing the writing of this book {in November 2004}, although there is no guarantee that they will be extant later.

Since incinerators appear to be such a promising alternative for getting rid of solid wastes, the characteristics of the flue gas that is produced will call for some comment. A final subject for analysis in this chapter is the pressing problem of getting rid of used tires, along with the issue of worn-out tires that can be disposed of safely without using landfill space while also generating energy.

**Chapter 3** is called 'Sustainability in the built environment'. As its title suggests, it analyses sustainability in connection with different various scenarios and human activities: that of individuals and households, in commerce, industry and health-care systems, agriculture, transportation, education, and in the construction industry. These will be brief overviews, lest this book bloat to the proportions of an encyclopaedia. The reader will thereby gain a glimpse of what is generally involved, with a focus on two actual cases:

1. The matter of recycling used floor carpets — the USA.
2. The development of a rational transportation system, associated with sustainability in transportation — Brazil.

**Chapter 4** has to do with industry and its strategies to achieve sustainability. Industry is the main producer of waste and an inefficient user of raw materials, yet it stands as the area with the greatest potential for increased efficiencies and savings. Information will be provided about concepts such as industrial ecology, industrial metabolism, industrial integration, materials flow analysis, and dematerialization — all of which are of paramount importance for sustainable industry. Five cases studies will be presented in this chapter:

1. The Kalundborg complex is a paradigm case of vertical and horizontal industrial integration that has led to significant reductions in natural resource usage where the waste of one industry is used as the input for another — Denmark.
2. The case of a metallurgical firm where usually rejected by-products are recovered after treatment of the flue gas — South America.
3. A power plant that uses sugar cane wastes — Belize.
4. Bogotá's Transmilenio bus system and its social and environmental benefits — Colombia.
5. A holiday resort's treatment and re-use of scarce resources — Virgin Islands — the USA.

**Chapter 5** discusses the pressing problem of electricity sustainability. The technical aspects of conventional electrical generation will not be analysed, but rather the chapter's focus will be on the potential use of non-conventional sources that are sustainable due to being renewable and not amenable to depletion even with irrational levels of consumption. The chapter will also seek to confute the misleading notion that non-conventional electrical sources are more expensive than conventional ones, even today. Four cases will help to prove the point:

1. The use of a photovoltaics system in buildings — Denmark.
2. The use of solar collectors in a supermarket — Finland.
3. Tidal energy used for electric generation — France.
4. The use of sea waves — Scotland.

**Chapter 6** deals with the fundamental concept of the measurement of sustainability. To explain the point it is necessary to operate with indicators or measurements that are both qualitative and quantitative, to establish limits or thresholds for certain activities, productions, extractions, etc. and that link up to the fundamental concepts of carrying capacity and resilience. The breach of thresholds or carrying capacity of the environment are discussed through various case studies, such as:

1. The Ogallala case of water overuse — the USA.
2. Over-fishing — Peru.
3. Lack of food for elephants — Africa.
4. The carrying capacity of the Rhine River — Central Europe.

One additional case study will deal with indicators:

5. The selection of urban indicators in Guadalajara — Mexico.

**Chapter 7,** called 'Sustainable Impact Assessment', considers that all human actions produce a disturbance on the environment, and, in many cases, on the society where those actions take place. This makes it imperative to analyse the tools that are available to determine the impact produced by such actions. A methodological way to collect, analyse and process information will be discussed: this is paramount to any sustainability process, and also for determining, out of any given set of projects — all of which contribute to sustainability — which contribute the most to sustainability objectives. A hypothetical case — albeit based on a real situation — will be analysed, namely:
- Regarding the wishes of a community to exploit its renewable resources for electric generation.

**Chapter 8** is given over in its entirety to a case example on tourism, where most of the concepts outlined in the preceding chapters will be considered. This chapter will also analyse the fundamental concept of multipliers, in order to show the economic impact of this project, and to analyse the impacts on the different strata through a stepped-matrix analysis. This example shows all the steps that a community should take to become sustainable.

## 1.2 Defining sustainability

Many definitions have been proposed for sustainability, although one of the most widely accepted is that found in the Brundtland report (see Bibliography: World Commission on Environment and Development, 1987). It states that: Sustainable development is development that meets the needs of the present without compromising the ability of future generations to meet their own needs. This is a technical definition. Alan Fricker (see Internet references for Chapter 1), quotes a definition by Veiderman when he says that, "Veiderman may have come closest to a definition [by stating that] ... *sustainability is a vision of the future that provides us with a road map and helps us focus our attention on a set of values and ethical and moral principles by which to guide our actions"*. In the present author's view, this is an accurate explanation that clarifies the Brundtland definition by providing a methodology to follow to reach sustainability.

The main concepts in the Brundtland report that are relevant to our subject are: development, present, and future (i.e. time). Let us examine them in turn:

*Development:*

When the word 'development' is mentioned, one reflexively thinks of economic development. Yet when it comes to sustainability issues, development means advancement in every area, including:

- **Economic growth,** which involves economic progress.
- **Social progress,** which facilitates attaining social equity and equality of opportunities for everybody, without social discrimination and equal prospects to obtain shelter, education, health care, jobs, etc.
- **Environmental protection**, ensuring that resources are healthily recoverable, so they can also be enjoyed by coming generations.

***The Present:***
The present, in the Brundtland Report, refers to the need to act in the present with a view to achieving growth that comprises not just economic progress but also environmental and social advancement.

***The Future:***
The definition does not refer to the immediate future but to the long-term future inhabited by our descendants — that is, the children of our children's children. Because of the advancement of science it is almost impossible for the present generation to foresee our descendants' needs, the problems they will face, and the other resources they maybe able to exploit that present generations cannot even imagine. Consequently, it is impossible to establish a time-frame for the achievement of those needs.

To shed further light on this subject, let us imagine being in the late $19^{th}$ century and that the notion of sustainability had recently been introduced, perhaps under another name. To what might this concept of sustainability currently refer? That 'present' may have included:

- People distressed about the noise and fumes produced by smelly steam- and gasoline-powered vehicles, which had just started to share the streets with pedestrians and horse carriages.
- The spread of some diseases, and ignorance about their cures.
- Horse dung accumulating on the streets because of horse-driven transportation.
- Children working in sweatshops.
- A lack of health services.
- Etc.

If somebody in that era had thought about their future, it would have probably been modelled according to the notion of using far more advanced steam machines, better coal-heating systems, enhanced horse carriages for urban transport and improved railways for long-distance travel, advances in the production of food, cotton and wool, and improving long distance communications through the wire telegraph. They might even have thought of better long-distance communications by improving the recently-invented telephone, etc. In other words, they probably were thinking of improving what they had, since research into all kinds of what we now call advancements was not organized, let alone funded.

With the advantage of hindsight, it is clear that if plans had been drawn at that time for future generations based on such assumptions, they would have been almost worthless. Why?

Because the world did not evolve according to their expectations, and since at that time nobody could have imagined that the future would bring, amongst other 'novelties', these activities:

- The global economy and its consequences.
- People travelling by air and the demand for aluminium and other metals.
- The harnessing of atomic energy and the danger of radiation and nuclear waste disposal.
- Advances in women's and children's rights, as well as the absorption of women into the work force.
- The discovery of antibiotics as well as the development of machines for medical analysis.
- The outbreak of AIDS and its consequences.
- The discovery of plastics and the waste they produce, as well as of plastic textiles and their consumption.
- The invention of computers and the dissemination of information; and
- Global warming and the depletion of the ozone layer, produced by human activities.

All of these advancements, and many others, are familiar to us, but were completely unknown to that generation, just as the future is unknown to us.

The fact that the future is unknown to us makes establishing time scales and objectives a difficult task. Even so, it does not mean that time scales cannot be set up about various other things. Some of our current objectives can be counted on to still be common objectives tomorrow, including:

- Everyone's right to shelter, education and health-care.
- The enjoyment of equal opportunities and respect for all human beings, regardless of religion, skin color, or nationality.
- The protection of the environment.
- The right to work and earn a decent income.
- The right to live in a clean environment with access to basic infrastructure.
- The right to participate in the management of one's city.
- Etc.

Plans can be made on these and other subjects. For instance, in city assets management Gordon and Shore (see Internet references for Chapter 1), propose the following time scale and stages:

Operational phase (the present): 1 - 2 years.
Tactical (immediate future): 2 - 5 years.

Strategic (future): 5 + years.

Considering the three fundamental concepts of development, present, and future, it follows that sustainability is temporal in the sense that it is time-related. Taking into account all these relevant factors, it can be said that sustainability is a process to attain a goal embedded in a system to support it. The Dictionary (Merrian Webster's Collegiate Dictionary, 11$^{th}$ edition) provides a definition for both concepts:

***Process:*** A series of actions or operations conducting to an end.
***System:*** A regularly interacting or independent group of items forming a unified whole.

Sustainability is not a methodology; it is linked with **the will of people for a change,** and as Alan Fricker states (see Internet references for Chapter 1):

*The challenge of sustainability is neither wholly technical nor rational. It is one of change in attitude and behaviour. Sustainability therefore must include the social discourse where the fundamental issues are explored collaboratively within the groups or community concerned. We do not do that very well, partly because of increasing populations, complexity, distractions, and mobility, but more because of certain characteristics of the dominant paradigm that are seen as desirable.*

Sustainability is a process involving people, institutions, natural resources, and the environment. It is implemented collectively and really points to the future. It is a process that involves changes — most of the time, considerable ones — in behaviour, attitudes, consumption patterns, spending and purchasing habits, and how society perceives and values the environment. It also depends on policies being enacted and enforced, such as recycling and reusing. For example, industry might be compelled to treat and reuse its industrial wastewater, rather than discharging it and making up for their loss with clean water from city sources.

Water usage can also be limited (such as by installing domestic water meters) by promoting the production of less domestic garbage by recycling at origin, and by exacting a collection fee based on weight, etc. These types of measures, making people pay not a flat rate but a fee related to consumption and/or production, usually works because as somebody said *"the most sensible human organ is the wallet"*.
As a bottom line, sustainable development can be condensed in Quality of Life. Quality of life embodies the following features that the author feels are related to sustainability:

- Adequate shelter for every person.
- Education for everyone at all three levels (primary, secondary and university).
- The opportunity to develop any professional or artistic career.
- Access to a good health plan and health protection system.
- Safety, at least to a reasonable level.
- Equal employment opportunities without distinction of gender, race, religion or language.
- The right to elect governments.
- Protection of the environment so all can enjoy nature and its biodiversity.
- A right to abundant clean water and to breathe clean air.
- Good working conditions. Of course, nobody can affirm that a person will keep the job forever, but it is good to know that there are other job opportunities requiring the same or higher levels of skill or education, and also by diversification of jobs opportunities in the area.
- Representation in the local setting, not only from a democratic point of view (electing representatives), but also considering the degree in which the society participates in the elaboration of plans, their monitoring, the allocation of funds, and decisions that affect citizens such as new projects, regulations, bylaws, etc.
- Adequate infrastructure, including transportation, clean water, clean air, connection to the sewer network, pavements, etc.
- Opportunities for capacity development.
- Opportunities to participate in cultural activities and events.

Obviously, this list is incomplete since many more features can be considered.

Naturally, there are several definitions of 'quality of life' but there are common indicators that can be utilized, although most depend on what each individual community desires. (The Internet references for this chapter mention some websites from cities in the UK, USA, and New Zealand).
Once a community has reached an agreement about its vision and interpretation of quality of life, how do they measure it? They need to use indicators, as explained in Chapter 6. To illustrate, a few quality of life indicators are listed below:

- The percentage of people owning their homes. Naturally, this is an indication of social justice as well as of the economic conditions of an area.
- The number of new dwellings built in already developed land. This is important because it relates to the amount of new land taken for

housing, which in most cases occurs at the expense of agricultural land. When old houses using large lots are demolished and new dwellings are built on the vacant land, it is good for the environment since it recycles a resource by making better use of the land, and considering population density.
- The number of break-ins per 1,000 households. This is associated with safety in the area, its economic conditions, as well as the efficiency of the police force.
- The percentage of people receiving unemployment benefits, which is an indication of economic conditions as well as of education levels.
- The ratio of the average house price to average income, which is linked to social and economic conditions.
- The percentage of domestic waste that is recycled, relating to the care of the environment and social participation in decreasing the amount of waste dumped into landfills.
- The amount of waste per person — for the same reasons as above.
- The number of road accidents per 10,000 inhabitants, associated with safety.
- The percentage of low birth weights, providing revealing correlations with public health and economic conditions.
- The number of days each year with clear skies and low air pollution, which links up with public health and the environment.
- Etc.

## 1.3 Weak and strong sustainability

Two main theories have been developed about the utilization of natural resources. One says that natural resources are utilitarian and are there to support humankind: in other words they are seen as just another commodity, thereby as amenable, to a certain degree, to substitution.

The other approach is not utilitarian as it claims that resources should be used — but respecting their intrinsic qualities, enjoying the biodiversity they offer – .They must be used in more rational and restrained ways, since humankind cannot substitute most of them, so we must do this in ways that at the same time preserve this capital for future generations.
Probably there is a grain of truth in both of these approaches, since the first one advocates using resources today without caring for the future, which is somewhat egotistical since upcoming generations are not considered. The second approach is not realistic because society needs resources to survive.

As in many such cases, the truth probably lies halfway between these extremes. Therefore, why can humankind not live from its resources but

carefully enough to keep them for future generations? What is needed, then, is a disciplined and planned way to utilize the earth's resources so that they are always renewing themselves (Smith, 1991). This approach is sometimes called in the technical literature **living off the interest, and saving the capital**.
This book maintains that the second approach captures the essence of sustainability. So, what is in there to sustain? **To protect, maintain and improve present lifestyles and preserve them for future generations.**

The reader has probably come across the terms 'weak' and 'strong' sustainability, two expressions that embody the different approaches summarized above. The first belongs to traditional economics, which considers environmental resources as replaceable and interchangeable by labour and other production factors. The second approach — strong sustainability — denies this assumption and supports the need to preserve resources not only because of their material utility but also because they are fundamental to support life and because they also provide other benefits that are not interchangeable, such as landscapes, the beauty of a mountain, etc.

**1.4 Sustainable development vs. economic growth**

Many people think that the expression 'sustainable development' is in itself an oxymoron (see Glossary), and that the words contradict one another: that is, some think that sustainability and development cannot coexist, and that the expression is senseless. Many others believe that it is not an oxymoron as they consider that sustainable development (in its broadest meaning) and economic growth can indeed coexist, on condition that the latter does not involve taking indiscriminately whatever is needed from the environment and society.

To clarify this last point it is possible to establish a parallel between this and something that happened in the Middle Ages, in the early 14$^{th}$ century. Lords and knights ruled over individuals in a servile, feudal category, duty-bound and subject to the lords. These lords and their knights thought that they had the privilege of taking whatever they needed as some sort of divine right, including women (known as the 'right of pernada' or 'droit of seigneur'; see Glossary), money, food, horses, etc., for their own benefit.

In present-day society, economic interests work in a similar manner but perhaps more subtly. They take from the environment what they need, and as much as they need, and from society what they consider will make more money for them. This involves an indiscriminate use of natural resources, in some countries it makes for poor working and economic conditions for workers.

Considering this, everyone would probably agree that society should aim at economic growth that is not incurred at the expense of people and/or the depletion of their natural resources. **Economic growth does not necessarily mean a better living:** a society can have a huge rate of economic growth measured through its Gross Domestic Product (GDP), maybe a figure as high as 8 percent per year, because of very vigorous exports. However, that society might also have its own people living in poverty because the wages paid to its workers (and even children) make that export possible, while not covering their most elementary needs.

Besides, when talking about economic growth and sustainable development it is necessary to recognize that the former has, more often than not, a political connotation, while the latter has a broader meaning as well as a time-dependency for reaching its objective.

Evidently, there is some difficulty in explaining the meaning of these terms. However, they can be very well understood given the explanation by Vaughan *et al.* (Vaughan, 1981), who characterize them as follows:

- Economic growth is quantitative. Considering a baseline, it is an economic change or variation related to investment, output, income, and consumption.

- Sustainable development refers to a qualitative change. This means changes not only to the economy, but also institutional, social and environmental changes.

As Donella Meadows (Meadows, 1993) put it:

> *No one can lead the way to the sustainability revolution who doesn't start with his or her own life... It means living life for quality instead of quantity... It means living life for what is really worth living.*

If economic growth translates into a healthy input of hard currency, then a sustainable economy should divert some of these profits to provide decent wages, education, hospitals, housing and food that is available for everyone, and by making judicious use of its resources. What a shame, then, that such a country's products are no longer competitive with those of other producers! What then is the solution? The solution lies in **finding equilibrium**.

It is always possible to lower production costs without sacrificing wages if resources are used in a more efficient way, such as by employing fewer raw materials, less energy, less water, by recycling, etc. If society used resources thinking only of today's needs, it would resemble a person living on his or her savings until they are gone. The sensible thing for this person to do would be to put that capital to work, earning interest, and then live off the interest, without touching the capital. This analogy can probably be replicated in the actual world, in that capital should be preserved (land, forests, water, fish, and other resources) not only for the present but also for the future, while living only from the 'interest' produced by this natural capital.

This means that society should greatly improve the relationship between raw material inputs and the finished products made with them. Humankind needs to decrease the rate of consumption of non-essential items, decrease water use and recycle it many times, lessen the consumption of paper and board products, and use only fibre produced from trees in planted forests. When it is not possible to recycle water, as is the case of water for drinking and cooking, then, the use rate should equal the natural restocking rate, at most.

What about non-renewable raw materials? Obviously, society cannot live off of their 'interest', as it is not possible to 'replant' minerals, for instance. People should make a more efficient first use of these non-renewable materials, and also institute vigorous recycling programs of products made out of them. This applies to glass, aluminium, iron, plastics, electronic equipment, etc.

This last paragraph applies to minerals but, of course, not to fossil fuels used for combustion in cars and power plants, since there is no way to recycle them. While it is true that there has been improved efficiency of fuels in cars, boilers, aircrafts, and in industry, this is not enough; humankind is 'eating into its capital'. Society should establish a time limit for phasing out the main consumer of fossil fuels — that is, transportation —and cede to devices such as fuel cells. It would then be possible to make a better use of hydrocarbons, such as by producing recyclable plastics and to generate large amounts of electricity in power plants, until a better option is found, such as, perhaps, nuclear fusion.

An appeal should be made to wealthy nations, which, with only a fraction of the world population, are consuming a very large percentage of the world's resources while producing the greatest amounts of pollution on the planet. It is a fact that if each person in the world had the same natural resource consumption rate as the average US citizen, then the Earth could not sustain life, and would therefore be unsustainable. Yet if that nation's consumption fell to rational levels, the planet could probably not only sustain its actual population but it might even withstand a large demographic increase.

In sum, there are sound reasons supporting the view that there is no contest between sustainable development and economic growth: they both could co-exist provided that the existing capital is judiciously used and reused.

In 1968, Garret Hardin (Hardin, 1968) wrote a now-famous work analysing the use of resources that belong to everybody, such as air, water, land, minerals, etc., known collectively as 'the commons'. This concept is well summarized by Murdoch University (see Internet references for Chapter 7), which states that:

> ...a "commons" is any resource used as though it belongs to all. In other words, when anyone can use a shared resource simply because one wants or needs to use it, then one is using a commons. For example, all land is part of our commons because it is a component of our life support and social systems. A commons is destroyed by uncontrolled use – neither intent of the user, nor ownership are important. An example of uncontrolled use is when one can use land (part of our commons) any way one wants.

Hardin posed an example of a herdsman adding one sheep to his flock, reckoning that the land used and the grass consumed by the additional animal will be shared by all the local shepherds. His gain of an additional animal means the loss of only a fraction of the resources. In analogy with our modern world, and paraphrasing Hardin, one may think of a man buying a second car for his family, thereby 'gaining' more comfort, and perhaps more mobility, and maybe increased status in the neighbourhood. But what are his 'losses'? He reckons that the new car will contaminate the air, use raw materials, and contribute to road congestion, so it will use the commons. However, these losses, affecting the commons 'owned' by everyone — such as the air, raw materials and space — would be shared with them, and hence his losses for him will turn out to be only a fraction of his gains.

The 'tragedy' is that everyone else can also think this way, and then each one will add a new car, buy a new TV set, etc., using what belongs to everybody, and in so doing depleting it.
Some great inventions have brought large benefits to a sector. For instance, the development of the steam engine and its application to transportation (railways and ships) benefited both the social and the economic sectors with the transportation of people and merchandise. Its application to industry favoured the rapid economic development of some industries, such as textiles. The steam engine brought genuine progress, both from the economic point of view and sometimes also from the social point of view, especially as regards transportation.

Old photographs and other pictures show how common it was, in the last decades towards the end of the 19$^{th}$ century and the first decades of the 20$^{th}$ century, to depict progress through smokestacks ejecting dark and thick smoke. The same applies with large ships, with all of their chimneys belching smoke; had they had indicators these would have shown a positive trend associating people's movement and economic growth. However, at that time no one thought about the consequences of casting into the air such large amounts of contaminants; that is, no one thought about the third component in sustainability: the environment, a commons.

Other more recent examples of this irresponsible action are:

- The daily dumping of hundreds of domestic solid waste into the sea, because it was more 'economical' to do so that to dispose of the garbage in a landfill. True, it was more economical because nobody took into account the environmental cost to the marine ecosystem of such actions. This waste produced turbidity in the water, very likely altering the penetration of sunrays into the water, and consuming the oxygen needed by sea creatures. Its consequences are still to be assessed.

- Connected to the above was the fact that dangerous chemicals found their way into the sea, especially mercury. Some species ingested this chemical, in turn feeding higher species, and finally being fed to people. Thus, there was a boomerang effect with what was deposited in the sea: the poison came back to hurt man.

- Modern chemistry has produced excellent products that have served certain purposes very well indeed. That is so with compounds such as PCBs (see Glossary), which have been used as an insulator in electric transformers, and Freon gas (see Glossary), used as a refrigerant fluid with industrial applications and in millions of fridges and air conditioning units.
  What are their effects? When combined with air in the higher layers of the atmosphere, they break down the ozone layer that protects us from the dangerous ultraviolet rays. The consequences? Skin cancer.

This shows how actions in one sphere, for instance in economics, can influence other areas, such as the social and environmental spheres, whereby in some cases substances like mercury in the sea can alter the food chain and rebound on us. These examples and comments seek to show that in taking an action it is necessary to consider not only its direct consequences, both the positive and the negative, but also how it interacts with the rest of the economic, social, and ecological world. This is the whole concept behind sustainable development: society seeks to improve its economic growth, but it must at the same time advance social progress and environmental protection.

As mentioned before, some people think that economic growth is not possible at the same time as sustainable development, because one is done at the expense of the other. In other words, economic growth needs land, raw materials, fuel and, in general, natural resources, so humankind must use them in order to secure economic growth. On the other hand, economic growth cannot be stopped just to keep resources, and since these two concepts appear to be opposed or in conflict, what is the solution?

Some consider that the solution lies in finding a balance between these two objectives, in other words, in living in a sustainable way. Sustainability as a process often involves making an analysis to determine the best course of action when several projects, plans, programs, and options are considered.

## 1.5 People's participation

Nothing can be done regarding sustainability without people actively participating in the process. People's involvement means conveying people's needs and wishes, collaboration with local authorities in defining plans and policies, taking part in the monitoring process, and contributing with ideas and analysis. Most of the time, people are not consulted; maybe they are perfunctorily asked about something, but merely with the actual purpose of complying with regulations about people's participation; after that, however, decisions are taken without really considering such input. And yet citizens can talk very clearly, make their resolve known, and force change in decisions already taken. See case studies in sections 1.5.1 and 1.5.2. Lindseth (2001) characterizes communication as:

1. *A process with its basis in indicators.*
   This means that any analysis of issues and decisions about them has to be made based on *data* supplied by indicators.
2. *A consultation process,* involving the abovementioned discussions that include the interested parties.
3. *Process contribution,* implying that everyone will contribute to the process, not merely act as observers but as achievers.

The Brundtland report (section 1.2), is very specific regarding public participation when it mentions:

> *Progress will also be facilitated by recognition of, for example, the right of individuals to know and have access to current information on the state of the environment and natural resources, the right to be consulted and to participate in decision making on activities likely to have a significant effect on the environment, and the right to legal remedies and redress for those whose health or environment has been or may be seriously affected.*

In some countries, the process of involving the general public in the decision making process has reached a state where people make decisions about the allocation of municipal funds, and participate in the selection and determination of priority investments. This is called 'participatory budgeting'. It began in 1989 in the city of Porto Alegre, Brazil, and has now been extended to 180 other Brazilian municipalities and many other countries, especially in South America but also in Europe and in South Africa. The system received an award from the United Nations Organization and recognition from the World Bank as an effective public administration tool.

The mechanics of this system of public involvement are well described in a paper written by the Inter-American Development Bank. (See Internet references for Chapter 1.)

---

*1.5.1 Case study: Community participation in Albertslund - Denmark*

*Albertslund is a community of about 30,000, located 15 km west of Copenhagen. It boasts about having some of these quality of life indicators:*

- *Unemployment: 4.6%.*
- *Average class size in schools: 20 students.*
- *Population density 1,270 /km$^2$.*
- *Consumption of electricity for domestic use has decreased by 15%.*
- *$CO_2$ production fell because of a new electrical power plant.*
- *The community produces a report that outlines its progress with sustainability.*

*In the 1980s, together with some other organizations the municipality of Albertslund established a 'User's Group' that represents the welfare of the people in their communities. Its initial purpose was to obtain its citizens' input about the rates to be charged for the heat and electricity generated by a municipal waste incinerator. That initial participation led the Group has had a deeper involvement in other urban issues related to water, sewers and waste management. It is now very active in promoting new ideas and themes, as well as maintaining connections with the Local Agenda 21 activities.*

*Their Green Report, which was begun in 1992, indicates the gains or losses they have achieved in terms of sustainable measures the community has set. For instance, it records their energy-related consumption of natural gas, oil, electricity, etc., breaking the figures down in terms of households, commerce and public*

> buildings. Also, at the request of the community, it even includes the levels of $SO_2$ and other gas emissions. There are both short- and long-term targets.
>
> (For more information, consult Long-term Plan of Action - An agenda for sustainable development - Albertslund, Denmark: Internet references for Chapter 1).

> ### 1.5.2 Case study - The will of a town - People defending their environment and health
>
> *Esquel, a small town located in Patagonia, southwest Argentina, is the gateway to a paradisiacal area, with lakes, forests, majestic mountains and a relaxed atmosphere in a temperate rain forest. Reflecting the economy of the country, the city has high unemployment rates where the inhabitants survive as best as they can with fine fruit production, small-scale industries, and tourism.*
> *A multinational firm intended to mine a gold-bearing local mineral by using dangerous chemical compounds to obtain the precious metal. There was the danger of groundwater contamination due to the tailings that would result from the mining process.*
>
> *In 2003, people stormed the municipality under the slogan "water is more precious that gold",* **forcing the local municipal council to call for a non-binding referendum on the construction of the mining project.** *There was a 75% attendance, and 81% of participants voted against the project. They said very clearly that Esquel's citizens did not want to start a US$100-million open-pit gold mine project to the north of, and just 7 km. upstream from, their town.*
> *The mine's closeness had raised the townsfolk's concern about contaminating their site's pristine beauty. Had the project proceeded, a great deal of dust would have been raised as a result of the blasting of 42,000 tons of rock per day, and with the subsequent grinding process that was also required.*
> *There is no doubt about the economic benefits that this mine would have brought to Esquel and the entire province of Chubut, since such a massive investment would have created 300 direct jobs and about 1,300 indirect ones. Even so, the* **people considered their social and health development was far more important than economic gain.**

> *The developer, a large Canadian transnational, had prepared reports about the safety of their methods, which included the use of 180 tons of cyanide per month to leach gold from the ore. Of course, the townspeople's main concern was that aquifers might somehow be contaminated with this poison, and spoil the town's water source.*
> *The developer also claimed that it would use a new method whereby cyanide is not impounded but treated and mixed with unwanted rock and buried in the void left by the extracted ore.*
>
> *Nevertheless, Esquel's people remained unconvinced, and have since hired an independent US mining consultant.*
> *The referendum has indefinitely postponed the project for the developer. Its future is uncertain, as the developer owns the land and has exploitation rights, although the people's response to the project shows their will to fight the project and halt it forever.*
>
> *This example shows the will and participation of citizens successfully fighting something that is against their interests.*
> *Obviously, this book is not the place to take sides in the confrontation, but the author would like to add something very important to this account. Whether they are right or wrong, despite their meagre economic situation the Esquel people chose to **forego the economic benefits that this project would have brought, traded those off in order to preserve the place as it is for generations to come. This is the very essence of sustainability.***
>
> This case resembles a project (listed in the Note in section 7.7) that led to a disaster in Guyana due to leaking cyanide.

## 1.6 The ecological footprint

The Gross Domestic Product (GDP) is an indicator measuring the economic progress of a country considering the sale of its products, the value added, etc. However, it does not take into account any environmental losses that this can produce — such as the loss of nutrients, erosion due to logging of forests, or contamination due to mining operations. In fact, it cannot consider any of the social implications.

This raises a question for the reader to consider: how much and how many resources does a human being utilize to live?

Let us do a little retroactive history:

- The bread, pizza, and pasta that a person eats comes from crops that use some tract of land; the same is true about the mill producing flour, and the bakery making the goods, baking, cooking and storing them, etc.
- That person's meat originates in cattle that need to be reared in some area, and this also is true about the industrial plant that will slaughter and process the meat, about the supermarket that sells it, including the lot needed for customer parking.
- The daily paper that that person reads, as well as the books and stationery, all come from trees that occupy space in forests, and must be processed in pulp and paper plants that also require space.
- Clothing requirements come from cotton plantations that of course take space, as do the mills that weave it, the manufacturers that make the clothing, and the store in the shopping center where people buy it. The same applies to the tract of land utilized by sheep producing wool, and all those corresponding manufacturing processes.
- People sleep in beds that come from trees growing in forests, which naturally occupy space and consume nutrients.
- Refineries produce the fuel a car uses. These chemical plants take plenty of space (and cause a great deal of pollution), while cars in a city need a fleet of trucks to transport the fuel they consume, which is transported on highways that also are taking up a lot of space.
- Artificial lakes utilize hundreds of hectares. The water that a population drinks and uses comes from these lakes, and more land is needed for water treatment plants to purify the crude water; ending this process, the wastewater produced by human beings also takes space to treat and purify.
- It is also necessary to consider energy consumption, coming mainly from fossil fuels burnt to generate electricity for homes, offices, and factories. However, in this case how is a link established with land use? Because the '$CO_2$ sinking' concept.
It refers to the amount of space that forests need to sequester enough $CO_2$ from those generation plants, thereby preventing an increase of this gas in the atmosphere.
- Etc.

To give an order of magnitude to these considerations about space: Rees (see Internet references for Chapter 1) estimates that in the USA each person needs about 4.5 hectares of land!

Perhaps it would be useful for the reader to calculate how much space the whole population in the United States needs, or, on a more modest scale, how much is needed by the inhabitants of the city where he or she lives.

This measure of how much area a person needs is called the 'ecological footprint'. It was developed by Wackernagel and Rees (1996), and has been described by its creators as *"...the land area necessary to sustain current levels of resource consumption and waste discharge by a given population"*. When one links this value with another fundamental indicator called the **carrying capacity** (section 6.12), it is possible to calculate the stress that is being put on the environment.

This is a very rational indicator that dramatically shows how much a society is consuming. It is a plain fact that **there are currently no sustainable cities in the world because none is able to live from its own resources.** Cities mostly import water, produce, construction material, electricity, and many other inputs from beyond their geographical limits. They also discharge their sewage and solid waste beyond their urban borders. Consequently, it makes sense to determine the extent of this geographical area, i.e. the 'footprint', which a city needs for functioning.

As expected, a city in one country has a different footprint than another's in another country. Thus, US cities have a footprint several times larger than that of cities in Latin America or India, and that is also larger than the footprint of European cities. This corroborates the widespread understanding that the US consumes a disproportionately larger portion of the earth's resources, considering its size and population, than other countries. It is also evident that if the whole world used that American footprint, **planet earth would be unsustainable.** Since the **carrying capacity** of the planet is limited, this clearly shows that the well-being of some countries is based on using the land and resources of others.

Chambers and Lewis (2001; Internet references for Chapter 1) show in two case studies how footprint analysis can help to reduce land use. In one of the cases regarding a business, BFF, they state: *"Changing to a renewable energy supplier and reductions in the need to travel reduced the company footprint from 1.72 to 0.61 hectares. Working at BFF initially accounted for 33% of the individual's average earth share – a figure which was reduced to 12% following energy and travel changes".*

This indicator, measuring the use (and abuse) of resources and their consequences, is very useful, although it does not take into account environmental issues such as **erosion** or **loss of nutrients**, to mention just two. By the same token, it fails to consider **social aspects** such as the adequacy of the payment received by workers extracting and processing these resources. For instance, when a country exports its grain, it would be necessary to consider the loss of nutrients resulting from harvesting its crops, or the investment in fertilizers needed to keep the soil productive.

Many other examples exist as a result of cities' appetite for resources that are produced far away from their borders. For instance, large expanses of virgin forest have been logged in Brazil to make room for soybeans, to be used for cattle feed; the meat produced thereby will go to feed people in cities. So while the footprint indicator can identify this problem, it still actually tells us nothing about the social disturbance caused to aboriginals living in the area, or about the damage to the habitat of countless wild species.

Dramatic examples abound that illustrate the concept of the ecological footprint especially as it relates to cities. For instance, the Global Development Research Center (GDRC — Internet references for Chapter 1) reports that the City of London has an ecological footprint 120 times the area of the city itself. The GDRC also reports that for the typical North American city (the author assumes that this would not include Mexican cities) of about 650,000 requires around 30,000 $km^2$ of land just to meet its domestic needs. By contrast, a similar city in India requires only 2,800 $km^2$.

Many researchers have investigated this issue and have arrived at a figure for the average footprint of the world of about 2.8 ha/capita, with lows and highs. Low values correspond to developing countries — such as Bangladesh, with a footprint of about 0.45 ha/capita; at the other end is the US with the mentioned value of about 9.8 ha/capita. This indicates that to maintain its life style the US population needs 3.5 times more land per capita that the world's average citizen. This shows that the US, with only 4.5 percent of the world's population, uses 25 percent of the world's resources! This percentage increases to 33 percent of the world's resources when the total emergy (section 1.8) is considered.

Dupont, one of the largest industrial corporations in the U.S.A. is quoted in *Business and sustainable development: A global guide* as saying:

> *Dupont's mission is to achieve "sustainable growth", a goal which is defined as "increasing shareholder and societal value while **decreasing the company's environmental footprint** along the value chain in which we operate". Dupont's own perception of sustainable business borrows from the Brundtland definition, and commits the company to "implement those strategies that build successful businesses and achieve the greatest benefit for all stakeholders without compromising the ability of future generations to meet their needs.*

Dupont has also formed partnerships with other big corporations, such as the Ford Motor Co. at its car assembly plant in Oakville, Ontario, Canada, and, as it also indicates in the above publication:

*Using a new contract based on the number of cars painted, rather than the quantity of paint consumed, Dupont helped Ford achieve significant savings. As a result, hydrocarbon emissions from the plant have dropped by 50 percent, and costs are down by a third*

However, no mention is made about how the changed contract effected the decreased emissions.
Figure 1.1 shows that a city imports raw materials from other areas and disposes of its wastes to other areas. At the same time, it transforms raw materials and materials and exports manufactured products.

**This shaded square represents the land area equivalent to footprint**

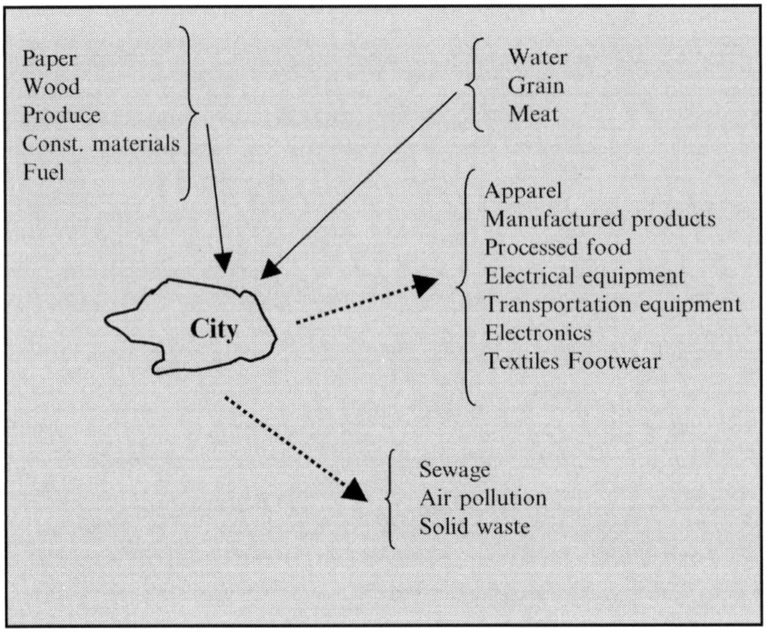

Paper
Wood
Produce
Const. materials
Fuel

Water
Grain
Meat

Apparel
Manufactured products
Processed food
Electrical equipment
Transportation equipment
Electronics
Textiles Footwear

City

Sewage
Air pollution
Solid waste

*Figure 1.1*     **Inputs and outputs in a city**

## 1.7 The ecological rucksack

Researchers at the Wuppertal Institute, Germany, developed the idea of the ecological rucksack. (see Internet references for Chapter 1). This indicator is similar to the ecological footprint but it relies on the material and energy flows measured by the material input per unit of service (MIPS). It measures in kilos the quantity of direct and indirect material necessary to produce a good, minus the mass of the product itself. The name 'rucksack' comes from the notion that the indicator measures the environmental impact or 'burden' that backs up a product or service, but that is not physically included in the final product.

Thus, to build a product like a passenger railway car, for instance, this sequence is followed:

- Produce an inventory of all materials in its manufacture, including:
    o steel beams for the chassis;
    o forged steel for wheels and bogies;
    o aluminium beams for the body and in sheets for the roof and sidings;
    o window frames and glasses;
    o metal doors;
    o sinks and toilets;
    o aluminium sheet for the floor;
    o carpets for the floor;
    o brake mechanism;
    o electric wiring; and
    o etc.

- Tracing processes backwards, calculate the weight in kg. of each **direct** component.
  Thus, for steel one must go back to obtaining the raw material and equipment needed to mine the iron ore, its transportation, processing in the blast furnaces, Bessemer converters (see Glossary), rolling steel sheets and profiles, etc.
  The same is required for the aluminium products and all the other components.

- Tracing processes backwards, calculate the weight in kg. for each **indirect** component.
  This accounting includes all the elements not embodied in the car but which were required by or that otherwise participated in its construction, such as:
    o electric energy;
    o water;
    o fuel and lubricants;
    o packaging;
    o spare parts for the transportation equipment such as trucks;

o etc.

- Capital costs and labour are not included.

Therefore, the MIPS calculates the material inputs for all assembly levels and for all activities, including extraction, transportation, distribution, utilization, and final disposal, and then deducts the materials content of the car itself. It is a similar calculation to the Life Cycle Assessment (LCA; see Glossary and Appendix section A.5), but it is based on material flow instead of environmental impacts. The ecological rucksack is therefore the sum of all the materials that are not physically included in the final product but that have been used to build it. In order to allow for comparisons, this calculation is based in five standard areas, including renewable and non-renewable resources, as well as land, air, water and biota (see Glossary).

Total Materials Requirements (TMR), based on the MIPS, is another tool developed to monitor materials flow at a regional level. It calculates all the materials obtained at a global level, including their ecological rucksacks, to support the country's economic activities. This allows determining a quantity, measured in tons, of resources used per capita on a yearly basis, which then permits instituting comparisons between countries. It is similar to the GDP, although TMR is considered more realistic since it measures economic health but is based on physical factors.

As an example, a national policy in a country granting economic incentives for the installation of wind farms (see section 5.3.1) will most probably provoke a decrease in the country's TMR per person, because less carbon, oil or gas (i.e. resources) would need to be extracted to generate energy, reflecting gains for the environment. TMR's values are around 70 tons/capita of natural resource use in industrialized countries. (Figure 1, reference cited above).

## 1.8 Emergy accounting

Odum (1996, and Internet references for Chapter 1), devised another measure that can be used to take into account externalities (section 1.11). It is called E**m**ergy Accounting, and introduces concepts such as **emergy** and **emdollar**. The word 'emergy' was coined with the meaning of 'embodied energy', having the 'emdollar' as the economic equivalent to emergy. The concept is simple: It measures the worth of something not by its market value, but considering the amount of available energy that was used for its manufacture, production, marketing, etc.

Market values are not a good measure because usually they do not represent the true value of something. The reason for this is that a market can be distorted by subsides, monopolies and other issues. As an example, in many countries oil is under government control, raising or lowering the price of gasoline, according to political and or social issues, and usually not considering the demand or the supply.

On the other hand, according to Odum, goods and services should be appraised according to a general and universal measure, which is the amount of energy that is emergy used. As an example, assume the baking of bread made out of wheat. There is an effort in the bakery to produce bread from the raw material (flour) that can be measured in **em-joules** (which is the emergy unit). It is then necessary to appraise in this unit, the work of the baker, the content of heat needed for baking, the emergy needed for the oven construction, etc.
Let us now analyse just the raw material: flour

Now it is considered the effort made to grind the wheat to produce the flour, the emergy included in the personnel, construction of the mill, equipment, packing, transportation, etc. Let us now go to the wheat itself: To produce a grain of wheat there is need of sunlight (measured in emergy), in rain (measured as the amount of energy needed to evaporate water from the ocean and to create the winds that transport a cloud over the field where the crop grows). It is also necessary to consider the energy spent by the farmer, the energy used to fabricate the farming equipment, etc.

To consider the economic aspect of this example, one may assume that the wheat crop is purchased by another country. Odum maintains that the
*"...price paid by the purchaser takes many times more emergy from the seller that is in the buying power of the money they pay in exchange".*
Odum's work is cited to call to the reader's attention other, more rational systems than the GDP to measure a product's real value. This economic theory is of course more extensive and complex than the present outline. For this reason, interested readers are encouraged to consult the books and Internet reports cited at the end of this chapter.

A report released by the University of Michigan (see Internet references for Chapter 1) provides an illuminating example whereby the amount of energy invested to process and prepare a breakfast cereal is computed to be 15,675 kcal/kg, while that meal's energy content is only 3,600 kcal/kg. In another example, 2,200 calories is said to be required to produce a 12-ounce (341 millilitres) can of diet soda, which only **produces 1,000 calories of food energy**. Yet another disturbing fact from the same publication states that packaging is second to labour in the production cost, averaging fully **8.5 percent of the total cost**.

## 1.9 Resilience (social, economical, and political)

Before defining 'resilience' let us posit the following case: A region in a Country A is highly dependent on car manufacturing. Three assembly plants are built in the area for as many car manufacturers. There is also a vast array of suppliers of parts, sub-assemblies, and assemblies for this sector. One school in the regional university teaches car manufacturing, and technical trade schools offer courses on welding, stamping, painting, etc.
Aided by a favourable exchange rate and the excellent labour pool with decades of car-making experience, the industry is thriving in the region. Most production targets exports to neighbouring Country B, which has a huge population and a reasonable-sized middle class that buys the entire production of all three factories, since those models are not manufactured in Country B.

Unfortunately, Country B has many economic, social, and labour problems, and its currency has been declining over the last two years, and wages and salaries have increased at a much lower rate. What is the result?
People in Country B have discovered that they now have more important needs than buying a new car, so its domestic car market is declining at an extent that many domestic car plants are drastically reducing their personnel. As the reader can imagine, this situation will also affect Country A, and layoffs are underway. The question is: To what extent will the reduction of activity in this vital sector affect the economic and social base of Country A?

It is obvious that the decline of industrial production in Country A will substantially affect the economy and the social well-being of its population. Without a diversified economy, the consequences to the region, from both the economic and societal point of view, will be extremely serious since the population is not prepared to work in other activities, nor does the region have an industrial base that can absorb the unemployed.
In this case, it can be said that the **resilience**, that is the capacity of the region to react to this situation and recover, is very low. That resilience is low because the labour market is not in principle prepared to engage in other activities that are as well paid as they had been before the decline in car manufacturing, or the region lacks the commercial base needed to absorb the decrease in the population's purchasing power of, or, perhaps, maybe the resilience is poor on both counts.

Is it possible to have an indicator that can measure this dependency on one or a few industrial sectors? Yes, one indicator does consider the top five or ten industries and relates them to the total industrial base. Are there measures to gauge if there is also resilience in the fabric of the society? Yes, this can perhaps be measured by a series of indicators detailing the savings/capita ratio between household incomes and rent or mortgage instalment payments, percentage of people working per household, etc. It is also evident that if the savings/capita ratio has a high relative value, or if in a household there is a high proportion of people with stable jobs, then surely a large number of those who are laid off will be able to be out of work for a certain period. That is, these would indicate some resilience, or a capacity to react in this situation.

If the resilience, as an average is low, there will probably be serious social, administrative, and political problems. Social problems are likely to result because many people will become homeless if they can no longer pay their rent or mortgage, forcing them to leave their homes. Administrative problems may arise if the authorities give in to certain groups' demands for monetary help — thereby creating grievances from other sectors of the population, and perhaps even eroding the overall population's confidence in government's ability to solve problems fairly. There may also be problems arising from the political cost of a local government's ability to afford to alleviate the situation about how others will view its measures.

There is also the ability of the government to cope with this circumstance and with its capacity to help this people, counting with the human and economic resources the region possesses. This resilience is also linked to the infrastructure the City Hall possesses to promote new lines of action for jobless people, for creating the economic and tax conditions for encouraging the establishment of new types of industries or activities, for training programs for capacity building, etc.

## 1.10 Environmental resilience

In section 1.9, the concept of resilience was described and now the definition will be applied to the environment. The principle is always the same: resilience measures the capacity to absorb changes, to react and to recover, and it is some sort of the measure of the **elasticity** (see Glossary) of a system.
For instance, a car's aerial antenna bends when it travels on a tree-lined street and passes below low branches. Once the car clears them, the antenna returns to its original straightness. The antenna therefore has certain elasticity and a capacity to recover its original form without any damage.

Now assume that someone purposely bends the antenna such as to exceed a certain limit. The antenna will not recover its original straightness even if it did not break. Since the bending is proportional to the effort, it can be said that the resilience of the antenna is the amount of pressure it can withstand without destroying its elasticity.

Is there any equivalent to this action with the environment? Yes, and commercial fishing provides a good example. Tug nets in fishing boats are as big as a soccer field, and due to this size, commercial and non- commercial species are harvested together with aquatic plant species. As expected, the effect is a large catch, but the problem is that the catch trawls everything, even turtles. Naturally, fish reproduce, and by the next season there could be a similar amount as in the previous one; in other words, resilience by the aquatic population is required to cope with the damage and restore its levels. But it could happen that because of intense fishing, the number of fish that are able to reproduce diminishes, compounded by the disappearance of their food, whereby a point can be reached when no more fish are found in that area. In other words, the catch size can offset the resilience of the natural system.

For this reason, quotas have been established to capture certain species, such as seals, for instance. These quotas express a limit of how much can be captured, and are known as the **threshold.** Once these thresholds are breached, there is a danger of extinction. Sometimes after this has happened, the species may reappear in their usual grounds during the closed season, but at other times this will not happen. The same occurs as in the case of the antenna: once the breaking point or threshold has been surpassed, there is no recovery. For an actual example of this happening, see section 6.12.

## 1.11 Externalities

These actions are produced by activities that do not have a market value (they are external to the market), but that affect people and the environment. A typical example is air pollution. Smokestacks in chemical and power plants produce sulphurous gases, particulates, and they generate acid rain. These pollutants affect people, animals, plants and structures, yet the price of the product that the chemical plant demand or the rate charged by the electric company per kWh do not reflect these damages.

In all likelihood, these entities do their best to improve the quality of their emissions by installing more efficient filters in their smokestacks, or perhaps using a better quality of coal; however, they believe, or want to believe, that their responsibilities end at the top of their smokestacks.

In short, there is a failure in the market to reflect these externalities, but in a sustainable process these externalities must be identified, and devise mechanisms to consider them. The way to compute an economic value for these externalities lies in using contingent valuation (see Glossary).

## 1.12 Capital

Consider a certain region. What is its capital from the point of view of sustainability? It will involve capital of various kinds, namely:

Human capital, that is:

- Their degree of education, preparedness, and professionalism.
- The number of its research institutes and universities.
- Its economic and industrial activity.

Human-made capital, that is:
- Products than have a market value.

Natural capital, that is:

- Non-renewable resources, such as minerals, oil and gas.
- Renewable resources, such as forests, fish, and drinkable water.
- Non-market values assets, such as lakes, beaches, scenery, etc.
- The richness of its soil and climate, for crops and grazing.

Cultural capital, that is:

- Traditions and beliefs.
- Their art (music, literature, plastics, etc.).
- Social, political, and legal institutions.

Perhaps this classification does not fit the traditional definition of capital by economists, but one is not interested here in Economics but in sustainability, and from this point of view a kind of capital that is related to social, economic and environmental issues should be defined. This appears to be a daunting task, although this elemental classification can usefully relate human activity to sustainable issues. For instance, an exodus of professionals from one country to other countries will decrease its human capital, since the original country is losing people who were trained in its universities and research centers.

Another example lies in the argument that a country exporting its industrial products gains economically. While indeed it does, one ought also to analyse how much it costs that country to produce those goods from the environmental and social point of view. If they were produced in sweatshops or places that employ child labour, that gain was made with a **decrease of the**

**quality of life of its people**, denoting a lack of social justice, at least in that area.

That same industrial activity is at the same time probably causing some contamination in the exporting country, and decreasing its renewable and non-renewable resources. If the country exports agricultural products, there will probably be a decrease in its natural capital due to the degradation of its soil because of salinity, as well as due to the use of herbicides, fertilizers, etc. As commented previously, the classic economic indicator for measuring the economic growth of a country, its GNP, does not reflect what the society as a whole pays for that growth and how the environment is being damaged. Different measures have been proposed to correct the GNP, but that subject is beyond the scope of this book.

A very important point to keep in mind is that natural capital is something to be considered as a stock, a reservoir that is able to maintain a **flow of resources.** Section 1.3 remarked that to keep economies sustainable, humankind should live off the 'interest' and keep the 'capital' intact; that is, the economy should utilize this flow of resources rather than deplete the stock. Reed (1996) puts this in the right perspective when he says, *"A fundamental question for ecological economics is whether remaining stocks of natural capital are adequate to sustain the anticipated load of the human economy into the next century".*

## 1.13  Local Agenda 21

This powerful and seminal document was drafted in 1991 by the International Council for Local Environmental Initiatives (ICLEI), and was presented and approved at the Earth Summit of 1992 (see Glossary) as Chapter 28. It launched various guidelines for municipalities to start communicating and working collaboratively with urban inhabitants in order to promote a sustainable environment. The ICLEI also proposed a research and development project, called the 'Local Model Communities Programme' (LA21 MCP), to work with selected municipalities to test and evaluate different structures for sustainable development.

The results of this experience were published as 'ICLEI Local Agenda 21 Planning Guide: An Introduction to Sustainable Development Planning'. It was released in three languages and can be purchased at http://www.iclei.org/ICLEI/la21.htm At this electronic address, the ICLEI establishes what they call 'LA21 Campaign Milestones', whereby five actions determine an excellent guide to achieve this purpose.

## 1.14 The Bellagio principles

In 1996, an international group of researchers met in Bellagio, Italy, to analyse progress made on sustainable issues. That meeting produced the Bellagio Principles, which are now recognized worldwide. These principles are (see Bellagio Principles, in Internet references for Chapter 1):

1. Guiding vision and goals;
2. holistic perspective;
3. essential elements;
4. adequate scope;
5. practical focus;
6. openness;
7. effective communication;
8. broad participation;
9. ongoing assessment; and
10. institutional capacity.

The enunciation of these principles is clear enough, and provides a good example of the steps that have to be taken in sustainability undertakings.

## Internet references for Chapter 1

**Author:** Peter H. Raven (2000)
Title: *Foreword*
Address:
http://www.ourplanet.com/aaas/pages/foreword01.html

**Authors:** by K. Bidwell and P. A. Quinby (1994)
Title: *Sustainability and the value of ancient forest landscapes*
Comment: Supplies impressive figures about the devastation that humankind has caused the planet, and the rate at which this has been done.
Recommended reading.
Address:
http://www.ancientforest.org/rr5.html

**Authors:** A.R. Gordon and K.R. Shore
Title: *Life Cycle renewal as a business process*
Address:
http://irc.nrc-cnrc.gc.ca/fulltext/apwa/apwalifecycle.pdf

**Author:** Alexa Stanard, Associated Press (2003)
Title: *South America's wild Patagonian glaciers are melting faster than in previous years, say scientists*
Address:
http://www.enn.com/news/2003-11-19/s_10511.asp

**Author**: Bilal M. Ayyub (2000)
Title*: Methods of expert-opinion elicitation of probabilities and consequences for Corps facilities*
Comment: This document thoroughly reviews the mechanics of the expert opinion elicitation process within its 120 pages, including the presentation of their results – subjects seldom found in the available literature. It also includes an extensive Bibliography.
Address:
http://www.iwr.usace.army.mil/iwr/pdf/MethodsforEEfinal1.PDF

**Author**: Alan Fricker (2001)
Title: *Measuring up to sustainability – Sustainable Futures Trust*
Comment: Very good analysis of the concept of sustainability. This refreshing paper has managed to condense all the difficulties inherent in sustainability, while at the same time providing a general overview about indicators.
Address:
http://www.metafuture.org/articlesbycolleagues/AlanFricker/Measuring%20up%20to%20Sustainability.htm

Public participation
**Source**: The Danish Government (2002)
Title*: A shared future - balanced development 16. Public participation and Local Agenda 21*
Address:
http://www.mst.dk/udgiv/publications/2002/87-7972-279-2/html/kap16_eng.htm

**Source**: I.C.L.E.I. (2004)
Title: *What is Local Agenda 21? A short introduction*
Address:
http://www.iclei.org/europe/la21/la21.htm

**Source**: TDM Encyclopedia -Victoria Transport Policy Institute (2003)
Title: *Evaluating transportation resilience -Evaluating the transportation system's ability to accommodate diverse, variable, and unexpected demands with minimal risk*
Comment: This source states that "…resilience reflects *uncertainty*, our inability to know what combination of conditions will occur in the future."

This comprehensive paper makes a good attempt to define resilience, and how to evaluate it. Recommended reading.
Address:
http://www.vtpi.org/tdm/tdm88.htm

**Authors:** C.S. Holling and Brain Walker (2003)
Title: *Resilience defined*
Comment: Interesting paper on social resilience. Explains the difference in resilience between an ecosystem and society's adaptive capacity.
Address:
http://www.ecoeco.org/publica/encyc_entries/Resilience.pdf

**Source**: TDM Encyclopedia - Victoria Transport Policy Institute (2003) Sustainable Transportation and TDM
Title: *Planning that balances economic, social, and ecological objectives*
Comment: Regarding sustainability in transportation activities. Provides good definitions and comments about sustainability principles and indicators.

Address:
http://www.vtpi.org/tdm/tdm67.htm

**Authors**: Jeff Tryens and Bob Silverman (2000)
Title: *The Oregon benchmarks as a measurement system for sustainability*
Comment: Good overview of available frameworks, as well as of the Bellagio Principles.
Address:
http://www.oregonsolutions.com/A_govt/pb_whitepaper.cfm

**Source**: International Institute for Sustainable Development (1997)
Title: *Bellagio Principles – Guidelines for the practical assessment of progress towards sustainable development*
Comment: Description of each one of the principles by the organization that "...brought together an international group of measurement practitioners and researchers to review progress to date and to synthesize insights from practical ongoing efforts".
Address:
http://www.iisd.org/measure/principles/1.htm

**Author:** William E. Rees, University of British Columbia
Title: *Revisiting carrying capacity: Area-based indicators of sustainability*
Comment: Very good definition and analysis of this subject. Includes topics such as 'The ecological argument', 'Why Economics cannot cope', 'Technology and trade: No boon to carrying capacity', 'Appropriated carrying

capacity and ecological footprints', 'Footprinting' the 'Human Economy'. Visiting this Website is highly recommended.
Address:
http://dieoff.org/page110.htm

**Author:** Virginia Abernethy (1993)
Title: *Population politics: The choices that shape our future "The carrying capacity of the United States"*
Comment: Excellent analysis of this subject, and includes physical and cultural carrying capacity.
Address:
http://dieoff.org/page58.htm

**Author**: Howard T. Odum (2000)
Title: *Emergy accounting*
Comment: In this 20 pages paper the author expresses the main concepts behind emergy, and presents a case study in Ecuador with a detailed explanation and example of the mathematics involved, as well as an energy system diagram.
Address:
http://dieoff.org/page232.pdf

**Authors**: Nicky Chambers and Kevin Lewis (2001)
Title: *Ecological footprinting analysis: Towards a sustainability indicator for business ACCA Research Report No. 65*
Comment: The authors mention a number of approaches that can be used by businesspeople wishing to assess the environmental impact of their goods and services.
Address:
http://www.accaglobal.com/research/summaries/23915

**Source**: The Global Development Research Center (2004)
Title: *Urban environmental management*
Address:
http://www.gdrc.org/uem/index.html

**Source**: University of Michigan — Centre for Sustainable Systems (2004)
Title: *Life cycle-based sustainability indicators for assessment of the U.S. food system*
Comment: Comprehensive analysis of projects for buildings, agriculture and food systems, transportation, renewable energy, and packaging.
Address:
http://css.snre.umich.edu/

**Source**: City Councils of the eight largest cities in New Zealand (2003)
Title: *Quality of life in New Zealand's eight largest cities*
Address:
http://www.bigcities.govt.nz/indicators.htm

**Source**: City of Glendale, California (2002)
Title: *Quality of life indicators report*
Address:
http://www.ci.glendale.ca.us/government/cdh/ns/qol_indicators.pdf

**Source**: East Lindsey District Council, UK (2003)
Title: *Quality of life indicators*
Comment: Excellent information about trends in indicators.
Address:
http://www.e-lindsey.gov.uk/documents/QOL_Indicators.pdf

**Source**: Calvert on line (2004)
Title: *Calvert-Henderson - Quality of life indicators*
Address:
http://www.calvertgroup.com/aboutindex_2951.html

**Source:** California State University – Monterey Bay (2001)
Title: *Section IV: Quality of life indicators*
Address:
http://hhspp.csumb.edu/html/community/tellus96/s4/

**Source**: The European Commission – Directorate General (1997)
Title: *Long-term plan of action - An agenda for sustainable development Albertslund, Denmark*
Address:
http://europa.eu.int/comm/urban/casestudies/c035_en.htm

**Authors**: Signe Gilson and Richard Gelb (2004)
Title: *The influence of development characteristics on the ecological footprint of an urban household*
Comment: Very good information for footprint calculations including land-use related footprint categories. As the authors point out: "*The purpose of this paper is to summarize research findings on the relationship between urban development characteristics and household resource consumption using the concept of ecological footprint*".
Address:
http://www.cityofseattle.net/environment/OSE%20Footprint%20Report%203-19-04.pdf

**Source**: Office of Sustainability and Environment – City of Seattle (2004)
Title: *What's new?*
Comment: Seattle is one of the key cities in the world about sustainability issues and their implementation. This Website is the gate to a series of articles related with the practical implementation of sustainable measures, such as buildings, urban household, hotel water conservation, 'smart energy', etc.
Address:
http://www.seattle.gov/environment/whats_new.htm

**Source:** Wuppertal Institute for Environment Climate Energy – Department for Material Flows and Structural Change (1999)
**Authors:** J.H. Spangenberg, F. Hinterberger, and S. Moll and H. Schütz
Title: *Material flow analysis, TMR and the mips concept: A contribution to the development of indicators for measuring changes in consumption and production patterns*
Comment: The reading of this 14-page paper is strongly recommended.
http://www.seri.at/Data/data/publications/fh/material_flow_analysis_1999.pdf

**Source:** Inter-American Development Bank (2003)
Title: *Assessment of participatory budgeting in Brazil*
Address:
http://www.iadb.org/sds/doc/ParticipatoryBudget.pdf

# CHAPTER 2 – THE CULTURE OF WASTE

## 2.1 Introduction

Sustainability relates with Economics, Society, and Environment. However, one common fact that links them all is the **generation of waste.**
This chapter is divided into two: The first part analyses the **current generation of waste as well as its treatment.** The second intends **to establish policies for the future treatment of waste** — or, better yet, for ceasing the generation of waste. This first part begins by raising some capital questions:

- *What* is waste?
- *Which* are the components of waste?
- *Where* is waste generated? and
- *How* is waste treated?

The second part will deal with:

- *Why* is waste produced? and
- *What* is society doing to correct this problem?

## 2.2 First part: Current generation and treatment of waste

### 2.2.1 What is waste?

The dictionary defines 'waste' as something useless, unwanted, or defective and the word 'by-product' as something produced in an industrial or biological process in addition to the principal product.
From the point of view of sustainability, the word 'waste' does not have that meaning as, though it may be unwanted, it is not something useless and is certainly not defective. Even if in a manufacturing process a product or part of it does not conform to the manufacturer's quality specs, it does not thereby become waste, but is, rather scrap material that is usually brought back to its original state and then processed again.

The Indigo Development Corporation (see Internet References at the end of this chapter) has a good definition of waste: they call waste a 'dissipative use of natural resources', as indeed it is, because if released into the air, soil or water there will actually be an unrecoverable dissipation of a natural resource. Therefore, it is believed that the kind of 'waste' referred to here could be better called a by-product, and, as a consequence, having some economic value. The logging industry provides a very good example: A tree is sawed and transported to a sawmill, where it is de-barked — the bark being the first by-product, which used then in the sawmill as fuel for a boiler or for producing ethanol (section 5.3.4.2).

The de-barked timber is then sawed into raw wood products, such as saw logs, veneer, building materials, pulpwood, etc., all of which could as easily be regarded as the main product. In the process sawdust is generated, which is another by-product normally utilized to produce particleboards, and also as biomass (section 5.3.4). The branches of trees, cut into small pieces, are also used as mulch in gardens to retain humidity around plants.

Raw wood products have many different uses, such as for boards and structural units in house construction, doors, windows, cabinets, furniture, etc. At the end of their life, these products are considered waste, and this is wrong since many beams, floors, windows can be reused for the same purpose, or taken advantage of the beauty of their grain to make useful and beautiful things. Clearly, the destiny of these wood products is not in a landfill.

Sometimes, on the other hand, these remains cannot be used for any of these purposes. But then it is always possible to grind them down into biomass fuel, or into mulch, which is a protective covering (like sawdust, compost, or paper) that can be spread out or left on the ground to reduce evaporation, maintain an even soil temperature, to prevent erosion, control weeds, enrich soil, or to keep fruit (such as strawberries) clean.

If logs are destined for producing pulp and paper, after its primary use as newspaper, stationery, books, etc., the used paper can be recycled and converted into cardboard, or as new writing material. Paper bags used for the collection of tree leaves in the autumn are generally made of this recycled paper; at the end, both container and contents end up being used as mulch. Therefore, this very important sector of the industry, if properly managed can comply with the closed flow that is commonly found in nature (section 2.7).

Other items, such as cars for instance, are recycled by being sent back to the steel industry, which feeds them to electric furnaces as steel scrap for re-processing. In many cities, scrap yards permit people to purchase at low prices used spare parts for their own cars, thereby reusing many parts.

These are two examples of the correct treatment course for intermediate by-products and for items that are no longer wanted; they are raised due to their importance in the waste stream as a result of their volume. However, there is no reason why similar procedures cannot be applied to many other items.

Archaeologists often remove garbage from the past with their hands and fine brushes. Their inspection and study of the ceramic shards of cultures long dead enables them to envision the lifestyle of those people, their progress, and the art they developed. By inspecting their food leftovers, they learn about their dietary habits, and perhaps even some things about the diseases that afflicted them. This can be very useful, since archaeologists are often thereby able to give society a glimpse of very ancient cultures that were hardly known.

What would an archaeologist 1,500 years from now find — say, in the $36^{th}$ century? We can easily imagine them digging in the garbage left by their $20^{th}$ and $21^{st}$ century ancestors. Their inventory would probably include plastics in a myriad of different forms. Some plastic boxes with glass they might run into were called computers and TV sets. Millions of black, donut-shaped things the archaeologist would probably discover were called tires, used in something known as a car; heavy metals such as chrome and nickel would be found used, together with iron, to make many different things; also, millions of small plastic disks that were once used to play music; glass in many different forms might also come to light, and so on.

It is likely that at that time all oil and coal will be gone, and forests may have disappeared. Given the lack of forests, people may live in an arid and eroded world, with insufficient natural food and water, and only a few animals. Due to the scarcity of food and space, populations might be kept steady at 16,000 million people, and colonizing other planets may have become a key for survival. How would they react upon realizing that most of the planet's resources, resources that they would have been entitled to share, are gone and turned into garbage? Just as different stages in human history have come to be widely known as the Middle Ages, the Renaissance, the Modern Age, the Atomic Age, etc., they will no doubt call ours the Waste Age.

From the point of view of sustainability, waste is of paramount importance. Every publication on the subject provides guidance about how to dispose of it. This therefore involves discussions about land filling, incineration, vitrification, and recycling methods, and about how to eliminate or reduce the consequences of waste, such as emissions from incineration plants. Also subject to scrutiny are various sophisticated ways of getting rid of radioactive waste.
Unfortunately, the world is losing the battle. Adequate land for landfills near large cities is become more and more scarce, which means than daily garbage has to be hauled to sites far from the cities that produce them, implying an increase in transportation costs, an increment in the number of trucks with

corresponding growths in their consumption of fuel, and consequently more waste (air pollution) added to the environment.
What then is the solution?

The answer is twofold: The human race has to develop good means of **waste management,** the analysis of which is the purpose of the first part of the current chapter, as well as to **reduce the production of waste,** which is treated in this chapter's second part. Many people will argue that waste management is somehow being undertaken properly now, through recycling. Well, this not really the case: what is being done is to put waste to new uses, which of course is a very good endeavour, **but it would be better not to produce that waste in the first place.** That would stop requiring anyone to get rid of or recycle it.

### 2.2.2 Which are the components of waste?

A very large proportion of domestic, municipal, institutional, and construction waste, and a sizeable part of industrial waste, ends up in landfills. Somebody examining the composition of the waste in a landfill can formulate a reasonable idea of the society that generated it from the amount and type of that waste. A walk through the streets of any city in the U.S.A. or Canada with an eye on the type of waste left at the curbside to be picked up for the landfill, shows the remains of still usable furniture, mattresses in a very good condition and often even without stains, kitchen utensils which, if repaired, could again be in good working order, etc. The amounts and quality of such waste originates, without a doubt, in a prosperous society.
Figure 2.1 shows the distribution of domestic waste. A research comparison of landfills in different cities in the US, Canada, the EU, Latin America, Asia and Oceania, broke down the composition of waste as indicated in Table 2.1.

This table only provides a rough idea, because many factors make it very difficult if not impossible to chart the refuse of all countries, let alone to find an average. The reasons for this are:

- Not all the agencies (City Hall, Statistics Departments, private firms, etc.) consider the same factors. For instance, in North America floor carpets are a very important item because the cold weather calls for their utilization in practically all houses, office buildings, airports, shopping centers, etc. This is not the case in many other countries.
- There is no consensus about what the 'paper' label covers. Does it include packing, cardboard, special papers?

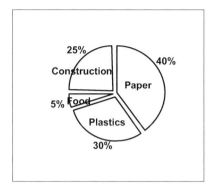

*Figure 2.1*  **Distribution of domestic waste**

- Some items such as 'garden refuse' may be common in some countries but non-existent in others, and its inclusion of course reduces the percentage assigned to other items.
- Some countries class 'food' within the 'organic' realm, along with other perishables, as the mentioned garden refuse, or some textiles.
- The EU's figures that appear in Table 2.1 comprise all its constituent countries, but not all these countries have the same economic level, customs, habits, climate, etc. A consolidation of criteria among the member countries has presumably been undertaken to establish their various types of waste, after which the results indicate that paper leads the ranking, followed by organics and plastics.

However, in the balance of the countries shown in Table 2.1, figures corresponding to 'organics' might include something other than food (although not medical wastes or dead animals, since these are not supposed to be dumped in a landfill); as mentioned above, this could include garden refuse and textiles. In many large cities around the world, there is no room for gardens or backyards, so there is no garden refuse; for this reason, one should discard these values as non-comparables. In sum, statistics show that in most countries three items lead the pack: paper, construction and demolition materials, and plastics.

Given these considerations, it would appear that more measures should be taken to recycle a larger amount of these components of municipal waste.

*Table 2.1*

Composition of waste

| | Paper | Plastics | Metal | Glass | Organics | Construction and demolition | Waste causing problems | Hazardous waste | Floor carpets | Garden refuse | Textiles |
|---|---|---|---|---|---|---|---|---|---|---|---|
| Four states in the United States | 26.1 | 6 | 6.2 | 2.5 | 28.9 | 18.1 | 5 | 0.6 | 5.2 | | |
| One city in Canada | 64 | 17 | 7.8 | 11.4 | | | | | | | |
| European Union | 18/38 | 7/10 | 3/10 | 2/5 | 14/38 | | | | | | |
| Two cities in New Zealand | 10.5 | 7.5 | 7 | 3.5 | 11.5 | 15.5 | | | | | |
| One city in the Philippines | 25 | 13 | 4 | 4 | 27 | | | | | | |
| Two cities in South America | 25 | 18.3 | 6.9 | 22.3 | | | | | | | |
| One city in Japan | 44 | 24 | 3 | 5 | 16 | | | | | | |
| One city in Australia | 16 | 7 | 3 | 4 | 31 | | | | | | |
| One city in the Gaza strip | 10.6 | 10 | 1 | 23 | 70 | | | 1.5 | | | 2.7 |
| One city in Armenia | 18 | 1 | 3.5 | 4 | 30 | | | | | | |

### 2.2.3 Where is waste generated?

Table 2.2 depicts eight general origins for wastes, while Table 2.3 shows the different types of wastes and their treatments.

*Table 2.2*     **Origins of waste**

| | |
|---|---|
| **Origins of waste** | Households |
| | Municipal |
| | Institutional |
| | Wastewater treatment plants |
| | Industry |
| | Hazardous |
| | Nuclear |
| | Construction |

*Table 2.3*     **Types of wastes and treatments**

| | | | Waste treatment | | | | |
|---|---|---|---|---|---|---|---|
| | | | Recycling | Landfilling | Incineration | Biological treatment. | Second use |
| **Type of waste** | Solid | Households | | | | | |
| | | Municipal | | | | | |
| | | Water treatment plants | | | | | |
| | | Institutional | | | | | |
| | | Industry | | | | | |
| | | Hazardous | | | | | |
| | | Nuclear | | | | | |
| | | Construction | | | | | |
| | Liquid | Households | | | | | |
| | | Municipal | | | | | |
| | | Water treatment plants | | | | | |
| | | Industrial | | | | | |
| | | Nuclear | | | | | |
| | Airborne | Municipal | | | | | |
| | | Industrial | | | | | |

Regarding disposal and/or treatment, wastes are dumped onto the ground (landfills), into bodies of water (rivers, lakes, oceans), and spewed into the atmosphere, incinerated or recycled, as shown in Table 2.3. All of these procedures will be analysed below.

Table 2.4 compiles data about the generation of waste for eight different sources, and for solid, liquid, and airborne wastes.

Table 2.4  Generation of waste

| Waste source | Solid waste | Liquid waste | Airborne waste |
|---|---|---|---|
| Domestic (house holds) | Human and animal waste | Wastewater From kitchen, laundry, toilets | |
| | Household: no longer wanted and reusable items Clothing, wire hangers, cutlery, dishes, kitchen hardware, kitchen appliances, camping equipment, bikes, shoes, computers, printers, fax machines, books, magazines, furniture, plastic boxes for storage, home-office furniture, unused paint cans, shop and garden tools, carpets | | |
| | Household: small items Paper, boards, plastics, glass, wood, clothing, aluminium, metal cans, metals, different fibres (from carpets, e.g.), small appliances (hair driers, electric shavers, kitchen appliances) Food scraps and diapers (probably the only waste that is worth bagging) | | |
| | Household: large items Mattresses, appliances (fridges, dishwashers, clothes washers, clothes dryers, stoves, bicycles, shop tools, garden tools, etc) | | |
| | Household: kitchen vegetable waste Any vegetal scrap, peelings, as well as eggshells | | |
| | Household: garden waste Cuttings, leaves, weeds, branches | | |
| | Household: other items Paints, pesticides, insecticides, cleaning and laundry chemicals, fluorescent lamps and ballast, small alkaline batteries (from music players, flashlights, remote controls, etc.) | | |
| | Household: hazardous waste Expired prescription items, swabs, paints, solvents, chemicals, stain removers, small batteries | | |

| | | | |
|---|---|---|---|
| Municipal | Car waste<br>Old cars, tires, batteries, motor oil, grease, carpets, motor coolants, windshield fluid, waxes | | Emissions from cars |
| | **Solid waste** | | |
| | Paper products in streets, sidewalks and parks<br>Plastic and glass bottles<br>Beverage cans<br>Refuse from storm drains | | |
| | **Wastewater** | | **Air contamination** |
| | Constituted by rainwater collected by storm drains. | | Air contamination from municipal trucks, road equipment and car fleets, as well as transportation system if run by the city |
| | **Solid waste** | **Liquid waste** | **Airborne waste** |
| Institutional waste | • Paper product from offices<br>• Food from restaurants<br>• Plastic from offices and restaurants | Waste water from kitchens, toilets | |
| Waste water treatment plants | Sludge (mixture of solid matter and water) produced after mechanical and biological oxidation process | Treated wastewater | Methane produced by digestion process |
| Industry | • Metal scrap from manufacturing processes<br>• Materials, products, original chemicals, etc., as non longer needed and in good condition that is surplus of manufacturing processes<br>• By-products of processes (ashes, $CaCO_3$, etc.)<br>• Packing material, such as wooden boxes, steel containers, drums, board, etc.<br>• Hazardous materials, usually chemicals<br>• Defective products<br>• Spent oil from vehicles and industrial equipment, as well as cooling oils from machine shops | • From processing plants<br>• Hot water | • Emissions from chimneys and furnaces<br>• Air emissions from paint cabins<br>• Emissions from leaks<br>• Steam<br>• Gas flaring in oil refineries |
| Hazardous waste | From hospitals, health centres, animal clinics, etc. | | |
| Nuclear waste | Spent rods | Hot water discharged into rivers | |
| Construction waste | Discarded shingles, timber, plastic floors, electric wire, electric gadgets (telephone and electric plugs), acrylic sheets, windows, doors, construction panels (gyproc) (see Glossary), | Construction water | Dust and particulate |

## 2.2.4 How is waste disposed of or treated?

Table 2.5 analyses the treatment of different types of waste, following the source and categorization of Table 2.4.

*Table 2.5*  **Treatment of wastes**

| Source | Treatment |
|---|---|
| **Domestic (house-holds)** | **Solid waste** |
| | **Recyclable items**<br>*House sorting and recycling*<br>City Hall can supply plastic boxes (made out of recycled plastic), usually called 'blue boxes'<br>Recyclables such as paper, plastic and glass are yard stored for further use to manufacture new paper and cardboard. Crushed bottles are sent to the glass industry for manufacturing new bottles. Aluminium cans, flattened and packed, are delivered to the aluminium industry for melting. Metal cans are also stored in the scrap yard for melting<br>Wire and plastic hangers can be donated to drycleaner stores<br>Electrical material such as copper and aluminium wire, outlets, electric boxes, etc., are also used and re-melted by the corresponding industry<br>Wood, beams and boards can be chipped and used as mulch, although treated wood is not mulched because of its constituent chemicals; for that reason, it is usually incinerated<br>Some items such as shoes, that are not apt for second-hand use, can be recycled. See 'Miscellaneous products' for recycling athletic shoes, in Internet references for Chapter 2 |
| | **Household not longer wanted and reusable items**<br>*One solution is house sorting and donations.* Several charitable organizations and churches are more than happy to receive and pick up donations for sale at their annual sales, or for sale in shrift shops<br>*Community street sale:* A nice practice in North America is when people in an area organize a community garage sale on several streets, whereby people sell their household items at a fraction of their cost |
| | **Composting**<br>If there is a backyard, composting at home is an easy task, since it means placing all vegetable peelings and organic waste in a bin, and occasionally adding water. Compost bins need no maintenance, just occasional watering. They produce no odour (even in the summer time), attract no cockroaches, and rodents cannot reach the waste. It delivers an excellent fertilizer for the garden and back yard, and saves a great deal of space in a landfill, while eliminating transportation costs |
| | **Appliances**<br>These items are sorted at the landfill and then sent to the manufacturers for parts recycling or melting |
| | **Cars**<br>As mentioned, the main option is to use them as scrap material for electric furnaces, melted and reused in a myriad of ways, or delivering them to scrap yards |

| | |
|---|---|
| | **Tires** |
| | When a person buys new tires, the old ones should be delivered to special centers, which often are the same shops |
| | This material can have many different uses, including to generate electricity by direct burning or by producing fuel (section 2.12.5) |
| | **Domestic hazardous waste** |
| | Many household items are considered hazardous. They cannot be placed on the curb-side but have to be delivered to specials depots. For instance, for paints: See 'Emerging products - Paints', in Internet references for Chapter 2 |
| | See also 'Electronics' in Internet references for Chapter 2 |
| | Expired medicines can be taken back to the local pharmacy to be destroyed |
| | **Liquid waste** |
| | **Human and animal wastes** |
| | Generated in households (domestic wastewater) and emanating from kitchens, laundry rooms, and toilets. This wastewater is piped to an underground sewer that runs below the street level in front of a house. It is then conducted to the wastewater treatment plant where is treated in different ways to be finally discharged into a river, or the sea. If properly treated, this can be used for irrigation purposes |
| | Sewer trunks and conduits do not take land space, since they run underground, but are very expensive to build. Water treatment plants are also expensive to build and maintain and use considerable land, but they are so important to human health in cities that the proportion of dwellings served by the system as well as the quality of treated water discharged are even used as indicators (see Chapter 6) to measure sustainability |
| | At present there is no other alternative, especially for large cities. However, in some circumstances it is possible to treat and reuse wastewater generated in large hotels or in large apartment buildings (section 4.5.1) |
| | Many large cities have entire areas not served by their sewer system. In such cases, each dwelling carries out its own treatment by means of cesspools. These eventually fill up and have to be drained by special trucks. These devices are not advisable because under certain circumstances they can severely contaminate the water table (see Glossary). Besides, during heavy rains the water table can rise enough to provoke the overflow of the cesspool, backing up into the dwelling |
| | On the other hand, sewers are also subject to problems, as, for instance, when some people make dangerous use of the concealed sewer system to drain unwanted solvents, gasoline and/or chemical products. It is possible that they don't know that these products are not discharged into the sewer system for several reasons: one is that they disturb the normal aerobic and anaerobic processes in wastewater treatment plants; another reason is that flammable products such as gasoline and motor oil can cause explosions in the sewer system with very serious consequences to personnel and installations |
| | **Spent cooking oil** |
| | A sustainable use of such oil is to save it for collection (from restaurants, hospitals, hotels, etc.) by special trucks for processing as a fuel for trucks |

|  | Solid waste | Liquid waste | Airborne waste |
|---|---|---|---|
| **Municipal waste** | Sorted and recycled at the landfill | **Rain water:** In some cities, storm drains share the sewer system, so rain and domestic water end up being mixed. This practice is inadvisable for various reasons, including that the wastewater treatment plants end up having a greater volume of water to treat in a brief period. Another reason is that during heavy rains, water treatment plants sometimes cannot cope with the sudden input of water, resulting in a discharge of wastewater and rainwater into a river or the sea. Some cities build large underground reservoirs and/or sometimes open water ponds to serve as temporary storage of storm-water. During a heavy rainfall, rainwater is conducted there, to be released later into a river, thus avoiding potential flooding. Water can also be used for irrigating parks, sweeping and washing streets or sidewalks, etc. Consequently, it is better to have separate systems for wastewater and rainwater | **Emissions:** Can be decreased by replacing trucks and buses with vehicles equipped with fuel cells (section 5.3.5) |
| **Waste from water treatment plants (WTP)** | **Treated water** | **Sludge (solid and liquid)** | **Methane** |
|  | Water following the biological oxidation process can be discharged into a body of water, used for irrigation, or reused to flush toilets in hotels and large apartment buildings | Raw sewage entering a WTP is more than 99% water. After a filtration and settling process, and through biological oxidation, sludge is produced. This product does not appear to have any commercial value; however, if properly treated in a digester (see Glossary) it can yield two very important products: nutrients and electricity<br><br>**Nutrients**<br>The human digestive process keeps nutrients in our waste. At the same time, with appropriate retention time and at the right temperature, as much as 90% of harmful organisms (pathogens) in the waste are destroyed (this destruction can reach 100% with further treatment). | **Electricity**<br>In venting the anaerobic digestion process, methane and carbon dioxide are released into the atmosphere. Methane can be utilized to produce clean energy. This utilization has a double advantage: it produce electricity and, by not releasing it, we prevent contaminating the |

|  |  | Consequently, the resultant slurry is an excellent product for use as a fertilizer in agriculture — thereby averting the need for a large scaled supply chain (see Glossary) to mine and process non-renewable minerals such as potassium and phosphorous | environment |
|---|---|---|---|
| **Industry** | **Solid items** | **Liquid discharges** | **Air emissions** |
|  | **Re-smelting**<br>Most of the scrap metal produced in the manufacturing process can be re-smelted.<br>**Wood**<br>Reutilized in the plant, or reduced to mulch.<br>**Steel containers**<br>Sold as metal scrap or to other manufacturers<br>**Drums**<br>Reutilized in the plant, sold to other industries, or sold as scrap<br>**Board**<br>Sent to pulp and board manufacturers to be recycled<br>**Packaging**<br>Processed to pack own products<br>**Waste brokerage** | **Fluids**<br>Most industrial plants use large quantities of oil for lubrication and for cooling the tools of automatic machines, such as lathes<br>These fluids are contaminants but can be used for other purposes, such as for fuel in cement kilns. This actually brings an economic benefit to a plant while also helping the environment, since less fossil fuels have to be burnt in the cement kilns, or in industrial and municipal incinerators |  |
|  | **By-products from processes**<br>For instance: Ashes, fly ash, chemicals | Hot water from steam condensers, scrubbers, etc., can be used for other purposes | Steam discharged from processes or turbines, can be further used as low-pressure steam |

| | **Hazardous wastes**<br>Vitrification (section 2.3.1) and burying | | |
| --- | --- | --- | --- |
| | **Flue gases**<br>Solids collected in filters and cyclones should be analysed to find other uses, but not to be buried. Material collected after filtering should be kept in tight storage because of the high content of heavy metals | **Slurry**<br>From scrubbing operations can be used for heating water | **Treated gas**<br>After filtration, released to atmosphere |
| | **Sludge**<br>Sludge from water treatment plants or other industrial processes can be very efficiently dewatered to produce pure water, and the dry sludge can then be incinerated or sold as fertilizer | **Wastewater from kitchens, showers and toilets**<br>If treatment is possible, this can be used for reprocessing in the plant or for irrigation. Residue should be incinerated. | |
| | **Swine farms**<br>Manure and urine from poultry and swine farms can be recovered from pig and poultry pens through grated floors, treated, and the solid residue can be used as a natural fertilizer | **Treated water**<br>Water can be separated from manure, treated, and used for flushing pigpens and for irrigation. See section 5.3.4.4 | **Methane**<br>Flammable methane gas is generated in the digestion process for manure, and can be used to generate electricity and heating. |

|  | See section 5.3.4.4. |  |  |
|---|---|---|---|
| **Hazardous waste from hospitals and health care centers** | Incinerate at high temperature. Ashes to be sent to dedicated landfills |  |  |
|  | **Solid items** ||||
| **Construction industry** | **Scrap bricks, concrete, asphalt, rubber, etc.**<br>Should be crushed and used as construction materials ||||
|  | **Lumber**<br>Either reused or mulched ||||
|  | **Steel, aluminium, zinc**<br>Sent to smelters ||||
|  | **Doors, windows, bathtubs, balcony grates, gargoyles, large lamps, etc.**<br>There is a good market for these items in the antiques business ||||
|  | **Electric wiring**<br>Sent to smelters ||||
|  | **Lead piping**<br>Sent to smelters ||||
|  | **Plastic piping**<br>If possible, palletize for recycling, although not all plastic is recyclable in that it is not always possible to return to the original components ||||

## 2.2.5 Waste and its effect on the environment

Why is humankind so concerned about the production and disposal of wastes? There are economic, social, and environmental reasons for this.
From an economic point of view, large amounts of money are spent hauling, transporting and disposing of waste, funds that could be used more efficiently in other activities; the generation of waste therefore involves a poor utilization of public funds.
From the social point of view, and especially in developing countries, waste is a health hazard. As for domestic waste, there is a potential danger of rodents and insects feeding from it to cause the propagation of serious infectious diseases. Industrial waste can also be a critical component in the deterioration of public health when one considers acid and alkaline discharges, particulates and other pollutants that contaminate the air, soil, and water.

Regarding the environment, waste produces a series of unwanted effects such as the leakage of heavy metals into the groundwater from landfills and from industries, the misuse of valuable land as landfills near cities, the contamination of rivers, the disappearance of aquatic life, the raw materials, fuels and contamination produced by transporting waste, etc.

It would be utopian to believe that society can function without producing waste or altering the environment. This is why levels or standards of contamination of the air, soil, and water exist and are used as yardsticks measuring maximum allowed pollution levels. Very well known standards refer to the amount of pollution measured in gram/litre (gm/l) or in 'parts per million' (ppm), or in 'parts per billion' (ppb), that as a maximum are considered sustainable, i.e. that can be permitted for long periods. International standards exist for volatile compounds, but their value often depends on the characteristics of the area where these measures are taken, the existence of winds, certain barometric pressures, etc. These indicators thereby reflect local values instead of global ones.

One of the most common standards used for water is the Biological Oxygen Demand ($BOD_5$), which measures the amount of oxygen needed to oxidize, burn or break down organic matter in five days. The higher the value, the more serious the problem, because this express that reduced amounts of dissolved oxygen are present in water, which is needed to support healthy aquatic life. This not only encourages fish migrations but the production of large amounts of gases.

Here too, the damage done to the ecology depends in part on the physical characteristics of the area. For instance, untreated sewage discharged into a narrow and swift mountain course of water has, on its way to the sea, more opportunities for oxygenation than sewage released into a river in a valley with low water speeds. The damage done will also obviously depend on the flow of the watercourse.

Table A.7.1 in section A.7 of the Appendix shows approximate allowed values for different discharges in water. Similar tables can be prepared for discharges into the atmosphere and soil; the Internet is an excellent source of information. When in a determined situation measures (thresholds) have been established, it is necessary to monitor the evolution of these values, because this is a dynamic procedure, not a static one.

Figure A.7.1 shows how the content of several contaminants have varied over a ten-year period. This underscores that, once thresholds are determined, it is absolutely necessary to monitor the evolution of contaminants, and to adopt required measures when these values diverge from those initially established. The ideal would, of course, be a declining curve for each contaminant, and the slope for each one would illustrate the effectiveness of the remediation measures adopted.

## 2.3 Hazardous waste

The Medical Dictionary Search Engine defines 'hazardous waste' as *"Waste products which, upon release into the atmosphere, water or soil, cause health risks to humans or animals through skin contact, inhalation or ingestion"*.

The problem people usually have is in identifying when a substance is a hazardous waste. It is a difficult issue, prompting a leading institution such as the United States Environmental Protection Agency (EPA) to provide extensive information for determining when a substance or compound is considered hazardous waste. In general, they define that a solid waste is hazardous if it shows one of the following characteristics: if it is ignitable, corrosive, reactive, or toxic (EPA, in Internet references for Chapter 2).

Probably the most famous (or infamous) hazardous compound created by man for industrial purposes, is Polychlorinated Biphenyls (PCB). This is a substance utilized since the 1930s for electrical equipment, especially transformers, and for hydraulic fluids, adhesives, and plasticizers, as well as many other uses; however, it was prohibited many years ago as it is a known carcinogenic. When deposited on the soil it makes its way to the water table, and when burned it produces dioxins and furans (section 2.5).
The chemical has also been banned in the EU and the year 2010 is a deadline to get rid of all PCBs in its diverse forms, and for decontaminating affected equipment. It appears that high temperature incineration is the method of choice for destroying PCBs; however, other alternatives also exist. In Canada, the compound is disposed of in two incineration plants, generating considerable outcry.

### 2.3.1 Vitrification

It is possible to vitrify some hazardous materials such as very contaminated soils and sludge, in a practice where these are subjected to a process at very high temperatures that destroys organic materials and keeps the melted remains inside a vitrified brick. This final product can have several uses or it can safely be dumped in a landfill.
The United States' EPA has published a report about this vitrification process that explains the particulars of the method (see 'waste vitrification' in Internet references for Chapter 2).

## 2.4 Recycling

This has tremendous momentum and is applied everywhere in the world: in households, factories, institutions, schools, etc. Its adoption is not uniform, however, since some countries have very much higher recycling rates than others. The main items normally recycled are paper and paper products, some kinds of plastic, aluminium cans and bottles. It is believed that at present in some countries recycled paper has reached 80 percent of total usage, however it is not possible to recycle 100 percent of paper because cellulose fibres lose their quality by shortening.

## 2.5 Incinerators

Incinerators are thermal devices for waste incineration. Incineration works at high temperatures and essentially converts waste into toxic gases, also producing toxic fly ash and ash containing different toxins. These devices can incinerate a broad range of municipal, medical and hazardous wastes; unfortunately, it is known that they generate a flue gas that is rich in particle matter, mercury, lead, dioxins, and furans.

Dioxins and furans are a family of chemical polychlorinated compounds created when there is an incomplete combustion of hydrocarbons in the presence of chlorine. Both are generally known under the general term of dioxins, and, as pollutants, they remain for a long time in the environment. Atmospheric currents transport and deposit them on plants eaten by cattle, where the compound accumulates within fatty tissues. Some companies have developed (see 'cleaning flue gas' in Internet references for Chapter 2) a sort of filter that treats this flue gas via 'adsorption' (see Glossary), and using inexpensive activated char.

For hospital waste, the incinerator comprises a series of operations that include a rotary kiln, a boiler (to recover heat), filters, a scrubber (see Glossary) and, finally, a charcoal filter to get rid of dioxins and furans.
Glass and metallic elements are usually recoverable from the ash of incinerator wastes after incineration, and these can be sent to appropriate smelters.
It would be ideal to produce enough heat from the incineration process to generate energy, but this is not always possible. Factors exist limiting this energy recovery, and one of them is the too large fractions of chlorine-containing compounds (Ron Zevenhoven and Loay Saeed, in Internet references for Chapter 2).

There is much discussion about the convenience of using incinerators. Some believe that incineration releases cancer-causing and other toxic chemicals from smoke stacks, including heavy metals (Rachel's Environment & Health Weekly # 592, April 2, 1998) (see Incinerators in Internet references for Chapter 2). Others, such as the New York Department of Environmental Conservation (see New York incinerators in Internet references for Chapter 2), which operates ten incinerators, are apparently satisfied with their results.

Incinerators produce ash that has to be eliminated, which can present some problems, mainly due to its metal content. Because at high temperatures ash is composed mainly of silicates, most of them insoluble although some are soluble. This means that when the ash is spread in a landfill leaching may occur. Rabl *et al.* (see Internet references for chapter 2), produced an extensive report about the installation of waste incinerators in six European countries. Switzerland has a law forbidding landfills, so many incineration plants have been built in this country.

As a bottom line and even considering its main drawbacks regarding emissions, it is safe to affirm that waste incinerators are making great strides in many countries, and will perhaps be the leading method for waste disposal. On the other hand, it seems that few options remain open since land for land filling is becoming very expensive; however, the location of waste incinerators in Europe appears to be a difficult proposition because of the high population densities and the closeness of incinerators to urban centres.
An inherent advantage of incinerators is that garbage volumes are reduced by 90 percent, and weights decrease by 75 percent; this is very important as it sharply reduces the need for land to build landfills, but naturally, provisions should be taken for the final and safe deposition of ashes. Another advantage is the potential utilization of waste heat to generate electricity and district heating.

Some industries produce highly contaminated liquid wastes. If their contents cannot be recovered then it is necessary to burn them in liquid waste incinerators, and in many cases, there is no need of fuel but for the initial start up. Dunkan Kimbro *et al.* from Franklin Engineering Group, and Phil Knisley *et al.* from Eastman Chemical Company, provide a good description of the particulars and operation of such incinerators (see Internet references for Chapter 2).
There is a viable alternative for liquid wastes, and it is their use in cement kilns. This is not a new idea, but some of its advantages are now emerging, and countries in Europe and in the Americas are considering it. Some advantages of using this system are:

- The high temperature needed for cement manufacturing, in excess of 1,400 degrees Centigrade. This makes for an almost complete destruction of organic components.

- Savings in fuel consumption, and recovery of the potential energy contained in the waste. According to Daniel Lemarchand (see Internet references for Chapter 2), 300,000 tons of fuel are saved annually in France because of wastes burnt in cement kilns.

- Many metals are embedded in the clinker (see Glossary).

- Because alkaline lime is a raw material for the production of clinker, its presence helps in acid neutralization.

- Cement kilns can accept liquid wastes such as solvents, paints, varnishes, etc.
  Because of their high temperature, cement kilns can be used to destroy hazardous wastes such as solvents and chemical compounds; several tests have shown that no significant hazardous emissions are found within the flue gas.
  However, it has also been found that cement kilns burning hazardous waste produce considerably more particles than kilns working without hazardous waste.
  Many cement kiln companies are interested in burning these residues due to their economic advantage, considering that 50 percent of operating costs for cement kilns are due to fuel. In some cases, there is the added benefit that cement companies are paid to burn these wastes. By the same token, some cement kilns burn tires.

As useful and convenient as the use of cement kilns for burning waste might appear, incinerators are believed to do a better job, as they are especially designed for the purpose, while cement kilns were not. Analysis of the flue gas from both cement kilns and incinerators shows that the former release larger amounts of hydrochloric acid, and of NOx, SOx and particulate matter, as compared to the latter. For more detailed information the reader might want to consult: http://www.geocities.com/~watchdogs/myths.html

This site refers to an EPA study called "*Myths and Facts About Protecting Human Health and the Environment: The Real Story About Burning Hazardous Waste in Cement Kilns.*" This is a Commentary on the Claims about Burning Hazardous Waste in Cement Kilns in the Publication entitled 'Protecting Human Health and the Environment', published by the Cement Kiln Recycling Coalition.

Section 2.5.1 illustrates a case of a waste incinerator that does not dissipate waste heat into the atmosphere, but uses it for the urban water heating system of the city of Göteborg, Sweden.

---

**2.5.1 Case study: Heat from incinerator and wind energy for Göteborg, Sweden**

Göteborg is Sweden's second city, with a population of about 500,000. It has an extensive district water-heating system that is more than 400 km long, which receives heat and electricity from a variety of conventional and non-conventional sources. These heating sources are all linked, and a management control system takes heat from where it is most convenient. A large proportion of the heat (60%) and energy needs is covered from waste heat and through energy from non-conventional sources. The balance is provided by hydropower and by boilers that are being converted to natural gas, with one of them able to burn biomass. The system's non-conventional sources of heat and energy are as follows:

- Heat from the **waste incineration** (section 2.5), plant in Sävenäs. This is the largest waste incineration plant in Europe, and heat is transferred through heat exchangers. There are also plans to use the flue gases from the incinerator.
- Heat from Shell **refineries.** This is waste heat from the processes that in the past released it to the atmosphere or the sea.
- Heat from **heat pumps** (section 5.2.8.1) at the Rya sewage treatment works. Four heat pumps extract heat from the sewage and deliver it to the district heating system.
- Heat from the **truck engine testing facility** at the Volvo Truck Manufacturing Plant.
- Heat from Chalmers University originated in their **research facility for fluidized bed** (section 3.5.1).
- **Energy and heat from bio fuels** (section 5.3.4).
- **Energy from wind turbines** (section 5.3.1).
- There are also two gas-fired plants whose main purpose is to produce hot water; however, these heat energy co-generation units (CHP) also generate electricity.

"Waste heat from the refineries and the incineration plant at

> *Sävenäs accounts for almost three-quarters of the district heating, together with heat from heat pumps at the Rya sewage treatment works".* (see the largest plant in Europe in Internet references for chapter 2).
>
> This undertaking is also a clear example of integration, and it is interesting to quote them stating that:
> *"The amount of energy supplied is about 10% less than the energy used. We use in other words more energy than what is supplied. This is possible through the reuse of energy". "Renova AB - the leading recycling company of northern Europe handles more than 600,000 tons of waste per year. More than 90 per cent of all the waste handled is recycled"*

## 2.6  Second part: Decreasing waste generation

The first part of this chapter was concerned with the present-day situation of waste generation and its disposal and treatment. This second part will refer to the future, where it is not a matter of finding ways to get rid of waste, but, rather, to not generate it. Accordingly, four actions are considered namely: reducing consumption, reusing, recovering, and recycling. The general idea is to try to mimic nature, as far as possible, in producing with a minimum of materials and reprocessing the wastes to use them again. For this reason, this section begins by comparing how nature proceeds, and how society behaves regarding waste. The comparison is useful as it shows society's weakness in its extraction-production-use and disposal systems, and points to potential ways of improving them. A road map will help readers to navigate through this section.

## 2.7  Nature's closed waste cycle

A closed cycle means that something is created, used, wasted, and then formed again in a perpetual and a kind of circular path along time. The most important closed cycles in nature are the 'carbon cycle', the 'nitrogen cycle', and the 'hydrological cycle'. The first one is responsible for the production of oxygen, the second for building organisms, and the third for the production of fresh water. Life, at least as humankind knows it, would be impossible without these three cycles.

Let us very succinctly analyse the carbon cycle.

The sun generates all the energy for the planet but man or animals cannot directly harness it; consequently, both use an intermediate: **Plants with green leaves.** Figure 5.2 in section 5.2.7 shows a very elemental diagram of this relationship (the photosynthesis process). Through their roots, plants take water from the ground and use their leaves to absorb carbon dioxide ($CO_2$) from the atmosphere. In a natural world, man and animals and even plants generate this $CO_2$ in the process of respiration.

Here is a summary description of this complex process: Green leaves possess a substance called 'chlorophyll' (see Glossary) that absorbs sunrays. Then, synthesis takes place with the production of carbohydrates (compounds of carbon, oxygen and hydrogen) to form sugar, starch and cellulose. Energy is stored in the plant in the form of chemical bonding between sugars and starches. As its 'waste', this whole process releases oxygen ($O_2$).
Sugar is converted to Adenosine Triphosphate (ATP), which the plants themselves use for growth and to store energy. In turn, humans and animals ingest these plants, taking advantage of that stored energy, and, together with ATP and inhaled $O_2$ from the air, produce water and carbon dioxide ($CO_2$), which is exhaled to the atmosphere and absorbed by the plants' leaves, closing the cycle.

Therefore, it can be seen how 'waste' from plants ($O_2$) is a fundamental input to humans and animals, allowing them to breathe, and how 'waste' from human and animals ($CO_2$) is also a fundamental input to plants, permitting them to use the sun energy's for their growth.

## 2.8 Society's open path for wastes

In the built environment — consider cars for instance — the engine takes oxygen from the air that oxidizes the fuel in the combustion process, releasing energy that is partially converted into mechanical movement, mobilizing the vehicle while releasing carbon monoxide (CO) and dioxide ($CO_2$), along with other gases as waste products. Green leaves can absorb the $CO_2$, closing the cycle, but CO and other gases are dumped into the atmosphere in a **linear or open way,** that is, in a circuit that is open since they are not recycled. Besides, as not all of the $CO_2$ released is absorbed by plants, it contributes to climate change by producing the greenhouse effect (sunrays that enter the Earth's atmosphere cannot escape from it), contributing to the warming of the planet. $CO_2$ is very efficient at absorbing infrared radiation from the sun, and then it prevents it from bouncing back into space; this process is helped by methane and water vapour.

Figure 2.2 sketches the mentioned linear path, or open circuit. Examples of how humankind is not following the closed cycle of nature abound. Consider these few cases:

- In mining minerals, a very large percentage of the ore is just waste, since metals usually constitute a small proportion of the material extracted. In the process, fuel is consumed, $CO_2$ is released to the atmosphere, water is contaminated, etc. These truly are different types of waste, since they are not reused, recovered or recycled.

- During manufacturing processes, considerable waste of raw materials, water and other vital elements usually takes place; although some are recoverable, until recently they were routinely released into the environment.

- After a product reaches the consumer, its life is usually shortened due to substitution for more updated and perhaps more efficient models; the product ends its life in a landfill, while there usually remains a lot of potential in it.

- Gases from processes such as those in coal-fired power plants release sulphur that combines with water in the atmosphere, generating sulphuric acid ($H_2SO_4$) — producing acid rain, with very severe effects on the environment. Since sulphur is not restored to its original state in coal, or used in other processes, this makes for another example of following a linear path.

These cases of dumping have a far larger effect that not merely closing a cycle. They often lead to very serious environmental and health consequences:

- Release by accident or leakage of large amounts of fluorocarbons originally used in air conditioning and cooling equipment. This affects the ozone ($O_3$) layer in the upper atmosphere, thereby destroying part of the natural shield against the sun's ultraviolet rays. Ultra violet rays are high-energy radiation that can penetrate the skin and alter their cellular metabolism, provoking skin cancer.
- When lead (Pb) in the air, in the form of lead tetraethyl, was used as an additive to gasoline (to prevent an effect known as knocking), the exhaust gases released into the environment that contained lead had very severe effects on several species. This led to the prohibition of the sale of leaded gasoline in many countries,

although unfortunately it is still available in many developing countries.

Compare the above examples with what happens in the natural world. Because of their metabolism, animals produce waste, which ends up as droppings on the ground; nature immediately starts a mechanism to use this waste. How?
Certain bacteria and other organisms begin a process of breaking down the organic matter, for instance by converting the organic ammonia in the waste into nitrates and other elements. Why is this conversion important? Because plants cannot absorb organic ammonia, so it needs to be decomposed into other chemical inorganic constituents, such as nitrates that can be used by vegetation. Therefore, the **waste turns into a nutrient**.
In turn, animals feed from plants, absorbing this nitrogen — which is essential for their growth — and completing the cycle. The nitrogen cycle in this very basic explanation elucidates why manure is useful as a crop fertilizer. Obviously, man is far from developing a mechanism that is identical, let alone one that is similar.

Figure 2.2 schematically shows that people extract and process raw materials, and then dump any remaining burden (in mining, agriculture, industry, fishing, etc). These raw materials are developed into manufactured products, and the surplus is then dumped (for instance in food processing). Finally, the user dumps the product at the end of its life (such as a TV set), unless some recycling takes place. It is obvious for everybody that **this procedure is not sustainable,** since there is no return to an original element.

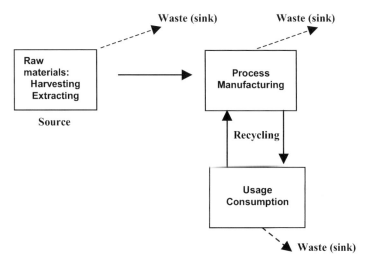

*Figure 2.2*      **The linear path of wastes**

## 2.9 Ecosystem metabolism and metabolism in society

In the first part of this chapter, waste was analysed, and different measures were proposed to deal with it better. Thus, these questions (section 2.1) were addressed:

- *What* is waste?
- *Which* are the components of waste?
- *Where* is waste generated? and
- *How* is waste disposed or treated?

These questions clearly fail to express the most direct strategy, which is: the best or most efficient approach to fight waste is not to generate it. Obviously, it is not realistic to expect that no waste at all will be made, as it is unavoidable for some. So the statement should perhaps be modified to say: the most efficient approach is to generate the least possible amounts of waste. The logical step to take in launching into this complex problem is to ask:

## 2.10 Why is waste produced?

To address this query it will be useful to explore the notion of 'metabolism'. Just what is 'metabolism'?

*"Metabolism is the sum of chemical processes in a cell or organism, by which complex substances are synthesized and broken down, and growth and energy production sustained".* (Shorter Oxford English Dictionary Volume 1).

Any living organism abides by a metabolic course of action, in that an ingested substance follows a series of processes necessary for chemical changes to take place in its living cells. A community, a society as a whole, can also be considered as resembling a living organism, and is thereby subject to similar metabolic processes. That is, it 'ingests' goods and services, 'digests' them to sustain life for people and animals, and thereby produces other goods and services necessary for people to subsist, interact, and develop activities (such as transportation, education, work, entertainment, housing, etc.

Therefore, from the process point of view, a correspondence may be established between **metabolism in an ecosystem**, and what could perhaps be called the **metabolism of a society**. Table 2.6 shows that the dictionary's definition lends itself to drawing this loose comparison between a society's metabolism and that of an ecosystem.

> **2.10.1 Case study: Generating light out of garbage, Groton, the USA**
> Data taken from "Landfill methane outreach program" (see Internet references for chapter 2).
>
> The methane gas which consists of carbon and hydrogen ($CH_4$) produced by the city landfill in Groton, Connecticut, USA, is utilized to power about 100 houses. Usually, in this type of undertaking the gas is cleaned and used in diesel-generator units or gas-fired boilers to produce steam for turbine-driven electrical generators.
> In this case, the methane gas, is used to power a fuel cell that directly converts hydrogen into electricity, heat, and pure water (section 5.3.5). The system's secret is to remove contaminants in the gas, such as sulphur and halides, with a gas pre-treatment unit (GPU). The system was tested for six months in a landfill in Penrose, Sun Valley, California, and the plant was then relocated to Groton — with the cell producing 140 kW for a utility company.
> Besides the economic benefit of saleable electricity, this cell saves money because the municipality does not need to have installations to flare the landfill's escaping methane. It is known that methane is the main gas producing the greenhouse effect, so its consumption reduces this danger into the bargain.
> The Landfill Methane Outreach Program (LMOP) can be applied to hundreds of municipalities anywhere in the world, and the EPA has developed fact sheets and manuals to assist with that aim.
> For more information, consult:
> http://www.epa.gov/lmop/products/fuelcell.htm

*Table 2.6*     **Metabolism in an ecosystem and in society**

| Metabolism in an ecosystem | Metabolism in a society | Comments |
|---|---|---|
| Sum of chemical processes | Sum of chemical, social and environmental processes | The etymology of the word 'metabolism' is 'change', and, indeed, a change applies to both nature and society |
| In a cell or organism (protoplasm) | In a cell or organism. (household, industry, school, health centre, commercial entity and industrial corporation, etc.) | In a similar way that changes take place in the protoplasm, changes occur in society's 'cellular' organisms |
| Complex substances: | Complex substances: | Notice that raw materials |

| | | |
|---|---|---|
| Ingested substances ($O_2$, food, water, raw materials) | Ingested substances ($O_2$, food, water, raw materials) | are also included in the natural world. Examples are birds, building their nests with twigs and brushwood |
| Substances are synthesized (production of sugars in plants) and broken down (that is cells destroyed) | Substances are synthesized or broken down (synthesis of chemicals, manufacture of products, etc.) and broken down (as in fuel combustion) | In the natural world, inputs are not only meant to sustain life; bees use raw materials (flowers' nectar) to manufacture something (honey), as in a human industry. The same applies, of course, to society. |
| Growth (animals, plants) and energy production sustained | Growth (economic, social growth) and energy production sustained (as in power plants) | Growth in the natural world involves animals and plants. Same in society, including in addition economic growth |

However, the resemblance ends here, because of differences in the way the processes are executed. In nature, everything is efficient and is used and consumed just as needed, and wastes by one organism are generally food or input for another, in a sort of continuous circular cycle. See figure 4.2.

It has been already mentioned in section 2.9 and Figure 2.2, that humankind does not follow this efficient process. In Figure 2.3 society can be seen as a sort of black box attracting raw materials and releasing wastes. A closed cycle is not achieved, as wastes are just dumped back to the environment, unless they are recycled — and not even then.

A question arising now is: how is this waste created? The answer to this can be found in the actions within the black box.

Let us analyse each of them:

- Unlike nature, humankind consumes many superfluous goods and services that are completely and utterly useless, to say the least. Weapons production, war gases, and military equipment are probably the most visible examples where the quantity of waste, and not only in people's efforts but also in raw materials, fuels, etc., beggar the imagination. There are also countless and as worthless gadgets, contraptions and products that consume valuable resources that people buy because they are attractively packaged and displayed on a shelf, and helped by advertising and marketing drives, but whose utility is nil. Of course, most of them end up in a landfill, a sink, and with them go the materials used for their manufacture, lost forever.

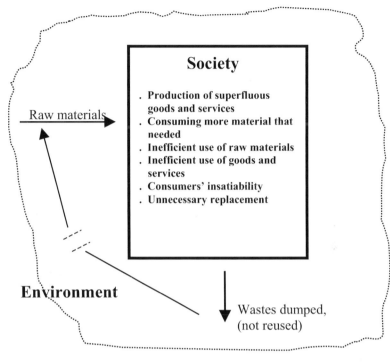

*Figure 2.3*  Metabolism in society

- In general, humankind uses many more goods and services than necessary for its functioning. It materializes in different quantities of inputs by different human nucleus. Some societies consume exactly what they need to survive (as in a lean human body), while others consume an excess (as in a stout human body). In other words, there are affluent societies with an excess of consumption, and poor societies with a deficit of goods and services. Of course, this excess of consumption translates into an inefficient and unnecessary use of resources.

- The way materials and energy flow from their extraction and generation until their final disposal are addressed in section 4.12. Among other concepts, the ratio between weight of raw materials and that of the final product come under scrutiny. Plenty of waste is produced here, and it is said that 90 percent of raw materials do not reach a product's final stages. Again, even if a product is made with the optimum ratio still does not mean that it is worth manufacturing, for humankind is still 'ingesting' many unnecessary things, or producing and consuming worthless products.

- The human race processes inputs in a very inefficient manner, without extracting the whole value of a product. One everyday example is found in many restaurant kitchens, where because of inefficiency there is a waste of vegetables, meat, milk, etc., due to poor food preparation procedures, creating mountains of perfectly good product that ends up in landfills. The utilization of fuels is another example, since man uses only a fraction of their available energy. A common example is the car, which only makes use of 30 percent of the caloric content of gasoline.

- Packing is another great culprit. Even when aluminium cans are recycled, there is a massive energetic waste when the amount invested to manufacture a can (the container) is compared to the energy provided by the product it contains (the soft drink) (section 1.8). Compare this with an apple, for instance, where almost 100 percent of the product can be consumed (pulp and skin) — that is, both container and content.

- Consumer's insatiability, fed by relentless advertisements, peer pressure, and lack of concern, also accounts for a large percentage of goods found in curbs — and, finally, in landfills — for items such as furniture, tools, TV sets, computers, etc., that the consumer society throws away because new models appear on the market, or with more power, or simply because people want something new. People do not bother to repair enough things. It is a fact that in developed countries it is hard to find repair shops for, say, a TV set, or a vacuum cleaner. These are scarce because people prefer to buy new ones, which is probably cheaper than repairing the old unit.

Clearly, none of this has a parallel in the natural world. All the above issues produce not only unnecessary waste, but, besides, items are seldom recycled and end up in landfills. **These examples address why waste is produced, but also provide a hope about solutions, since there are many measures that can be taken to avoid this waste.**
This raises the last question that needs to be addressed, namely:

## 2.11 What can be done to correct this situation?

This is hard to answer in that it calls for changing people's mentality, as it is a cultural issue whose deep roots also reveal their dependency of the economy. This is why this behaviour is called in this book the 'culture of waste'.

The economy of a society also determines this sort of performance, which is common in developed and affluent communities. Latin American countries, for instance, have completely different patterns of consumption and waste: in general, their products are used longer, through successive repairs and maintenance. Education (section 3.9) also has a prominent role in this issue.

Of course, it is impossible to think of a society without waste, as there will always be refuse, but it is one thing to have reasonable, normal, and unavoidable levels of garbage and quite another to have indiscriminate and unnecessary generation of waste. For instance, at the end of their life, the items that are really wasteful at the sorting point — such as inedible food, clothing, and bones — can be incinerated, and their ashes can be used as construction material. Thus, even if there is waste, its effects can be minimized, making a large landfill unnecessary and thereby preserving valuable land from frequent leakage of dangerous substances into aquifers. Yet the way to decrease the importance of this problem is by rationalizing production and reducing consumption, that is, by conserving resources.

## 2.12 Conservation of resources

Solutions to save resources — any kind of resources — for the future should contemplate the development of **policies** at a decision-making level, together with everyday **actions** at a personal level.

Figure 2.4 is a road-map to prompt understanding the measures that need to be adopted that consider both strata. Its information is based on issues discussed in:

- Sections from 2.12.1 to 2.12.6 (for policies, shown between brackets).
- Tables 2.7, 2.8, 3.3 and 3.4, show different strategies or measures suggested to reduce consumption of raw materials, water, energy, and reduce land use (as everyday, individual actions).
- Reengineering processes.
- Renewable electric sources.
- Table 2.5, includes measures to be taken for present-day recycling.

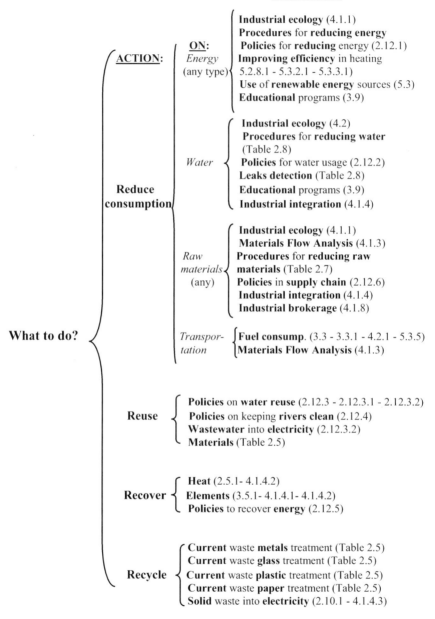

*Figure 2.4*     Roadmap for 'What to do?' procedures

The roadmap considers four different actions, so it is important to clarify the meaning and scope of each of them.

> **Note:**
> **Reduce:** Means manufacturing something, to reach the same or an improved final quality, while using less direct and indirect materials. For instance, laptops are much lighter now than years ago, because of the use of lighter components, more efficient and lighter batteries, etc.
>
> **Reuse:** This means that something is used again for the purpose it was built, and without chemical changes to its structure. For instance, a refrigerant works in a closed cycle as its chemical structure is always the same, but it changes its physical properties since in one moment it is a gas and immediately after it becomes a liquid. As a result, its usefulness as the same substance is infinite.
> There are plenty of examples, such as any second-hand items, from cars to houses.
>
> **Recover:** When something can be extracted from waste. For example, sulphur can be recovered from a flue gas. Of course, combustion makes impossible to reuse the fuel burnt, but some materials and elements contained in it can be extracted. Out of a worn tire it is also impossible to get all the elements that entered its making, such as pure hydrocarbons and sulphur, although the energy content of the original hydrocarbons can be recovered.
>
> **Recycle:** This occurs when the elements of something are broken to its elements yet they can be used again to manufacture the same items, or others. This is the case of aluminium, steel and metals in general; once the material is melted down, many different things can be produced with them. Some plastics, for instance, allow for obtaining their **original components** for any kind of use, yet this is not the case with other kinds of plastics, such as plastic supermarket bags, which can be recycled to render another product, but not a new plastic bag.

A question that this probably raises is: Which of these is the most noteworthy?
The list has been cast in decreasing order of importance, such that the most significant is Reduction, while the least is Recycling. Why?
As reduction involves the use of less material, indicating a lesser cost to the environment. Recycling, by contrast, while positive, always requires the consumption of resources in transportation, energy, etc.

## 2.12.1 Energy reduction

When energy in any form is lost without producing work or contributing to something, it vanishes and is thereby unrecoverable. Unfortunately, even the best equipment does not facilitate using or recovering a hundred percent of the energy contained in a fuel — not even the human body can do that — although many steps can be taken to minimize the waste. The following is a short list of policies at the decision level and for energy reduction. Note that noxious gases are not only harmful but also that they need energy to be produced, which is a complete energy waste.

**Suggested policies**

- Encourage research into the use of non-conventional energy sources; for instance, the potential conversion of electromagnetic radiation from the sun into usable energy.
- Support — through taxation or subsides, for instance — the production by communities, farms and rural homesteads of renewable energy sources, such as wind, photovoltaic, small hydro, biomass, fuel cells, etc.
- Promote research into fusion energy.
- Enact legislation for zero emissions in cars and trucks.
- Pass legislation for oil companies to modify the way gasoline prices are calculated, by incorporating into the economic costs and taxes the hidden costs due to damage to the environment. The cost per litre will probably more than double, but producers and consumers should share the added cost.
- Encourage car pooling and the construction of parking lots in mass transit stations.
- Increase taxes on drivers' licenses and car licenses.
- Stop construction of urban highways.
- Establish prohibited mobile traffic zones, save for taxis and buses, within downtown areas, as many cities around the world do.
- Dedicate one lane on highways for cars with more than two travellers — another common practice in some cities.
- Endorse the use of new types of lighting in households, and persuade people to switch off lights when a room is vacated.
- Urge people to make wise use of energy at home by planning their laundry and ironing times.
- Promote the use of methane from landfills for energy generation.
- Pass legislation about minimum levels of insulation in houses.
- Install solar panels in public buildings, and promote through taxation incentives their use in households, industries and commerce.
- Establish local bylaws prohibiting lights left on in public buildings during the night and on weekends.

## 2.12.2 Controlling water usage

**Suggested policies**

- Establish metered water systems for households, with differential prices for high consumption.
- Promote the need for people to save water.
- Establish maximum limits for water losses in the water network.
- Encourage the use of mulch in city parks to keep the ground humid, thereby decreasing the need for sprinkling.
- If open channels are used in a city's green belt for irrigation purposes, have them lined to avoid losses by seepage.
- Enact bylaws for water recycling for each industry, with a program to progressively restrict the amount of water drawn from the water network, wells, rivers, ponds, etc.
- By means of bylaws enforce the need to use treated water for flushing toilets in large buildings and hotels.
- Make the construction of water treatment plants mandatory for large industries, and the recycling of wastewater.
- Constitute an independent national body to determine the maximum allowable amounts of water to be extracted from all sources by each municipality — on an annual basis — in accordance with local characteristics. Instruct municipalities to plan building reservoirs to store rainfall water and its subsequent use for certain municipal activities. The fire hydrants network should be connected in parallel with these reservoirs.
- Encourage research to decrease water consumption in industrial activities. Industries should be convinced that, when the cost of purchasing raw water and its treatment and discharge are considered, they will find that using less water makes economic sense since it is profitable.
- Encourage municipalities and industries in the region to collaborate in the use and reuse of water and wastewater (see actual example in section 2.12.3).
- Some cities receive water from distant sources. The water can be transported using open channels or aqueducts or by pipeline. If an open channel is used, and if it traverses arid and dry zones, a part of the water will evaporate. This circumstance makes it advisable to use a pipeline system. Also, if there is a difference in altitude between the source and the city (water head), whatever system of transportation is used, that circumstance can be used to generate electric energy through a hydraulic turbine.

## 2.12.3 The use of water for industry and the reuse of wastewater

**Suggested policies**

Industrial wastewater has been one of the most damaging factors affecting the environment, and is usually a critical item as industry generally requires large amounts of fresh water.
Different methods can be used to save water (water efficiency), as for example in the following actual cases:

- **Reengineering a process** such that less water is utilized. For instance, one computer industry has replaced its water-cleaning process with another chemical procedure, saving more than 230,000 litres of water per day. It is assumed that this particular type of industry can save more than 40 percent of its water use with new processes, and when one considers that a large plant can require about 450 million of litres of water per month, this percentage assumes great importance.

  This need for reengineering is dramatically revealed by studying the amounts of water to produce similar products in developed and developing economies. In the latter, the amount of water is many times greater than what is needed in developed economies. One example cited by Don Hinrichsen *et al* (see Internet references for Chapter 2) reports that the amount of water to produce a ton of steel in China ranges from 23 to 56 $m^3$, as compared with 6 $m^3$ per ton in the US, Japan, and Germany. China also uses 450 $m^3$ of water per ton of paper, twice as much as European countries.

- **Reusing wastewater.** Equipment now exists to dry sludge and get pure water out of it. Industrial water contained in sludge and slurry can thereby be used many times over, making unnecessary any costing treatment, also raising the possibility of finding commercial uses for the dry matter (cake). Many industries can also recover wastewater using the reverse osmosis process (see Glossary).

  Sometimes industries make joint efforts with local municipalities to use and reuse water. The City of Harlingen, in the US, provides an example with an agreement of this kind between a large manufacturing plant and the municipality. The same amounts of water are used by the municipality and the plant, which now discharges it clean for re-use (see 'Texas Water Resource Institute', in Internet references for Chapter 2).

- **Sequential use.** Some industries produce steam and hot water, which can be used for other processes, for heating, etc., which saves water and energy.

- **Using water** only when necessary, through the use of sensors that stop flows when no product is present.

- Detecting and eliminating **losses** in pipes, tanks, and equipment.

- **Education programs** for personnel working in industries to save water, and rewarding ideas that conserve water. This should be mandatory, just as safety courses are in many industries

*2.12.3.1 Examples to follow in water reuse*

The EU is actively seeking out best practices to reduce or even eliminate wastewater in industrial production; for instance, in paper and board production. This industry uses large amounts of water, releasing processed water as effluents into watercourses, and compensating with the addition of fresh water into the production circuit. In other words, here is an example of a dissipative use of a natural resource, and another example of the linear flow instead of the circular flow model.

To solve the problem they have developed a methodology called the Paper Kidney Project, reusing the discharged water — thereby reverting to the closed-flow model — and avoiding the use of fresh water. The project claims increased productivity, as well as savings in energy. (See 'Towards zero effluent water systems in the paper and board industries' in 'The path towards sustainable industrial production....' in Internet references for Chapter 2.)

Another good example is posed by an IBM division that manufactures data storage systems for mainframe computers. In this case, wastewater is purified and used in cooling tower makeup (that is to compensate for water losses). Brown and Caldwell (1990; see 'Water reuse' in Internet references for Chapter 2) claim a reduction of 100 million of gallons per year in fresh water use requirements (or 378 million litres).

### 2.12.3.2 Case study: Wastewater contributes to maintain a renewable resource: The ingenuity of a town, Clearlake, the USA

*Mark Dellinger (see Internet references for chapter 2) most appropriately calls this project 'A rare example of a genuinely sustainable energy system', as indeed it is. People in this area of California put into practice the meaning of sustainability, namely, a joint effort involving everyone that addresses social, economic and environmental issues at the same time.*

<u>History:</u>
*The three small towns of Clearlake, Lower Lake, and Middletown are near the south shores of Clear Lake, the largest body of fresh water in California, northeast of San Francisco. The region is a tourist destination and its water keeps a temperature of about 25°C, which makes it ideal for nautical sports. Clearlake is the largest city in the Lake County, with a population of about 13,000.*

*This area has intensive geothermal activity, which is used to generate electricity, taking steam from geothermal 'reservoirs' in the earth's crust: at a depth of about 450 meters and a temperature of around 180°C. The steam drives steam turbines, which in turn run electric generators. The operation began in 1960, when the population was small and no problems existed with the supply and consumption of steam; i.e. the plant consumed less steam that the amount produced in the depth of the crust.*

*However, population growth demanded more energy, and consequently more steam. The result was a decline in steam production starting in about 1980, because the steam demand rate was greater than the water recharge rate in the underground reservoir or steamfield.*

*This natural system has a limit or threshold (section 6.11) measured in units for flow of steam, and while consumption was less than this limit, steam production was steady. When the consumption increased, there was a decrease in the carrying capacity of the steamfield (section 6.12), that is, its capacity to produce steam decreased due to a lack of water, so steam production began to decline. This is an excellent example of what happens when this threshold is breached (section 1.10).*

*This decline is a common phenomenon in this type of operation. The reason is that while the geothermal temperature remains constant, fuelled by the Earth's core, the quantity of water that comes in contact with these very hot rocks is not constant since it reaches the steamfield from fractures and cracks in rocks, and its volume depends on many factors.*

*With this problem in mind, would it not be natural to condense the steam after it has worked in the turbine, and then re-inject the resulting water into the ground in order to repeat the heating process? However, it is necessary to consider that condensation requires cold water, which is generally unavailable. Even if it comes from a large body of water, such as a lake, the consequences of discharging this hot water back into the source from where it was taken usually has severe effects on the ecosystem. The solution is to use air to condense the steam through cooling towers, although these devices provoke a huge loss of water by evaporation.*

*At the same time, the area had another problem that was completely unrelated to the decline in steam production. The natural population growth supplemented by increased tourism naturally produced more wastewater. The area had water treatment plants, but during the peak season its capacity was exceeded, and improperly treated water was being discharged into the lake — which of course was unacceptable.*

*The brilliant solution to these two problems was to treat the wastewater, although probably not at the formerly high levels, and then to use it for the injection process. This way the contamination of the Clear Lake by contaminated water ceased at the same time as the steamfield was recharged with this wastewater, ensuring the continuation of the process.*

*It can be appreciated that this solution is sustainable because as long as there is population, there will be energy consumption with all its benefits, social and economical. At the same time, as long as populations increase and demand more energy, more wastewater will be produced and re-injected into the steamfield.*

*There is no damage to the environment, and the energy generated continues to be renewable. Besides, the condensed hot water can be utilized of heat the area's houses, public buildings, hospitals, etc., producing an additional savings in electricity and in the consumption of fossil fuels.*

Figure 1 of the cited publication (see Internet references for Chapter 2), provides a very good synopsis of 'Suggested water recycling treatment and uses'. This paper also has information about wastewater usage in a large building complex that uses a double plumbing system. It is interesting to note that, according their figures, the cost of providing this dual system added only 9 percent to the cost of plumbing. There are also various actual examples on this subject in this publication.

### 2.12.4 Keeping rivers clean

**Suggested policies**

Humankind has managed very efficiently indeed to pollute rivers and underground water, the arteries and veins of the planet and a fundamental component of the hydrological cycle (see Glossary). However, keeping rivers clean should have the highest priority for many reasons, including that they are the usual source of fresh water for cities, industries and irrigation; they are the natural habitat of thousands of fish – which also have great ecological, commercial and sporting value – and they are also the habitat of birds and other species.

Many industrial plants discharge to a municipal sewer, but many others release almost untreated wastewater into surface waters. Because of this, many rivers have on their beds a thick sludge or mud produced by the deposition of organic matter and heavy chemicals, making dredging a priority, although some researchers claim that it would be dangerous for the environment to remove this type of sediment.

The great Rhine River, flowing through four countries with very heavy industrial activity and having at its basin a population of 50 million, was known in the 1980s as the sewer of Europe, as it was contaminated with chemical, industrial, and agricultural wastes. The water was undrinkable, swimming was very dangerous and the fish were gone. The four countries with shores on the river agreed to clean it and its tributaries, and the best indicator of cleanliness that they could think of was the **return of the salmon** (which had gone by 1958).

The target date was the year 2000, which is why the project was called Project Salmon 2000. The job is not finished yet since there remain large quantities of heavy metals, but not only have the salmon returned and are breeding, but another 40 species of fish are now also reproducing in the river. It is perhaps by chance that the city of Seattle, in the US, developed a similar project called 'Sustainable Seattle', and is using the same indicator, namely, the return of the salmon.

Is this a coincidence? Perhaps, but this is a true indicator of sustainability since the return of fish means that the water is clean **environmentally,** and, thus, as in a domino effect, this will facilitate **social** sporting activities and pose a lesser health risk to drinking water. At the same time, it also means **economic** benefits derived from the commercialization of the fish and related industrial activities. This also indicates, of course, that industries discharging their untreated wastes into the rivers were prevented from doing so.

This provides ample proof that very large undertakings, such as the Rhine River's 1,300 km, can be accomplished at a very high monetary cost indeed, although we would simply be undoing decades of mismanagement.

Since many industries discharge their wastewater in rivers, the lack of suitable treatment and especially the absence of environmental regulations and policies transformed rivers into open sewers.

Contamination of rivers comes, among other things, from:

- **Municipalities**
  Untreated or poorly treated wastewater from domestic sewage very often mixes with rainwater, which carries filth from the streets, and sometimes from industrial wastewater. The danger posed by untreated wastewater is measured by its $BOD_5$ (see Glossary). Because it uses the oxygen in the water, it deprives fish of this vital element. When the oxygen is consumed, the river dies and emanates odours and harmful gases. River contamination is also produced by leaching of noxious products from defective or damaged landfills that contaminate ground water by discharging into nearby rivers.

- **Chemical industry and pulp and paper industries**
  These plants usually draw water from a river and discharge improperly treated or non-treated wastewater and non-biodegradable wastes, while some industries such as tanning plants release large quantities of chromium and other heavy metals. An additional effect of industrial plants drawing water from rivers and lakes is the killing of fish sucked into the water intakes.

- **High-tech industries**
  These are also large consumers of high quality water, such as for instance the semi-conductor industry. Wastewater can end up loaded with heavy metals.

- **Food industry**
  Known to discharge sizeable contents of organic matter, for instance, by plants processing animal foods such as sausage manufacturers and meat-packing plants.

- **Mining**
  Especially damaging are operations of coal with high sulphur contents.

- **Agriculture**
  Produces thousands of tons of nitrogen, and phosphorous from herbicides, fertilizers, and liquid manure.

Although notable improvements have occurred in many countries, in others the industrial discharge problem persists mainly because of old installations, large concentrations of industries (as on the Vistula River, in Poland), lack of enforcement, and the large capital costs involved for water treatment plants. Waters are often so polluted that practically no fish can be found. Naturally, there is no easy remedy to this problem, since few industrial plants can afford the cost or have the room to build water treatment plants.
Another effect of industrial plants is the discharge of hot water into surface waters and lakes because of the use of river water to condense steam.

To prevent the contamination of rivers, these measures are recommended:

- Prohibit the discharge of domestic and municipal wastewater into rivers and oceans that does not meet stringent environmental regulations.
- Periodically monitor the number and species of fish in rivers and oceans, and their content for poisonous substances such as mercury.
- Establish thorough quality controls of river water in order to detect the origin of any eventual contamination.
- Make the construction of water treatment plants mandatory for large industries, as well as the installation of adequate devices to control the wastewater discharged by small industries.
- In large chemical plants built along rivers, establish safeguards to avoid the accidental spillage of chemicals into those rivers.
- Make certain that any type of commercial, industrial or leisure navigation traffic has adequate tanks for storing their wastes (not so long ago, kitchen as well as toilet wastes were discharged into the water).
- Enact bylaws prohibiting navigation of oil tankers not built with double hulls (to avoid disasters such as those of the Exxon Valdez, Prestige, and others).
- Approve only shipbuilding construction plans for vessels with double hulls.
- Prohibit the washing of oil tankers and the discharge of water into a river or the ocean (this, incredibly, was a common practice), and do the same as regards the cleaning of bilges.

- Keep a severe periodic control of industrial plants and check the quality of their discharged water.
- Limit the amount of $BOD_5$ for discharges into water sources.
- Limit the amounts of phosphorous and nitrogen in wastewater, and regulate temperature thresholds.
- Prohibit the discharge of water containing even small traces of chloride from pulp and paper plants.
- Survey channels and creeks on agricultural land than may be used to provide runoff of water loaded with contaminants such as pesticides and fertilizers.

### 2.12.5 Recovering energy from tires

**Suggested policies**

Why is an entire section of this book devoted to 'waste' from vehicle tires? Because used tires still account for a large amount of the waste now held in landfills. In the EU, about 46 percent of used tires end up in landfills, and the figure for the US is 38 percent (Kurst Reschner, Internet references for Chapter 2).
There are various ways to get rid of scrap tires, other than placing them in landfills — **an utterly senseless thing to do.**
Why?

Because tires are made out of rubber (from a tree) and petroleum compounds such as butadiene; they are artificial products, and have the distinction of having two lives.
The first life is to fulfill the purpose for which they were built, namely, serving as wheels for a vehicle to roll on. When a tire wears out to the extent that it no longer serves the purpose of driving safely, it can start on its second life, which is to return the hydrocarbons (fuel) that were used for its manufacture. This is done by burning the tire to generate steam and then electricity. That is to say, a tire's second life is to be a fuel.

Think of it this way: money and effort goes into extracting oil and purifying it, and then more goes into using that to make a tire. Then, when the tire is dumped into a landfill, it is much like burying again those hydrocarbons that were extracted as crude oil. Does this make any sense? Most certainly not. Tires have a high calorific content, measured in BTUs (see Glossary): higher, in fact, that some kinds of coal, so it is well-suited for use as a fuel through a controlled process called TDF (Tires-Derived Fuel). Tires can replace thousands of tons of fossil fuel when burned in cement kilns or in special boilers to generate steam in large installations, such as pulp and paper mills. The recycling of tires is possible, but the recycled material can participate only to a small extent in the manufacture of new tires.

Added to this is the hazardous nature of tires, for when they are stored in large piles they can generate devastating fires that can take many months to extinguish, and that can cause immense damage to the environment through their fumes.

Many fires have begun due to tires. One of the most serious happened near Westley, California, when a site holding about 5,000,000 tires caught fire due to lighting striking on September 22, 1999. It took almost a month for the firefighters, using very modern equipment, to put out the burning.

In short, scrap tires are not waste. They can be used as input to generate electricity as well as for other uses, including as filler for road construction, and as elastic media between a pavement and manholes and railway tracks embedded in it.

### 2.12.6 *Savings in the supply chain*

**Suggested policies**

The only way to save raw products and materials in a supply chain is to reuse or recycle them. This is not a new idea, but its implementation is far from easy since the problem is to develop a practical and economic methodology to do it. The term 'cradle to grave' policy is apt in this connection, since it involves manufacturers being responsible for the **whole life of the product** manufactured by their company; this confers responsibility to the company that does not end with the product's sale. There is also an approach that says manufacturers retain ownership of a product or its materials, while consumers have it only on a sort of lease.

Richard L. Ottinger and Mindy Jayne (see Internet references for Chapter 2), comment about this, saying that, *"Interface, the world's largest carpet-tile maker, estimates it cuts its materials flow by about tenfold by leasing floor-covering services instead of selling carpet and by remanufacturing old carpet"*.

The cradle-to-grave notion probably derives from the fact that human activities inevitably go from their cradle to the grave with no return; therefore, on a human scale, the obvious linear path begins on a birthday and ends with death.

This expression is common in the technical literature, and no doubt the reader has already come across it. However, in this author's opinion, it is a misnomer, because when it comes to products it refers to precisely the opposite of what humankind should be doing. That is to say, society does not want the 'life' of a product to finish when it is no longer needed, or if it becomes obsolete or damaged beyond repair — which, by the way, is what people have been doing for decades. Thus, resources should not be 'born' (extracted or harvested), 'live' (used) as products, and then 'die', (i.e. dumped in a landfill). Quite the opposite, they should be used repeatedly in a closed cycle. Perhaps the expression 'cradle-to-cradle' would be more appropriate.

Irrespective of the choice of words for this cradle-to-grave concept, the principle behind it is sound, and the purpose of this section is to analyse how this principle, that in this book is called 'back to the consumer' policy, ought to work.
In fact, this concept has been around for many years with very good results for some items. Take, for instance, car batteries, which are made of lead, sulphuric acid, plastic separators, and a plastic casing. It is a very simple and reliable device used every day by millions of cars, trucks, farm equipment, etc. It is also accounts for more than 60 percent of the usage of lead.

When, after years of service a battery 'dies' (i.e., its electrochemical properties not longer hold), a car owner needs to purchase a new one, and usually the old one is kept by the garage doing the change. Because of this, the garage collects a large number of batteries that are then sent to a place where they are stripped and their elements recovered. That is, the lead is re-smelted, the casing is used to make new casings or other products, as with the plastic separators, and the acid is sent to a chemical company for its recovery. Even in cases where owners themselves change the battery, it can only be disposed of by transporting it to an appropriate site — just as the garage would — since, if left on the curb, it would not be collected by garbage trucks.
For this reason — and note that there is a regulation prohibiting the curb-side collection of batteries — recycling rates are very high, probably better than 90 percent, making batteries the most recycled item in society.

The reader will notice that this procedure allows for the indefinite use of some resources, such as the lead and plastic, although not 100 percent of them, as there are always losses that have to be compensated for with virgin materials. Even so, lead, plastic, and acid are used again and return to the consumer. This is also the idea behind the cradle-to-grave concept, so a computer manufacturer has to take back the computer it made years before, to try to recycle its components (the case, plastic, circuits, glass, wiring, fans, etc.) in the best possible way. The direct consequence of this policy is for manufacturer to take pains in developing products so they will be easily disassembled, thereby allowing for upgrades.

There is no doubt about the usefulness of the concept, nor that it is not an easy procedure to implement, especially with complicated products where many different products and suppliers intervene in what is called the 'supply chain', which can be very long and complicated.
For example, Figure 2.5 sketches the general idea behind the 'Back to the consumer' concept through a simple post-consumer item like an electric blender used for home food processing.

This example starts with a **customer** purchasing a new blender, but getting a discount by delivering the old one, or any other small appliance. The **business** receives and transports the article to a depot maintained by a small **appliances manufacturer**, who breaks down the unit and separates its main components. On the assumption that nothing can be salvaged for reuse, that manufacturer sends the electric motor and its shaft to the **electric-goods company** that made it, and the plastic casing to the **plastic moulding company** that had supplied that element.
The electrical manufacturer would break down the electric motor into its components, electric coils and the shaft, and sends the first to the **copper wire manufacturer** who supplied the copper wire, and the shaft to the **machine shop** than had supplied that component.
The copper wire company would then deliver the material to a **copper smelter**, the machine shop sends the shaft to a **steel smelter**, and the plastic moulding company shreds, grinds and washes the casing and sends the resulting granules to a **plastic recycler**. Of course, an economic gain occurs in each of these transactions, as well as an economic gain for the smelters also, since they would be purchasing these materials at a lower price than virgin materials. Naturally, some virgin material might be needed to improve the performance of the molten materials in order to render a fine new product.

When the two smelters and the plastic recycler finish their work on the primary components, the process goes into reverse, with a new blender ending up in the hands of a **new customer**. Naturally, this example is an elementary one, and many more things need to be ironed out for it to work; but when it does, the elements that made up the old blender will be back with the consumer, decreasing the rate of extraction of raw materials and using up less landfill space.
This policy can be also adapted to more complex and diverse products, such as cars. Until now only the EU has made it mandatory for manufacturers to take back cars at the end of their life. Regarding tires, the issue has been already examined in section 2.12.5.

An EPA bulletin ('Vehicles', in Internet references for Chapter 2), mentions that DaimlerChrysler's CARE Car II Project is focused on achieving the goal of having vehicles that are **95 percent recyclable by 2005.** It also reports that, "*...in an earlier demonstration, the company worked with 27 suppliers to retrofit two Jeep® Grand Cherokees with 54 recycled plastic parts*". The reading of this item is recommended, as it also mentions efforts made by large companies such as Goodyear, Ford, General Motors, and Bridgestone/Firestone to increase the utilization of vehicles' used parts, including plastics, mats, foam, etc. GM claims that its EV1 electric car has an aluminium structure which is not only light but also 100 percent recyclable.

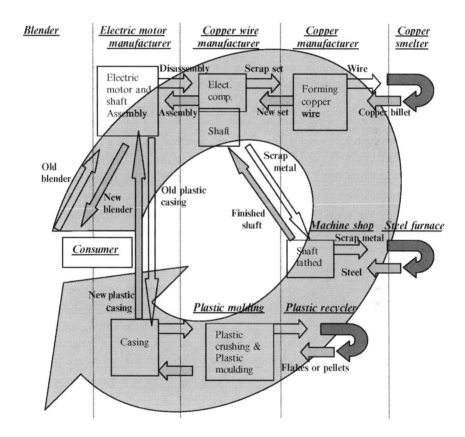

*Figure 2.5*     **Optimizing the supply chain for a house food processor, exemplifying the 'back to the consumer policy'**

Many other products can certainly follow the back-to-the-consumer policy for manufactured items, including computers, TV sets, fridges, washers, etc. Of course, it would not be reasonable to expect that manufacturers utilize in a new model the used parts from old equipment. While this is unfortunate, it is quite understandable, not only from the technical point of view (considering factors such as stress, improvements, new materials, and probably even new designs), but also since the labour, cleaning and repair costs would most likely be more expensive than in a newly developed component, even one that is identical. However, the overall policy whereby each company works closely with suppliers, as sketched out in Figure 2.5, is feasible, and could lead to major breakthroughs towards a more rational utilization of natural resources.

## 2.13 Actions to reduce consumption

This section proposes actions that involve people and organizations. Table 2.7 shows some of these suggestions for reducing the consumption of raw materials. Needless to say, many more exist, although this listing can help readers, and encourage them to seek more initiatives to limit consumption for their places of work.

*Table* 2.7   Suggested procedures to reduce consumption of raw materials

|  | Procedures | What is thereby reduced? | Additional information in: |
|---|---|---|---|
| In the household |  |  |  |
|  | Compost | **Mining of potassium and phosphorous,** used to make fertilizers | Section 3.2.2 |
|  | Buy grocery products in bulk (cereal, grains, sugar, flour, cookies, etc.) | **Logging of trees** for paper and cardboard, and **production of chemicals** for their manufacture | Section 3.10 |
|  | Demand less packaging in all consumer products — including food, computers, hammers, mattresses, medicines, household items, etc. | Ditto | Sections 1.8-3.10 - 4-13 - Table 4.5 |
|  | Replace plastic bags in supermarkets with your own plastic containers, woven baskets, linen bags, etc. | **Utilization of hydrocarbons** to manufacture plastic bags | Sections 3.1 - 3.10 and actual examples in section 4.2 |
|  | Promote with your local government representatives a ban on plastic supermarket bagging | Less land required for landfills |  |

|  | Donate to charitable organizations unwanted items | The **purchase of new items** by other people | Table 2.5 |
|---|---|---|---|
|  | Buy second-hand items at garage and garden sales | Ditto |  |
|  | Deliver tires to certified receptors for disposal | **Utilization of carbon/oil/gas for power generation,** when used as a fuel | Section 2.5 - 2.12.5 |
|  | Recycle paints. | <ul><li>Other people's **need to purchase** more paints</li><li>**Fuel,** when used in cement kilns and incinerators</li></ul> | See Internet references for chapter 2 Title: *Emerging products - Paints* Sections 4.1.2 - 2.5 |
|  | Do not put carpets out for garbage collection. They can be recycled | **The use of hydrocarbons** to produce more plastics (the raw material for carpets) | Actual example in section 3.2.2.1 |
|  | Donate electrical equipment (tools) | Other people's need **to buy new items,** and the necessity **to use more land for land filling** |  |
|  | Donate computers, printers, telephones, etc. | The **need to buy** new items |  |
|  | Donate furniture, kitchen and garden ware | The **need to buy** new items and the requirement for more land for land fills | Sections 3.1 - 3.2.2 |
|  | Have working fluids from appliances extracted by professionals. In some cases, solid waste involves the removal of liquids or gases: fridges, freezers, car and household air conditioners use as a refrigerant a substance that is very harmful for the environment; this gas has to be extracted before any recycling. The same also happens with transformers, which might contain PCBs (see Glossary) | The **danger of leaks of harmful gases into the atmosphere or fluids into the soil** In many cases, the by-product has a dissipative nature, such as chlorine, ammonia, sulphur, etc. |  |
| **Municipal level** |  |  |  |

|   | | | |
|---|---|---|---|
| | Curb-side recycling | **Mining, processing, harvesting of vast quantities of raw materials, such as aluminium, wood, glass, tin,** etc. | |
| | Make paper recycling in all government offices mandatory | **Logging trees** | |
| | Make recycling at landfills mandatory | **Use of valuable land**, as well as the use **of land filling equipment** and **fuel** | |
| | Develop a paid-by-weight system for domestic and industrial waste. Each household will thereby pay for what it produces; the same should be the case for industry | **Waste**, since people will be careful about what they refuse because every gram will count | Section 3.2.2 |
| | Develop systems of domestic garbage collection, such as a centralized vacuum system | **Purchasing of very expensive hauling systems, fuels, oils and tires** | Section 3.2.2 |
| | Provide for the imposition of heavy fines for people who litter the streets | The **number of people needed** to pick up litter and the resources to purchase sweeping equipment. At the same time, the need to **hire manpower** to clean storm grates and conduits will **decrease** | Such policies have been in effect for years on highways, and are enforced in some cities, such as Singapore |
| | Prohibit the use on non-degradable bags in supermarkets and other retail units | **Space in landfills**, as well as the **need to use hydrocarbons** to manufacture bags | Section 4.2 |
| | Encourage the use of waste from certain industries, such as from pharmaceutical plants (for fertilizers), ashes from power plants (for Portland cement and road filling), etc. | **Using raw materials such as phosphates**, as well as land | Actual example in section 4.6.1 |
| | Compost garden refuse | **Land space** in land fills, and the **purchase of fertilizers** | |
| Government | | | |
| | Provide for heavily taxing chemicals and inks used for printing commercial packaging; | **Mining of heavy metals** | |

|  | | | |
|---|---|---|---|
| | Make the manufacturing industry responsible for disposal of the items it produces, the cradle-to-grave policy (see Glossary) | **Purchase of raw materials, as many of them can be recovered from waste**. This also reduces **energy**, since it is generally cheaper to recycle something than to obtain it from raw material | Section 2.12.6 |
| | Use taxation to encourage the discovery of new uses for waste, striving to mimic nature by converting 'open' systems into 'closed' ones | **The use of raw materials such as minerals, metals and chemical products** | |
| **Industries** | | | |
| | Establish waste brokerage | **The need to purchase raw materials**, sometimes with specifications of minimum quantities | Actual example in section 4.6.3 |
| | Reduce waste by reengineering processes | **Weight of materials used**, as well as **thermal** and **electrical energy** | |
| | Reduce the need for pesticides and fertilizers | **Mining of potassium and phosphorous** | Section 4.9 Sustainable agriculture can use natural materials for these purposes |
| | Reengineer processes to use recyclable materials | **Extraction, and purchasing of raw materials, and manufacturing of chemicals** | |
| | Reduce weight ratio between input and finished product | ▪ **Need to re-smelt**<br>▪ **Energy** | |
| | Reuse scrap materials | **Use of virgin materials** | |
| | Establish programs with rewards for personnel with new ideas about manufacturing | **Waste** and **raw material** input in weight | |
| | Treat all cooling oils for reuse | **Fossil fuels**, using these products in cement kilns | |
| | Extend the life of durables, such as cars, fridges, TV sets and computers | ▪ **Purchasing rotation and a better use of raw materials and materials**<br>▪ **Electrical and heat consumption** | Section 3.10 |
| | Forge agreements with other industries to use waste from one industry for input to an other | **Use of raw materials as well as energy** | |

| | | | |
|---|---|---|---|
| | Reuse packaging from raw materials, products, sets components, etc. | **Use of new packaging and corresponding logging of trees** | |
| | Reduce packaging for final products | **Use of new packaging and corresponding logging of trees** | |
| | Recycle. For instance, a railway can renew the sleepers in its tracks and process them to make wood chips | **Land space in railway yards** and **landfills** | |
| | Find non-traditional uses for wastes | **Use of raw materials** | In Germany, gypsum is made from sulphur recovered from a coal-fired power plant |
| | Research to substitute materials. For instance, the replacement of steel used to make a device with steel alloys that can result in significant weight reductions | **Weight** in steel consumption | This policy has been widely used, for instance, in car manufacturing, sharply reducing weights by substituting steel with hard plastic |

The publication entitled 'Note to the reader from the UNO', listed at the end of this chapter under 'Internet References', contains in its Table 3 some very useful information on the dissipative substances and the products from which they originate, as well as the global output of these materials.

## 2.14 Working together

As an example of a community/government partnership in sustainability, the city of Ottawa-Carleton in Canada has a program called *'Take it Back!'* involving participating retailers in returning discarded items for recycling or to be taken back to the initial vendors.

One result of this program is that certain items are not picked up by the Ottawa-Carleton garbage collection service; these are:

- Oil filters
- Car batteries
- Antifreeze
- Motor oil
- Car parts

- Tires
- Gasoline
- Propane tanks
- Photocopier cartridges
- Computers
- Monitors
- Laser toner cartridges
- Non laser cartridges
- Printers
- Keyboard and mouse
- Electrical devices (Breakers, light switches, wiring, etc.)
- Televisions, VCRs, stereos
- Typewriters
- Flower pots
- Plastic flats
- Styrofoam flats
- Expired medications
- Needles and syringes
- Oil-based paints
- Solvents
- Fire extinguishers
- Pool chemicals
- Insecticides
- Wood preservatives
- Barbecue starters
- Disinfectants
- Appliances
- Bicycles and parts
- Brita water filters
- Building materials
- Camping gas cartridges
- Clothes hangers
- Dry cleaners bags
- Fluorescent lights and ballast tubes
- Fur coats
- Lead acid batteries
- Alkaline batteries
- Paint thinner and reducer
- Propane tanks
- Latex paint

Other programs dedicated to recycling deal with:

- o Automobiles, to recycle used cars.
- o Batteries (Ni-Cd) (See also 'Batteries' in Internet references for Chapter 2).
- o Scrap metal.

Information about these programs can be found at:
http://www.grrn.org/resources/ottawa_trish.html

Some countries have construction surplus programs that are very similar in conception to 'waste exchanges' (section 4.1.8). They involve non-profit organizations that receive donations of construction materials — such as surplus from construction companies, demolitions, material considered second-class because of minor defects, etc. These recipient organizations pick up donations of such varied materials as bricks, toilets, ceramic tiles, wood, stoves, windows, doors, kitchen cabinets, sand, Portland cement, etc. Their destination becomes low-income people wanting to build their houses, and who may buy these items at very low prices. The plan has a high social-welfare component, as well as an environmental component, since many materials considered as 'waste' are reused, rather than dumped into landfills. Table 2.8 shows some suggested actions to reduce water consumption.

*Table* 2.8  Suggested procedures to reduce water consumption

|  | Procedures | What is thereby reduced? | Additional information in: |
|---|---|---|---|
| **In the household** | | | |
| | Fix all leaks from faucets | **Water consumption,** which would cease to be wasted | |
| | Use small amounts of water, and if possible collect rainwater for irrigating the garden<br>Use mulch from chipped wood, grass, etc. to increase water retention | ▪ **Water use**<br>▪ **Water evaporation** | Actual example in section 4.11.2 |
| | Organize laundry for maximum amounts of clothing per load | ▪ **Water use**<br>▪ **Energy**<br>▪ **Laundry detergent and its chemicals**<br>▪ **Load in Wastewater Treatment Plant (WTP)** | |

| | | | |
|---|---|---|---|
| | | • Chemicals for potable water treatment | |
| | Use water-saving devices for shower heads and faucets | • Water use<br>• Energy<br>• Load in WTP<br>• Chemicals needed for potable water treatment | |
| | Organize dishwashing to use it at maximum capacity | • Water use<br>• Energy<br>• Dish detergent and its chemicals<br>• Load in WTP<br>• Chemicals for potable water treatment | |
| | Replace toilets with new models, which use far less water | Potable or flush water, as well as load in WTP | Section 3.2.1 |
| | In large buildings, recycle wastewater and use it for flushing toilets and for gardening | Potable or flush water, as well as load in the | Section 4.11 |
| Municipal level | | | |
| | Fix leaks in the water mains | Loss of treated water | |
| | Store storm water in underground reservoirs | Using fresh and treated water for wetting streets, sidewalks, parks, etc | Table 2.5 |
| | Install spring-loaded faucets in all public buildings. For instance, Beijing's City Hall installed thousands of new water faucets to avoid dripping | Treated water | Actual example in section 4.11.2 |
| | Replace toilets with new models, which use much less water. | Water and load in WTP | Section 3.2.1 |
| | Instruct personnel to save water | Water misuse | |
| | Install water meters in households | Water consumption | |
| | Establish a report system to inform on monthly progress in water consumption | The probability of more water consumption when personnel are aware that its use is being monitored | |
| Industries | | | |
| | Adopt new technologies to decrease the specific consumption of water per unit of final product | Water consumption. For instance, crude steel production requires about 11 m$^3$/ton in some countries, yet only about 5.5 m$^3$/ton in others | |
| | Research new technologies | Water consumption and | Section 4.1.3 |

|  | Measure | Impact | Reference |
| --- | --- | --- | --- |
|  | that use less process water | the production of more wastewater |  |
|  | Treat and reuse water in a closed circuit | Ditto |  |
|  | Promote sequential use | **Water**, since the fluid is utilized many times by several users | Section 2.12.3 |
|  | Eliminate leaks in the process, in tanks and in piping | **Water consumption** | Section 2.12.3 |
|  | Install spring-loaded faucets | Ditto | Actual example section 4.11.2 |
|  | Replace toilets with new models that use far less water. | Ditto | Section 3.2.1 |
|  | Collect rainwater for use in flushing toilets and for landscaping | **Water consumption** |  |
|  | Provide on water- saving courses for personnel | **Water misuse** | Section 2.12.3 |
|  | Establish programs with rewards for personnel contributing new ideas to achieve water savings | **Water consumption** through incentives that are generally less expensive that the water misused | The flow of new ideas that often come with this program can be amazing — quite apart from making sense economically, and ecologically. |
|  | Use waste hot water as an input for other industries | **Thermal energy** and **purchasing of heating equipment** | Actual example in section 4.6.1 |
|  | Establish partnerships between industries and the municipality | **Water** and **energy consumption** | Actual example in section 2.12.3 |

## Internet references for Chapter 2

**Authors**: Robert U. Ayres and Udo E. Simonis (1994)
Title: *Industrial Metabolism. Restructuring for sustainable development*
Comment: A complete description of the notion of 'industrial metabolism', with cases studies and methodologies from Sweden, Germany, and the US.
Address:
http://www.unu.edu/unupress/unupbooks/80841e/80841E00.htm

***Cleaning flue gas***
**Source**: Japanese Advanced Environment Equipment (2001)
Title: *Dry type desulphurization equipment*
Comment: Supplies technical information about a method to remove all kind of toxic substances.
Address:
http://nett21.gec.jp/JSIM_DATA/AIR/AIR_2/html/Doc_075.html

**Authors**: Ron Zevenhoven and Loay Saeed (2000)
Title: *Two-stage combustion of high-PVC solid waste with HCL recovery*
Comment: Discusses a waste-to-energy process.
Address:
http://eny.hut.fi/research/combustion_waste/publications/non-referee/torontoE.pdf

Title: *Incinerators make waste more toxic and…*(2002).
Comment: Useful information about operation of incinerators, with comments from different experts about emissions.
Address:
http://www.zerowasteamerica.org/Incinerators.htm

***New York incinerators***
**Source:** New York department of Environmental Conservation (2000)
Title: *Solid waste incinerators, refuse-derived fuel processing facilities, and solid waste pyrolisis units*
Comment: They state that 'Waste-to-Energy (WTE) is defined as a solid waste management strategy that combusts wastes to generate steam or electricity and reduces by approximately 90 percent the volume (not weight) of waste that would otherwise need to be disposed'. No mention is made about danger of emissions. Details given on facilities during 2002.
Address:
http://www.dec.state.ny.us/website/dshm/sldwaste/facilities/wte.htm

**Source**: Göteborg Stad (2003)
Title: *The largest waste incineration plant in Northern Europe*
Comment: Interesting short paper entitled: 'Sewage heats the city'.

Address:
http://www.goteborg.se/prod/sk/goteborg.nsf/1/press,mediaservice_(english),the_book_of_records,the_largest_waste_incineration_plant_in_northern_europe?OpenDocument

**Authors**: Rabl *et al.* (1999)
Title: *Impact assessment and authorization procedure for installations with mayor environmental risks*
Comment: Very well-documented report about the installation of waste incinerators in six European countries, with case studies in France, Austria, Belgium, Denmark, Spain, Switzerland and U.K.
Address:
http://www-cenerg.ensmp.fr/rabl/pdf/FullReport.pdf

Title: *Guidelines for a special waste incinerator* (2000)
Comment: Appendix 1 of this publication provides a very useful list of emission limits for contaminants in the flue gas expressed in $mg/Nm^3$, in accordance with current practices in the U.K., The Netherlands, Sweden, and Germany. Appendix 2 provides information about special emission standards for waste incinerators, and in Appendix 3 there is a recommended accepted criterion for suitability of industrial wastes for landfill disposal.
Address:
http://www.nea.gov.sg/cms/pcd/guideline_waste_incinerator_2001.pdf

**Authors**: Dunkan Kimbro *et al.* and Phil Knisley *et al.* (2000)
Title: *Emissions performance results for a liquid waste incinerator upgraded to meet the HWC MACT standards*
Comment: Technical discussion on the operation of three hazardous waste incinerators in the US.
Address:
http://www.croll.com/croll/ca/pdf/test_results.pdf

**Source**: Environmental Protection Agency (EPA).
Title: *Definition of hazardous waste*
Comment: Technical discussion on hazardous waste.
Address:
http://www.ehs.ohio-state.edu/docs/envaff/cmguide/chapter3.pdf

**Source**: Sustainable Cleveland Partnership (1998)
Title: *Medical waste incineration*
Comment: Due to its special characteristics, hospital wastes are burned rather that buried in landfills. However, incinerators also have risks, as mentioned in this report, not only for the potential production of dioxins and furans but also for the high content of mercury in the garbage. According to this publication, medical waste incinerators are considered the second most important source of mercury released into the atmosphere. It also offers some alternatives for reducing material to be burnt.
Address:
http://www.nhlink.net/enviro/scp/medical.html

**Authors:** Michelle Allsopp, Pat Costner and Paul Johnston – Greenpeace Research Laboratories – University of Exeter, UK (2001)
Title: *Incineration and human health*
Comment: In this large report (84 pages), the authors make a deep study of the environmental and especially the health impact of populations living near incinerators.
A visit to this site is recommended.
Address:
http://archive.greenpeace.org/toxics/reports/euincin.pdf

**Source**: Environmental Protection Agency (EPA) (1995)
Title: *Waste vitrification through electric melting, 1995 – EPA – Emerging Technology Bulletin EPA/540/F-95/503*
Comment: Bulletin of this prestigious US agency. Technology description, including diagrams, waste applicability, and test results.
Address:
http://www.epa.gov/ORD/SITE/reports/540_F-95_503.pdf

**Author**: Daniel Lemarchand (2000)
Title: *Burning issues, International Cement Review, February 2000*
Comment: Treats incineration in a cement kiln.
Address:
http://www.lenoco.com/burningssues.pdf

**Author**: Kurt Reschner (2003)
Title: *Scrap tire recycling – A summary of prevalent scrap tire recycling methods*
Comment: Methods associated with means of tire disposal and related statistics. Also details potential applications. See, in Figure 1, the dramatic picture of a tire fire in Stanislaus Co., California.
Address:
http://home.snafu.de/kurtr/str/en.html

**Author**: Forbes R. McDougall and Peter R. White (1998)
Title: *The use of lifecycle inventory to optimize integrated solid waste management systems: a review of case studies*
Comment: Life Cycle Inventory in five cities in France, Spain, UK, and Canada.

Address:
http://www.entek.chalmers.se/~josu/art-fmc.htm

**Source**: Luxembourg: Office for Official Publications of the European Communities (2003)

Title: *Waste generated and treated in Europe*
Address:
http://www.eu-datashop.de/download/EN/inhaltsv/thema8/waste.pdf

**Authors:** Richard L. Ottinger & Mindy Jayne (2000)
Title: *Global climate change — Kyoto Protocol implementation: legal frameworks for implementing clean energy solutions - Energy efficiency alternatives*
Address:
http://www.solutions-site.org/special_reports/sr_global_climate_change_1.htm

**Author**: Paul Hawken (1997)
Title: *Natural capitalism*
Comment: This paper starts with the sentence "Industry has always sought to increase the productivity of workers, not resources", and finishes with "The future belongs to those who understand that doing more with less is compassionate, prosperous, and enduring, and thus more intelligent, even competitive". Between them are medullar thoughts on economic concepts.
Address:
http://csf.colorado.edu/authors/Agerley.Harald/hawken.html

**Author**: Mark Dellinger - Lake County California Sanitation District (2004)
Title: *The geysers pipeline project*
Comment: This paper details the partnership that facilitated building the first wastewater-to-electricity system.
Address:
http://geoheat.oit.edu/bulletin/bull18-1/art37.htm

**Source**: U.S. Environmental Protection Agency (2004)
Title: *Landfill methane outreach program — Fuel cells — Demonstrating the power of landfill gas*
Comment: Example of utilization of landfill gas to power 100 houses using fuel cells.
Address:
http://www.epa.gov/lmop/products/fuelcell.htm

**Source**: U.S. Environmental Protection Agency (2004)
Title: *Batteries*
Address:
http://www.epa.gov/epr/products/batteries.html

**Source**: U.S. Environmental Protection Agency (2004)
Title: *Vehicles- Industry initiatives*

Comment: This paper features recycling efforts by Bridgestone/Firestone, Daimler Chrysler, Ford, and General Motors.
Address:
http://www.epa.gov/epr/products/vindust.html#chry

Source: U.S. Environmental Protection Agency (2004)
Title: *Emerging products - Paints*
Address:
http://www.epa.gov/epr/products/emerging.html#paint

Source: U.S. Environmental Protection Agency (2004)
Title: *Miscellaneous products – 'Nike's reuse-a- shoe' program*
Address:
http://www.epa.gov/epr/products/emerging.html#paint

Source: U.S. Environmental Protection Agency (2004)
Title: *Electronics - Multi-stakeholder initiatives*
Address:
http://www.epa.gov/epr/products/emulti.html

Author: Jean A. Bowman - Texas Water Resource Institute (1994)
Title: *Saving water in Texas industries*
Comment: Excellent information about water saving measures in the high-tech industry.
Address:
http://twri.tamu.edu/newsletters/TexasWaterResources/twr-v20n1.pdf

Authors: Don Hinrichsen *et al.* (1998)
Title: *Water conservation and management*
Address:
http://www.infoforhealth.org/pr/m14/m14chap6.shtml#top

Source: IEA Annex 33 – ALEP Guidebook – Internal Website for the working group
Title: *1. Description of the present situation, energy balance and emissions*
Comment: Valuable information, with charts and diagrams, about energy usage of different cities in Europe. Also a very interesting application of GIS (see Glossary) to develop heat maps.

Address:
http://www.profu.se/guidb1.htm

Source: US Environmental Protection Agency (2004)
Title: *Water reuse*

Comment: Water reuse at an IBM facility in California
Address:
http://www.epa.gov/watrhome/you/reuse1.html

**Source:** European Commission – Community Research
Title: *The path towards sustainable industrial production…*
Address:
http://europa.eu.int/comm/research/growth/pdf/nanotechnology02-conference/presspacklyngby-9a-leaflet-sustainable-development-version.pdf

**Source:** Devon Authorities Recycling Partnership (2003)
Title: *It's time to sort it out*
Address:
http://www.recycledevon.org/pages/index02.asp

Source: U.S. Environmental Impact Agency (2004)
Title: *Region 9: Water programs - Water recycling and reuse: The environmental benefits*
Comment: Visiting this Website is highly recommended for an idea of actual large-scale uses of recycled wastewater.
Address:
http://www.epa.gov/region9/water/recycling/

# CHAPTER 3 – SUSTAINABILITY IN THE BUILT ENVIRONMENT

The purpose of this chapter is to provide the reader with a glimpse of the **problems** affecting each area and **actions** that can be taken. Again, it is only an idea and general knowledge, and, as such, it does not pretend to cover even minimally the extent of each subject. The current purpose is only to make the reader aware of the different problems, solutions, inconveniences, advantages and disadvantages of each area considered.

## 3.1 Sustainability at the individual level

Sustainability is not merely the jurisdiction of government, or something to be left to local authorities. People have to think in a certain way while putting a little extra effort to sort out domestic items for recycling, or when deciding to ride the bus instead of driving downtown on an errand. Sustainability at the individual level means thinking about how each one within our own sphere can contribute to improving society's welfare, to take a series of actions, mostly on a daily basis, such as:

- Thinking twice before deciding to get rid of a three-year-old family car to get a new model. Is the change really worth it? What added features and advantages can one get from a new model?
- Consuming only the water needed, not wasting it. After all, who needs fixtures with four or five showerheads in the shower?
- Programming the laundry in order to save electricity, wastewater, and detergent going into the environment.
- Attending meetings relating to problems and solutions for the neighbourhood.
- Making the effort to compost at home.
- Helping educational authorities prepare the curricula of activities for children.
- Not littering the streets, beaches, roads, or recreational areas.
- Buying in bulk, thereby avoiding paying extra for fancy packages.

- Donating unwanted clothing, kitchenware, furniture, electrical equipment, bikes, etc., to charitable organizations and churches, remembering that what one person's junk can be a treasure for someone else.
- Not leaving lights on unnecessarily.
- Retaining the lawn's humidity, thereby saving water, by mulching.
- Going up and down flights of stairs, instead of using the elevator. This is good for health and the environment.
- Consider biking or walking to work.
- Sorting unusable stuff made out of metal and taking it to a junkyard for recycling.
- Designating containers at home and in the office for recycling paper, aluminium cans, and bottles.
- Teaching one's children about environmental measures.
- Using as scrap paper sheets that have been used on one side.
- Using a time-programmed car block heater in winter. This measure not only provides a quick engine start and heats up the car's interior, but eliminates the need for a warm-up period for the engine. During the 5 to 10 minutes generally used as a warm-up, exhausts can pollute the air considerably due to incomplete combustion produced by the car's still-cold engine.
- Delivering old tires to a tire-disposal facility.
- Buying and reusing fabric bags when shopping, instead of the plastic bags provided by supermarkets.

Many reason that they are not going to make a difference to the environment by following these and other suggestions. While this may be true, if millions of people followed these simple measures, along with others, they could greatly improve the planet's health. Besides, most of these measures make good economic sense. For instance, as when calculating the cost of using one's car just to transport the driver. Its operating cost — measured in insurance and the chance of accidents, wear and tear, gas, oil, parking fees, damage caused by vandalism, traffic delays, etc. — far offsets what one would pay in fares on public transit.

## 3.2 Sustainability in the household

A very high percentage of the world's population is urban. In light of this, individual households are evidently very important for the consumption of resources. Water, energy, land, sewage generation, energy for cooking, etc., take place at home, and for this reason, there is a special interest in controlling

## 3.2.1 Water use in the household

An international indicator determines the average amount of water that a person needs for hygiene, cooking, washing, etc. However, the fact that it is international does not mean that it can be applied everywhere, since that amount will also depend on local conditions. Obviously, countries such as Canada with huge reserves of fresh water cannot compare with other countries, even in the tropics, where water is at a premium.

Some measures to avoid excessive and useless water consumption include:

- Using special toilets with a better design that requires smaller amounts of water for flushing. In Europe, it is also common for a toilet to have two levers: one for flushing liquid waste and the other for solids. This makes a lot of sense since obviously the amount of flushing water required is different.
- Using special nozzles with different options in showerheads to reduce water consumption.
- Fixing low flow faucets and eliminating leaks.
- Using dry toilets is an option in some places; they use no water but, instead, degrade wastes biologically.
- Gathering rainwater for use to water the garden.
- Metering the water one consumes.

Water sustainability also means recovering all nutrients that are limited resources, and especially phosphorus. This cannot be done in the household but it is possible in wastewater treatment plants, with additional installations.

## 3.2.2 Solid waste in the household

Many communities around the world follow a few very simple principles that contribute to reduce waste that would end up in landfills. Such measures include:

- **Recycling waste**, such as bottles, aluminium and tin cans, plastic and paper, placed in containers supplied by City Hall (blue boxes), put once a week at the curb-side for collection by municipal trucks. Large apartment buildings also have special containers for these items, usually one for plastics, bottles and cans, and another for paper.

- **Composting.** Household kitchens generate much waste, which falls into two broad categories: what can be used for composting and what cannot. Vegetables, the skins of potatoes and onions, bread, eggshells, etc., are what make up most of the first kind of waste, and can be placed in inexpensive covered plastic containers to compost. Such containers, which are made of recycled plastics, do not allow rodents or other animals to come in touch with the compost. Consequently, not composted waste from the kitchen should amount to very little indeed, and be mainly constituted by bones, fat and other organic and inorganic products, making for the only waste suitable for a landfill.

- **Reusing.** Another type of unused items consists of furniture, appliances, bed frames, mattresses, etc., ought to be placed on the curb-side for collection.
  These discarded items should most definitely not be buried in landfills but recycled. Section 2.12.6 touched on the cradle to grave policy for goods that exists in some countries, where manufacturers of certain items are made responsible for their disposal.

- **Returning.** Large items, such as cars, batteries, motor oil, tires, etc., can be treated similarly, in terms of the manufacturers' responsibility, although some items like scrap tires (section 2.12.5) can be shredded at landfills and their components can be recovered, that is steel, polystyrene, and the rubber or synthetic rubber itself; each has different uses.

Regarding domestic waste collection, there are some initiatives that have already materialized, such as the pneumatic system for collecting garbage. This consists of pipes of large diameter that run underground with connections to a chute area within apartment buildings, usually in their basement, where several locked-door chutes in the building can be opened by each dweller who can use a coded magnetic card, resembling debit cards for bank automatic tellers.
When dwellers dispose of their garbage, that underground pipe network conveys it, using a vacuum — as in domestic vacuum cleaners — to a central garbage collection station where it is sorted and disposed of according to its characteristics. A large rotary fan in the central garbage station produces this

vacuum. To this author's best knowledge, these systems are operating in Stockholm and in parts of Copenhagen harbour; in this last case, it also serves a zone with many restaurants.

The advantages of this system are that it:

- Eliminates the need for garbage and waste containers and bags that must be left at the curbside for collection;
- does not provide food for rodents in the area;
- averts the issue of garbage odours, as there is no material to produce them;
- facilitates depositing garbage at any time, day or night;
- eliminates the need for garbage trucks and their operating personnel, leading to very large savings in investment and equipment maintenance;
- saves considerably on fuel, and prevents air pollution and noise, due to of this absence of trucks;
- eliminates the occasional problem of garbage spread on the sidewalk and streets due to damage to garbage bags caused by rodents, stray cats and dogs; and
- make for a rational, user-pay system, since the chutes can be equipped with scales that weigh the garbage deposited by each dwelling, so they can be billed from information on their cards.

Some disadvantages of this system are:

- The high initial cost for excavation, laying pipes and chute equipment, fans, etc.; and
- maintenance costs.

The advantages are felt to offset the disadvantages, especially considering the large investments needed for trucks and wages for garbage collectors and maintenance workers. The city of León, in Spain, also has this system installed, and in its case the network serves a population of 4,000 plus 150 bars and restaurants. Its production amounts to about 10,000 kg/day of organic refuse and 1,000 kg/day of glass. The cost of the system was 5.2 million euros, and maintenance costs are about 100,000 euros/year.

The reader might want to examine the Internet references for Chapter 3, on the León scheme, where more technical information is provided, along with photographs that show some of the system's components.

In most places, local authorities charge a flat rate for garbage collection, normally incorporated into the real estate tax. No doubt the system is unfair since it charges the same to both the big waste producers and small ones. A more rational system is the unit-based fee, where the user pays for the amount of waste produced. However, some practical difficulties arise in its implementation; for instance, a modern garbage truck equipped with automatic scales could do that job, yet many homeowners would then place their garbage in other lots.

---

*3.2.2.1 An example to follow: Recovery of carpets material*

*At least in the US and Canada, carpets are household items that are as much a part of any house or office building as any other construction material. For this reason, there is an enormous production of carpets of different materials, shapes and sizes, and there is also a very important replacement market for owners to change their floor coverings. Consequently, large amounts of discarded carpets end up in landfills, and as they are bulky, they take up considerable space.*

*Dupont, one of the world's largest chemical companies, began the Carpet Reclamation Program in 1991, which has already reclaimed about 30,000 metric tons of carpet (see 'Business and Sustainable Development: A Global Guide'). The company has expanded its recycling capacity in Georgia, US, and can now process about 23,000 metric tons per year, recovering clean nylon 6,6 resin. (See information under the above-mentioned title in Internet references for chapter 3).*
*See also "Carpet" in Internet references for chapter 3.)*

---

Another system involves the sale by municipalities, via supermarkets, of large paper waste bags (made from recycled paper, of course). To discourage people from producing a lot of garbage, these bags can be sold for a high price (this is sometimes called the Pigouvian tax, after the British economist who maintained that pollution could be addressed by exacting taxes). However, this raises another problem, which is illegal burning or dumping of waste. There is, however, some disagreement about this method; researchers H. Bartelings *et al.* (see Internet references for Chapter 3), for instance, think that this procedure may have undesirable environmental consequences.

*3.2.3 Energy uses in the household*

Energy in its different forms is a basic input to a household's well-being. It lights the house, cooks meals, heats water for kitchen and hygiene uses, provides heating in winter and air conditioning in summer, and has countless another uses. In high-rises, it powers elevators and water pumps, as well as runs safety systems. Energy savings in a household may be broken down into two broad areas:

- Permanent fixtures:
  Such as insulation installed during construction and involving walls and roof; usually builders follow strict regulations in this regard. In older houses, improvements are possible by installing efficient furnaces, through changes to the insulation type and to more efficient windows that can greatly decrease thermal loses.

- Measures taken to save energy, classifiable as temporary and ongoing, are probably where most savings can be realized, since these are based on the owner's will and interest in saving energy.

  Examples of such temporary actions are:
  - Eliminating drafts by covering windows with plastic sheets.
  - Setting up thermostats for heat and hot water.
  - Weather stripping doors and windows.
  - At night lowering the settings for heating at night.

  Examples of ongoing actions include:
  - Planning the laundry, dishwashing and ironing for the maximum loads.
  - Shutting off heating dampers in unused rooms.
  - Changing furnace filters more often.
  - Changing electric lights to low-wattage bulbs.
  - Not leaving lights on unnecessarily.
  - Making sure that fridges and freezers have the correct temperature settings, in accordance with one's needs, and that doors close tightly. Not putting hot liquids or food in these appliances. When something needs to be defrosted, not leaving it on the kitchen counter but inside the fridge; this way part of the energy previously spent on freezing the item is recovered in the fridge.
  - Buying energy-efficient appliances and equipment.
    Computers have added greatly to energy consumption, and the fact that many people work at home has

compounded this. Aside from using the computer less intensively, conservation measures include buying units that have earned a seal for minimal consumption.

### 3.2.4 Land use for the household

How can we measure the importance of the household in land use? International household standards suggest a minimum floor space per person, which is an indicator generally used when dealing with low-income areas, and naturally, it is also linked with land use in the city. Here there has been a shift in land use, most especially in many downtown areas that have been degraded because people have emigrated to suburbs, courtesy of the car and the construction of highways. The reason behind this exodus can be found in lack of parking space, increased crime, air pollution, etc. Because of these urban dynamics, cities have expanded towards suburbia, eating up land once devoted to agriculture and forests.

This, in turn, has provoked the need to extend necessary infrastructure services, such as transportation, water, sewage, and energy, with corresponding high costs. Consequently, and because of this migration, many central areas of the city are almost vacant, or with very little population, creating inner-city decline. The **population density,** which is the number of people per square kilometre, is then too low considered as an average of the whole city. This indicator (section 6.1) can be used to measure this **centrifugal effect,** that is: of people moving from the CBD (Central Business District) to suburbia.

For this reason, many City Halls are issuing regulations and bylaws to establish certain population densities for each area of the city; from the point of view of sustainability there must be — as always — a balance or equilibrium between land and people. Many think that the perfect solution would be the construction of tall high-rises, a measure proposed many decades ago by the famous Swiss architect Charles Edouard Jeanneret (Le Corbusier).

It is true that the construction of high-rises increases density, but from the point of view of sustainability density is not the only parameter to consider. It is necessary to take into account that the cost in energy needed to operate high-speed elevators, the necessary water pumps, and the lighting, which can put a severe toll on energy availability. The matter of the adequate height of these buildings has been and remains a topic currently analysed by architects, engineers and city planners around the world but it appears that there is not yet a solution or compromise.

## 3.3 Urban transportation

Urban transportation has had a very strong influence on city sustainability. Actually, one of the benefits than citizens should expect is to have reliable, cheap, safe, and abundant transportation. This is fine from the social point of view, but it can provoke havoc from the economic and the environmental standpoints.

As for the economy, there are problems because high frequency services mean many more trams or buses on the streets, involving a huge investment, maintenance, and payroll costs. As for the environment, it is obvious that the more buses the greater will be the resulting pollution. Therefore, again, the dichotomy faced is that what is good for one area, the social effect of transportation, could be bad for the other two areas: economics and environment.

Public transportation is a very difficult issue since it also relates to many others, such as:

- The need to keep streets in good repair.
- Land use allocation. This happens when a city builds dedicated highways for buses, as in Ottawa, Canada, or dedicated land for streetcar tracks, such as in Den Haag, Vienna, and many other European cities. Land use is also affected because of the need to build parking lots for people to use 'park-and-ride' systems.
- Fare structures.
- Social equity issues.
- The coordination of services and fares, if the system is run by private companies.
- The city and its different population density.
  If the bus route begins in a densely populated area of the city, it will quite probably have a large ridership. However, if between that point and its main destination — usually downtown — there are not enough patrons, which are related to population densities along the route, the operation will probably be unprofitable. Transportation companies make money when the same seat in used several times over one trip, since many more tickets are being sold per kilometre.
- The allocation of routes when the system is run privately, since a company will have some routes that are more profitable than others.
- The administrative efficiencies, if the system is run by City Hall.
- The city and its connectedness.
  In some cities, different areas are poorly connected and most trips go via downtown. This causes unnecessary private and bus traffic in an area that is usually already congested, as well as travel-time delays, longer trips, and more pollution. This fact is not simple to solve,

since if the public transit system is operated by private companies it could happen that not enough ridership will render profitable an inter-areas operation. If City Hall operated the system, these routes would also be unprofitable, and they would have to be subsidized.

Besides, improving this lack of connectedness usually involves building road infrastructures such as opening new roads, which can sometimes affect private properties, or the construction of bridges over courses of water or rail tracks, the construction of tunnels, etc.

- In some parts of a city, public transit is not available due to very poor road conditions or the sheer inexistence of streets — usually found in slums (section 3.4). Coincidentally, these are the very areas probably most in need of public transport because, obviously, their inhabitants seldom have cars.

Section 3.3.1 discusses the transportation system designed for the city of Curitiba, Brazil, which involves considerations of density and re-structuring the transportation system, which is both private and public. This system has been very successful and has been replicated in other cities.

---

### 3.3.1 Case study: The role of transportation in sustainable Curitiba, Brazil

*The following comments are based on information taken from this Web source, which the reader is encouraged to consult to learn more about this amazing city:*
*http://www3.iclei.org/localstrategies/summary/curitiba2.html*

*Curitiba is a large (1.6 million) city in Brazil. Dubbed the country's ecological capital, its sustainable progress has been hailed as an example for the world. This began in the 1970s, when the city decided that, since transportation and land use are closely related, transportation would form the nucleus of this project. The City Hall designed five main axes for the city, to be used as transportation corridors. These axes each have one two-way road complemented on either side with two additional corridors.*

*The main corridor is a dedicated two-way road for buses. The first adjacent road serves as an access road for cars. The next adjacent corridor serves for both cars and buses on a one-way basis for each side.*

*Figure 3.1*  **Bus and bus stop in Curitiba**

*This way, buses can travel fast and safely because they do not share the road with anything else; also, as they are articulated units, they can transport large numbers of people. Access to buses is through innovative tubular stations where the fare is paid before boarding, so drivers are free from fare duties. (See Figure 3.1.)*

*This efficient system has taken many private cars off the streets, it being so much more convenient and cheaper to use the privately-managed public transit. Of course, this has led to sharp decreases in air pollution, helped by the fact that the buses are diesel-operated and produce less pollution than gasoline-driven cars.*

*As mentioned, appropriate land use is fundamental for this scheme's provision of the ridership needed to make the system profitable. In this regard, Curitiba's City Hall issued bylaws on the types and heights of buildings along the routes, and their distances from them, thereby creating an increase in density along these routes. This is exactly what was needed to keep the system cost-effective.*

*A fleet of small buses interconnects the five main corridors that act as feeders. In this author's opinion, in the future these expressways will probably be used for a Light Rapid Transit*

*(LRT) system of light rail cars. In this way, the dedicated bus route can be seen as the 'seed' to increase population density along it, paving the way for more massive transportation systems, such as the LRT or a subway.*

*This system is proving to be very effective in building up density before the far more expensive LRT is built. An example of this practice in the city of Buenos Aires involves the terminal of one of the city's streetcar lines that was connected a decade or so ago to a huge urban development in the outskirts of the city, and it is appropriately called the pre-metro line. Investments in underground systems are so large that it makes sense to 'pave the way' for their construction that can only take place in many years to come.*

*Transportation is only one aspect of the paradigm that is Curitiba. Many other features make this city unique, including:*

The city was subject to periodic flooding from a river that traverses the urban area. The solution was to create storage ponds in city parks. Some secondary benefits include a cessation to erosion along riverbanks, as well as preventing illegal squatter occupations, a common issue in many Latin American cities.

- *Another known feature of many cities is that poor areas — which are actually slums — have no garbage collection services due to the absence of appropriate streets and the cost of such collection, since the local people cannot afford it. In Curitiba, people in poor areas bag their sorted garbage into garbage bags and take them to collection centres, where these are exchanged for bus tickets and food bags. Even children are encouraged to do this, and are rewarded with toys and candies.*

- *In many cities, domestic garbage collects daily, producing enormous costs in wages, the rapid deterioration of garbage trucks, high fuel consumption and the corresponding air pollution. In Curitiba, this problem was reduced many times over by weekly collections that pick up garbage that has been previously sorted by city dwellers. An exceptional 70% of this city's garbage is recycled, probably placing Curitiba far ahead of any city in the world (compare it with London, for instance, which recycles only about a 4%) The proceeds from the sale of usable garbage are spent on social programs.*

> *From the point of view of preparedness or capacity-building for diversification, the city has created centers to educate people in different trades and professions, such as marketing, finance, as well as mechanics, electricians and other trades.*
> *The publication mentioned above also provides details on 'Key Replication Factors', which can be very valuable for city planners.*
>
> *Another very important information source can be found in ICLEI (International Council for Local Environmental Initiatives; see Internet references for chapter 3).*

## 3.4 Upgrading slums in cities

Sustainability means equal social rights and opportunities for everybody, although more often that not a large proportion of city dwellers lives in very precarious conditions as, unfortunately, thousands of slums are abundant in cities not only in the developing world but also in the developed world. In many countries worldwide, there are shantytowns with deplorable social conditions, including 'houses' made of zinc sheets, wooden boards and cardboard. Economic problems, unemployment, lack of social plans, lack of working skills, are factors that have thrown millions of people living, or, more accurately, barely surviving in such conditions. These exist everywhere, although mainly in Latin America, Africa, Asia, Indonesia and some parts of Europe.

If a society wants a sustainable community, these conditions need to be, if not eliminated, at least improved. It is extremely difficult to eradicate these slums, because even if local governments build new houses for them in dedicated areas, they are solving the consequence but not addressing the root of the problem; so, after a number of years, the new settlement will be very similar to the one from where the people had come.
Traditionally this problem has been approached in two ways:

    A. Building new houses in another part of the city, and relocating people there. Table 3.1 condenses the advantages and disadvantages of this approach.

*Table 3.1*                    **Population relocation**

| Option A | Advantages | Disadvantages |
|---|---|---|
|  | People will have access to new dwellings, built with durable materials and with utilities connected | People are transported from where they have lived perhaps for decades, and will be cut off from familiar places, relatives, etc. |
|  | Dwellers will enjoy, perhaps for the first time in their lives, such 'luxuries' as piped drinking water, sewer services and lighting | Children have to start in new schools with new demands, new teachers and in a new environment |
|  | The new settlement will have streets that facilitate communication, and perhaps public transportation, as well as garbage collection | It may be impossible to reach agreement among the slum's dwellers about the financial investments that each inhabitant can afford, since what is good for one might be impossible for another |
|  | The local government will finance the new dwellings, with low rates of interest and a fixed repayment period | In large slums, it could be impossible to relocate everyone at the same time, which could cause friction among its dwellers. Besides, vacant shacks could tempt people from other areas to squat in them, not solving problems at all. |
|  | In many cities, these slums are obviously not in the best areas of a city, given that poor people generally occupy areas that nobody wants — such as flooded valleys, steep riverbanks or hills, and no longer used railways yards. There is always a risk that heavy rainfall will flood these areas and/or produce mudslides, leading to the loss of human lives. Unfortunately, this is common; from this point of view, relocating these people will improve their condition | From the social point of view, this option usually means that people are probably relocated on the outskirts of town, where land has less value; in so doing, they are separated from the city's social fabric. They therefore may feel that they are not integrated into the urban whole |
|  |  | Because of the new location, it is likely that existing infrastructure services have to be expanded to reach the new settlement. This means the construction or extension of water mains, sewer trunks, street lighting and transportation, and of course this also expands the city's area, reducing its population density |

B. Keeping slum dwellers in the same place while improving urban conditions is a second way of upgrading slums. This involves building basic infrastructure (water, sewage, energy and paving) in a first stage, and schools, health centers, etc. in a second stage. Table 3.2 condenses some of the concomitant advantages and disadvantages.

*Table 3.2*  Slum improvement

| Option B | Advantages | Disadvantages |
|---|---|---|
| In this option, basic infrastructure is supplied. It means the electrical connection to each house, building communal latrines, and installing communal faucets | Of course, from the social point of view, the advantages of this policy are the opposite of the disadvantages of the first, but it is necessary to recognize that people will probably be happier staying in their old places, with relatives and friends nearby, that moving to a new one | In the case of poor geographical conditions, such as flooded valleys and steep banks, works will need to be undertaken to remediate the problems by building containment walls to avoid flooding, or lagoons, or defence works to protect again mudslides |
| | From the economic point of view, this option is considerably better, since it is cheaper to provide basic infrastructure services, and for the people it is easier to pay for them than to have a new dwelling | It can perhaps be argued that no matter how much an area is improved, it will always be some sort of eyesore, especially if located in a heavily-populated urban area. On the other hand, it is necessary to remember that these people, however poor, are inhabitants of the city and have the same rights as other citizens |
| | In many cases, these slums are located very close to the downtown area, so that all services — such as transportation, water, sewage, and energy networks — pass very close by, although they do not reach the slum. This fact makes the provision of these services easier than in option A | |
| | The fact that the city is likely to encircle the slum makes the slum dwellers feel they are part of the city in which they live, and they are probably close to nearby schools and health centers | |

Examples are plentiful for both policy options, although it is believed, in line with experience gathered in Brazil, Argentina, Mexico, Indonesia, India, and the Philippines, that the second option is more favourable. It also gives its inhabitants the feeling that they are doing something for themselves, instead of expecting things from local governments. Besides, whatever the approach, it is also necessary to remember that these people are usually squatters in public and/or private lands, so they do not have tenure of the land they occupy. This circumstance poses a serious problem for the city, which has to decide about policy on this, and to consider that granting property title to people without payment establishes a dangerous precedent and creates inequality with the balance of citizens.

## 3.5 Environmental sustainability

### 3.5.1 Air in a sustainable environment

Air pollution has been in the minds (and lungs) of many people for many years. It is probably the type of contamination which, although not a blot on the landscape such as water and soil contamination, actually affects communities the most. According to the World Health Organization, 3 million people die every year due to problems related to air pollution. The Global Development Research Center (see Internet references for Chapter 3) reports that New Delhi's air pollution is so heavy that is comparable to smoking 10 to 20 cigarettes a day. This is so important as to influence climate, because of global warming, and to affect human health for people who are now exposed to more ultra-violet radiation than ever before.

Air pollution damages and destroys works of art, such the Taj Mahal and the Parthenon, produces the defoliation of forests, and contaminates lakes. Unfortunately, nature sometimes worsens these grim scenarios, as with gigantic volcano eruptions such as from Krakatoa, Mount St. Helen's, and Pinatubo, all of which have spewed enormous quantities of ashes into the atmosphere as well as sulphur gases.

Air is a renewable resource because green plants using solar energy (section 5.2.7 and Figure 5.2) produce one of its components, oxygen, although it also has a threshold, a limit, beyond which it deteriorates.
Among diverse examples of how the human race has managed to downgrade this fundamental component of life one could list:

- Smog: The word comes from the combination of the words 'smoke' and 'fog', which is produced by incomplete combustion from automobiles. Its most widely known example is probably the city of Los Angeles, although it is a common phenomenon in many cities around the world.

- Emissions from smokestacks from electric power plants and other industries, which makes for acid rain, a combination of sulphur and water vapour that generates sulphuric acid ($H_2SO_4$).
- The imperfect combustion of coal-burning plants.
- The burning of low-quality coal usually makes for sulphurous air. In the Katowice area of Poland this has had devastating consequences on the environment, affecting people, animals and forests.
- The release of an organochloride such as Freon — a gas used as refrigerant — is blamed, together with methane and $CO_2$, for the reduction of the ozone layer.
- Nuclear radiation, the most unfortunate example of which was the Chernobyl explosion.
- Production of $CO_2$ and methane gas ($CH_4$), not only from automobiles, power plants and factories, but also from farm operations. These gases increase the 'greenhouse' effect, and the temperature of the planet, and producing the melting of glaciers all around the world, as well as the creation of enormous icebergs in the Antarctic, broken from glaciers, because of the warmer weather.
- The incomplete burning of garbage, producing lethal furans and dioxins (section 2.5).
- The accidental release of strong chemicals by derailed trains.

What about smells? Many settlements have to live with odours from chemical plants, oil refineries, even pleasant aromas (for a while) from a chocolate manufacturing plant.

What is regrettable, to say the least, is that all of these effects are produced by humankind, and, because of it, their consequences return on us with a 'boomerang' effect. Pulmonary diseases, skin cancer, cancer produced by the air's lead content, terrible wounds due to nuclear radiation, and triggered natural disasters — all of these are the price the human race is paying for its insensitive handling of its natural environment.

Considering all of the above, one wonders if there exists intra-generation sustainability when all of these effects are present. Some measures have been taken, such us:

- The veto to use coal for heating in London.
- The mandatory replacement in Santiago, Chile, of oil-burning furnaces by gas-fired units.
- The prohibition against domestically incinerating garbage in apartment buildings in Buenos Aires.

- The ban from using leaded gasoline in most developed countries.
- Stiff penalties imposed on industries in the Ruhr area in Germany.
- The construction of subway urban lines to get cars off urban streets.
- New advances in coal-fired boilers using a process called 'fluidization'. When hot air is blown upwards at high pressure from the bottom of a layered 'bed' of coal, sand, and limestone ($CaCO_3$), the mixture of air and solids seems to boil as would a fluid under heat, producing a very good contact between air and fuel. This methodology is interesting in that it eliminates pollutants practically immediately. For instance, sulphur, present in coal, which in a normal combustion process produces emissions that lead to acid rain, combines with the sorbent, $CaCO_3$, and produces $CaSO_4$, (gypsum), which is then unloaded from the bed.

However, these measures are not good enough. Public pressure is needed to force authorities to ban activities that may hurt some economic interests while benefiting most of the population — including those responsible for producing the damaging effects, since air pollution is almost everywhere and affects everyone in the same way.
From this perspective, society and environmental interests must prevail over economic ones, although sometimes this appears not to be possible. A good example is the US's refusal to adhere to the Kyoto Protocol (see Glossary) for reducing acid rain.

Now, does sustainability relate to this problem of air pollution? Yes, it does. Not even considering the health approach, there is a social aspect, translated into a lack of social equity.
Why is this?
Affluent people can afford to move to higher areas of a city, or perhaps build their houses in beach areas where the dilution provided by prevailing winds can ameliorate the effect of air pollution, which is not affordable to other sectors of the population.
Is there anything that citizens can do to correct the environmental damage? Yes, there is.

In a democratic society, one has a right to force local authorities to take action. In some cities around the world, citizens have sued their cities, and forced City Hall to take measures. Agra, in India, and home of the Taj Mahal, is one of them. (See Bathia, in Internet references for Chapter 3).

There is also the issue of the quantity of emissions. According to Mail and Guardian (see Internet references for chapter 3), *"In Scandinavian countries, a typical oil refinery will produce at the most two tons of sulphur dioxide gas in a day, while in South Africa oil refineries produce as much as 82 tons of the noxious gas every day."*

What about particle pollution? It is difficult to take notice of this, since some particles are smaller than a human hair, but this does not mean that they do not damage the lungs. In fact, they do, and are responsible for cardiovascular ailments. Any medical lung association will bear witness to this.
Naturally, this topic and its relationship to sustainability have just been alluded to in this book. Sustainability in this context means that a population has the right to clean air to breathe, both outdoors and indoors; they have the right to demand that offending industries take measures or be closed down, and this is only possible when elected officials and decision-makers, if they are not already aware of the problem, are pressed by popular concern to undertake actions to reduce or eliminate air pollution. It takes a lot of coordinated effort, patience, and will to reach this objective, which will sometimes meet strong opposition due to resistance from political and economic interests.

Humankind and animals are constantly producing $CO_2$ through breathing, and natural processes also release it into the atmosphere. Therefore, the emission of this gas is linked to life. However, it is also considered as one of the main culprits of climate change, and the way to stop its harmful effects is to produce less and/or to find some other way to absorb the gas. Plants of course do the main job of this, by absorbing the gas and producing oxygen through the process of photosynthesis (section 2.7). However, this is obviously not enough, and one wonders if society has already altered the balance of the carbon cycle (see Glossary) by producing too much $CO_2$, to such an extent that plants are not able to cope with the increasing quantities of this gas present in the atmosphere.

During the last couple of decades, the concept of **'carbon sequestration'** has emerged as a way to reduce the concentration of $CO_2$ into the atmosphere. The idea is simple, consisting of dumping it into another medium, and that medium is soil. It is said that agriculture has a good potential for this carbon storage, with the use of diverse techniques. The American Farmland Trust (see Internet references for Chapter 3) reports that between 120 and 270 million of metric tons of $CO_2$ can be sequestered every year in the United

States, with cropland and using Best Management Practices, and from acreage conversion and bio-fuels. Another benefit would be to boost the farming economy by paying farmers for sequestration.

Another form of sequestration has also received considerable attention: it consists of giving credits to industries that spew $CO_2$ into one region, if they build a $CO_2$ sink elsewhere, even thousands of kilometres away, by re planting logged forests or by afforestation. This way, remote forests can absorb similar amounts of $CO_2$ as the industry produces. This is also called **'$CO_2$ sinking'**.

An interesting piece of software that is available can calculate emissions produced by cars and appliances in an area. It applies to the Japanese environment, but it can be used anywhere. This $CO_2$ calculator can be found at *http://www.gdrc.org/uem/co2-cal/co2-calculator.html*

### 3.5.2 *Water in a sustainable environment*

Fresh water is probably the scarcest resource humankind has on earth. It is considered a renewable resource as it comes from the evaporation of the oceans, the condensation of clouds, and the melting of snow. However, it is also necessary to take into account that not all the water that people consume is replenished in this hydrological cycle (see Glossary) — at least not on a human scale — since the rate of replenishment for rivers, lakes and aquifers is very broad, ranging from a few days to millennia.

The problem is that society is spending too much water, and as the population increases this consumption also rises. Besides, not all countries have the same availability of water; in fact, only a handful has abundant reserves. Even countries as large as the USA sometimes import water, mainly from Canada. Population Action International (see Internet references for Chapter 3) reports that, according to United Nations, by 2025 there will be between 817 million and 1,079 million people living in countries where water is scarce.

Water covers 70 percent of the Earth's surface but, unfortunately, this is not fresh water, which only covers 1 percent in the form of rivers, lakes and underground watercourses. Since this amount of water is somehow fixed, and because of increasing populations, there is no doubt that the resource is becoming scarcer.

To complicate the problem further, some of the planet's groundwater sources for drinking water are contaminated with chemicals from pesticides and herbicides, rendering this water unfit for human consumption. Underground water can also be contaminated with chemicals from mining operations, from landfills, and from chemicals from such animal waste as manure.

Because of its scarcity, used water — wastewater — should be reused many times.

The purification on wastewater from cities is performed in water treatment plants (WTP), usually in a three-step process. After that, it is disinfected using chlorine or ultra-violet rays, and then discharged into rivers and oceans, or used to irrigate certain crops.

In reality, WTPs mimic the action that takes place naturally in rivers and ponds, speed being the main difference. For that reason, and considering the cost of building WTPs, many communities are using a combination of meadows, marshes, and ponds to purify their water. This has the big advantage of not using chemicals, being economical, and providing areas suitable for wildfowl. Their disadvantage is mainly about the space they occupy, yet in those localities where land availability is not a problem this is an excellent solution.

WTPs use a system whereby wastewater is treated in one, two, or three stages called primary, secondary, and tertiary treatments, as well as through a preliminary operation to remove coarse solid matter. The primary treatment involves a settling of organic and inorganic solids, and the removal by skimming of matter that floats. In this primary operation $BOD_5$ (see Glossary), suspended solids and oil and grease are removed.

In the secondary treatment, bacteria are put to work to decompose solid matter; there is further precipitation and sludge removal up to a very high percentage. Also, most WTPs discharge this treated water into rivers or for irrigation after a further treatment with ultra-violet rays for disinfection.

However, the treated water still contains nitrogen and phosphorus, and these elements require a tertiary treatment to be recovered. Nitrogen is vented to the atmosphere, and phosphorous recovered for reuse. This is important, since not very many reserves have been detected of this element, which is a fundamental component for fertilizers. Therefore, if water is discharged without this tertiary treatment, phosphorus, a non-renewable resource, will be dumped.

In many cases, especially in rural areas, where isolation is a factor or when it is too costly to connect up a house's wastewater discharge to a city's sewage network, another system called a 'septic system' or 'individual home sewage system' can be used. Raw wastewater from the house is conducted via an underground piping to a concrete, plastic or fibreglass tank buried in the yard at a certain distance from the house. In this septic tank, solid matter deposits at the bottom as sludge, while lighter substances, such as grease, fat and soap suds, collect on the top as scum. Between these two layers, there is a zone of clarified water. Anaerobic bacteria decompose the solid matter and then the clarified, but still contaminated, liquid is sent through perforated piping, laid in trenches, to a drainfield that is also called an absorption bed or 'leachfield', composed of gravel, coarse sand, and other materials. Here, the water leaches to the ground and the soil filters it and completes the decomposition process. The size of the drainfield depends on the nature of the soil. For instance, some soils absorb water slowly, so the system needs a larger drainfield surface than those that do it faster.

Undigested sludge stays at the bottom of the tank, and it eventually has to be extracted; this is done with specialized trucks. There are tables to gauge the frequency of this operation, considering the number of people in the house and the septic tank's size. This system is efficient and safe if enough care is taken for its conservation, and usually it has a lifespan of between 20 to 30 years. Most cities and regions have bylaws establishing the distance of this septic tank or cesspool from the houses, water wells, large trees, etc. In cases where the household extracts drinking water from a water well, a chemical test of the purity of the well water is advisable periodically, to make sure that it is not contaminated due to fissures in the septic tank, or because of failure of the drainfield.

Also for this reason, it is necessary to consider that the drainfield has to be located at a certain vertical minimum distance from the water table below, to decrease the risk of contamination. This system is economical, since it avoids the need to connect to a municipal sewer system that might otherwise have to be extended: for instance, to catch wastewater from a new development. It also reduces the load of the WTP, although it does not recover some metals than can be collected from the sludge in a tertiary treatment of a WTP.

### 3.5.3 *Soil in a sustainable environment*

Soil contamination can be attributed to different causes, such as:
- Oil or chemical spills.
- Deposition on soil of dust removed from filters in smokestacks or petrochemical operations.
- Manure and urine from farm animals.
- Salt left by water extracted from an aquifer and evaporated by the sun.
- Phosphates, nitrogen and potassium from fertilizers.
- Chemical residues from herbicides used on crops.
- Contamination at the bottom of heavily polluted rivers, where it forms a thick sludge.
- Contamination produced by dumping mainly organic waste.
- Serious contamination in car scrap yards: after vehicles are crushed and flattened, large amounts of fluids — such as gasoline, oils, grease, brake and transmission fluid, windshield

washer, etc. — can end up in the soil without adequate safeguards.

A concern arises from the point of view of sustainability, because any of these contaminants can seriously diminish the soil's capacity to raise crops or to absorb water from rain. Besides, these may contaminate groundwater.
Affected soils can be recovered through a process called 'remediation'. Different techniques are applied to remediate soils, although all of them are costly.
One sad legacy of serious soil contamination occurred in Ukraine, where millions of hectares were polluted because of the Chernobyl disaster; the same happened to the radioactive contamination soil in Semipalatinsk, Kazakhstan, on a former nuclear test site.

There is a need to consider that everything in nature is related, and that any damage to one element can have consequences for many others. Thus, nitrogen and phosphate in soil usually leach to groundwater, or are carried by runoff to rivers, making for serious problems to human health, and that of fish. Runoff, in turn, is linked to soil erosion. Soil is a finite resource, and although it can have many different uses, a sustainable use of land has to consider the health of the soil, and its availability as a service-provider after one particular kind of use comes to an end.
The use of soil is related to water and air. In a landfill, for instance, the rational thing is to use the soil, but isolating it from the waste, especially at its bottom. To this end, in well-built landfills, waste is dumped on top of a thick plastic shield that rests over a compacted layer of clay. This way the liquid produced by rainwater after traveling through the waste (leaching), will be stopped from reaching the soil and the groundwater below it. The leach should be gathered through special conduits and treated the same as sewage.

Natural anaerobic (see Glossary) organisms decompose organic matter in the waste and produce methane, which usually escapes through appropriate vents. This methane is harmful to the environment because it is the main culprit for the depletion of the ozone layer. It also happens to be a good fuel; therefore, it can be cleaned and used in power plants to generate electricity (section 5.3.4.4).
Animal wastes are another serious source of soil contamination. When dropped on the ground, these wastes are decomposed by bacteria, leaching into the soil in the form of nitrates and nitrites, which are the nutrients that plants use. Yet sometimes certain amounts on these chemicals find their way into aquifers, especially nitrites due to their high solubility in water.

## 3.6 Team efforts toward sustainable environment

It has already been mentioned that for sustainability to be attained requires a team effort. That is, it is not a subject of engaging only the local authorities, but a matter to be addressed by the whole community; otherwise, the process will fail. Citizens need to collaborate with ideas, by monitoring, and with their taxes. Local governments of course have a lot of weight, since they involve policy-makers with the economic means and the necessary skills in their technical staff.
On the other hand, commerce and industry, and especially the latter, can play a fundamental role in diminishing their wastes, in using scarce resources better, and in working with the community to ameliorate things. The US Department of Energy's Energy Efficiency and Renewable Energy Network (EREN; see Internet references for Chapter 3), has developed a plan they call 'The Good Neighbour Project for Sustainable Industries' that offers very good information on sustainable business success stories about agreements between communities and their industries.

## 3.7 Sustainability in public administration and in urban life

Sustainability in public administration is associated with its efficiency, its handling of public monies, and its relations with the people they serve. The term is also about the way people interact in society, which is often promoted by municipalities, especially in areas connected with cultural activities. If, as in many cities, health and education also depend on City Hall, additional measures and circumstances have to be considered to make them sustainable.
It is customary to use indicators (section 6.1) to measure the quality of this service, such as:

- *Degree of preparedness of City Hall employees.*
  The higher the preparation the better the quality of service for the community.

- *Economic conditions.*
  Needless to say, a good administration will have a good and sound financial budget to address a city's different needs, which can be measured by some indicators, such as the percentage of taxes collected, justification in the allocation of public monies for different services, amount of money spent on public administration, etc.

- *Measure of decentralization of public offices in a city.*
  This refers to the creation of City Hall branches in certain areas of a city, when size justifies it, and normally including several adjacent

geographical areas. Citizens can normally visit the branch (in some cities they are called 'Centres of Community Participation'), where they can obtain any information regarding taxes and their payment, improvements in each area, available budgets, works to be done, etc. The population can also participate in meetings when important issues of their area are discussed, and such meetings are the place where people can submit their requests and meet with municipal technical representatives on issues such as pavements, water, sewage, etc.

This is a very important element related to the participation of citizens in handling municipal affairs, and for keeping citizens informed about what is going on in their area. For City Hall, they have the great advantage of channelling people's ideas, requirements, and necessities.

Another benefit is that citizens can obtain from each branch any documentation they need about birth certificates, fill complaints about the quality of services, pay municipal services, obtain marriage licenses, etc., without needing to go to a central, usually crowded downtown office, thus decreasing the need for transportation. Besides, branches can function as social centres where different cultural activities can be developed, thus contributing to the social development of the community they serve.

In some cities, City Hall has programs to train people — at municipal expense — to act as representatives of each of the different geographical groups that belong to a particular branch. Every area assigns its own representative, who, once trained, can sort out and screen problems raised by citizens, discuss them, and submit requests to City Hall. This is an efficient procedure because the technical offices in City Hall thereby receive screened information, with a minimum of background data and level of detail to be considered.

- *Number of meetings that public administrators hold with citizens.*
  This provides an idea of the degree of citizen participation in the governance of a city, and it is how grassroots information can be gathered.

- *Percentage of total City Hall budget allocated for salaries.*
  A useful indicator that can be compared with international standards, providing an idea of the size of the bureaucracy that citizens support to manage their city. Obviously, too large a value means that most tax monies collected are going to salaries, with little left for urban-related activities and works.

- *Percentage of employees per 1,000 inhabitants.*

Same as above.

- *Percentage of City Hall budget managed by citizens.*
  Provides an idea of the intensity of people's participation.
- *Efficiency and honesty in handling public monies.*
  No comment needed. People have a right to audit expenses by City Hall stuff in quantity and quality.
- *Percentage for cultural events.*
  This measures the efficiency and the municipal concern about citizens' cultural affairs.
- *Public facilities for sports.*
  This is important, since citizens need access to affordable swimming pools, tennis courts, soccer fields, athletic facilities, etc., as well as to other sports activities.
- *Number of libraries.*
  Another very important issue. The number and distribution of urban libraries enables people's access to Internet, serves as a meeting place for high school students to do their homework, and provides entertainment to citizens through a variety of books. Many libraries also lend music and movie CDs, and offer activities for children.
- *Public health in the city.*
  It is not necessary to elaborate on this issue, since its importance is paramount. Yet it is important for a city to provide easy access to hospitals and health centres, which must both be strategically distributed across the city, and in accordance with population density; this is especially so for hospitals. This category also has to do with special medical programs for the elderly, as well as facilities for transportation and travel for people in wheelchairs, etc.
- *Medical emergency services.*
  Associated, for instance, with the number of ambulances, the response time to a call, the quality of emergency wards, etc.
- *Emergency services.*
  These have to do with the delays in responding to citizens making telephone calls for suspected crimes, for social assistance, etc.
- *Relationships between children and schools.*
  Schools are not the only centres to prepare children intellectually, as they are also where their personality is formed socially. In light of

this, schools should make an additional effort to promote social interaction in children without distinction of race, color, or religion. They also have to promote cultural events for children to participate in, and for them to learn their vocations.
- *Safety*
  Indicators can be used to have a measure of safety in the city, such as crime per 1,000 inhabitants, size of the police force, percentage of crimes solved, percentage of road accidents, etc.

## 3.8 Sustainability in public health

Social sustainability involves access to good health care on an equitable basis for the entire population. This is a very difficult issue indeed, because of the costs inherent to providing health services; as a matter of fact few countries in the world offer 'free' and adequate government-based health services. Of course, it is never free since it comes from taxes paid by the population, yet it is free in that a person does not have to pay for access to an array of medical services, including surgery, while in other countries, private entities provide these at a cost, which is usually high. One indicator of this benefit could be, for instance, the percentage of the GDP that each country spends in health care; however, this does not measure the quality of the services provided.

It is necessary to consider that a healthy population — which mirrors a good and affordable health care program, as well as the access to adequate food and water — is in reality a natural capital or asset for a country. Therefore, the health care system is not a burden for the economy but a very sound and profitable investment. When health care works in an equitable basis for all citizens, without distinction of race, age, social position, etc., then it can be said that sustainability has been achieved in this sector.
Naturally, sustainability in health care links up with other issues. If the air in a city is highly polluted, its water quality is poor, or many vacant lots are infested with rodents, etc., obviously the connection between health care and environment is missing. The mission of a health care system is not only treat sick people, but also to make efforts to help people stay in good health; for that, it must work closely with local authorities in order to eliminate those circumstances or roots that can lead to a poor health.

From this perspective, health care systems must relate to the City Hall agency responsible for the infrastructure to build and adequately maintain hospitals, emergency services, and health care centres. It must also work with local authorities to eliminate potential focuses for diseases, such as cesspools, lack of adequate treatment for sewage, measures to secure clean water, air, road safety, etc. Of course, these sanitary measures give rise to considerable investments in infrastructure, elimination of slums (section 3.4), social programs, etc., in turn involving a necessary link with the economic sector.

There could be good indicators to express how much a country spends on public health, but that is not enough; they could merely reflect that the country has a costly health care program, perhaps with a large part spent on medical research, excellent hospitals, and very fast emergency services. Any indicators must be linked to others signalling the quality of the service, for instance:

- The percentage of people that, for economic reasons, cannot afford medical assistance.
- Children dying of malnutrition.
- Children with diarrhoea due to the quality of potable water.
- Average time people spend in hospitals waiting for surgery.
- Average time people spend in hospitals waiting for consultations with specialists.

If these indicators show high values, it is evident that the system is not working, no matter how high the percentage spent on health care.

Regarding public health, many cities have banned cigarette smoking indoors, that in restaurants, malls, offices, etc. This is a sustainability initiative promoted by the medical profession jointly with City Hall, and it is of course opposed for economic reasons by tobacco companies and many restaurant owners. This is also a social equity measure, and consequently leading to a sustainable gain as it guarantees the right of the vast majority of people to enjoy a meal without the smell and fumes of cigarette smoking. On top of that, it is probable to have other indicators linked with this banning, as for instance one showing the number of people affected with lung diseases provoked by smoking or second-hand smoking.

## 3.9 Sustainability in education

Education should, of course, be a fundamental part of the sustainability process, since it is the building block for a society to enrich its human capital. The success of an educational system depends to a great extent on the quality of training, instructor levels, and curricula. If the main activity of a certain region is, say, agriculture, it makes sense to emphasize careers related to this activity. This way, people graduating from the region's schools will stay in the area instead of migrating to other places looking to use the knowledge they gained.

From the social perspective, education is very often the key for people to elevate their standard of life, to improve employment opportunities, to get better access to new technologies, and a road to enhance their status in society. From the sustainable point of view, people's education is a very strong ingredient that is economically related to the region's prosperity, as it can promote the installation of industries wanting to take advantage of very valuable pools of people. Nowadays, many areas around the world have experienced the benefits of this policy, such as Bangalore, in India, which is one of the most important informatics centers in the world related to the software industry, due to its large cluster of experienced engineers. The same can be said for Singapore, Córdoba (Argentina), Krakow (Poland), and other places.

Education plays a very important role in the resilience (section 1.9) of a city or region. This is evident when a city does not have a diversified industrial base but depends on only a few select industries. When one of these industries collapses, for whatever reason, people with different abilities have more opportunities to find another job (section 6.12).

Sustainability also has to do with the physical installations dedicated to schooling. From this perspective, the use of very well known international indicators helps to measure the physical dimensions of rooms and laboratories, establishing the amount of floor space per student. One indicator also relates the number of students per teacher, in order to keep a good ratio between educator and students.

In many developing countries the reverse often occurs; that is, classes at every level are crowded with students, some do not even have chairs, and require sitting on the floor and taking notes in one's lap due to the lack of desks. It is obvious that the students need a minimum level of comfort for learning (section 7.8.5). It is also evident that a single teacher can hardly deal with maybe as many as 50 or 70 students, as there is no time to answer questions, to help students, mark exams, etc. For this reason, in the sustainability process one needs to work with these and other indicators to ensure quality of schooling.

Independently of what sustainability means in the educational sector by itself, as mentioned above, another fundamental and very important role is the responsibility of educators and adequate curricula. It is here, in the school, where the principles of sustainability are to be implanted in the minds of children. It is here where they learn to appreciate nature, to respect wildlife, and to care for their city's cleanliness. It is here where they find out how to save and reduce their consumption of water, energy, and raw materials. These are the people who will manage the world in a not-too distant future; consequently, if plans are drawn up for development that is sustainable, one of the best investments society can make is to educate the people who will be responsible for implementing it and keeping the measures outlined at present.

## 3.10 Sustainability in commerce

This author spent his childhood in what is now called a 'developing country'. At that time, FM radios, colour TV, DVDs, faxes, and computers did not exist, and for instance one had to call a phone operator to place a long distance call even within the country.
Of course, that was another world with different values and customs; different, but probably no better than we have now. Actually, many people believe that in many ways a person can now enjoy many more opportunities, especially women and young people, that society is much more free of prejudices, and that visiting other countries is no longer for the few but possible for many.
At that time, everything was bought in bulk at the corner store, including sugar, flour, beans, or cheese: the grocer just weighed it, wrapped it up, and collected his or her money in cash, since credit and debit cards did not exist. What does society offer today?

A supermarket uses an immense amount of space, with thousands of items for the customer to simply grab, in hopes that people will impulsively obey the drive to purchase, acquiring items they had no intention of getting and which they do not really need. This is fine, however, since everyone can make their own decisions about whether or not to buy something.
Obviously, the ways that people shop have also changed: instead of buying in bulk, as previously, a person now has dozens of different items that come in myriad sizes, wrapped in very expensive and glazed paper, printed in six different colors, and detailing the dietary content of the product (despite this information, in affluent countries people ingest more calories than ever, and obesity is becoming a serious problem).

Would not it be better to have everything in bulk, in different bins where people can help themselves, as we find in many bulk stores today?
Of course, everybody realizes that when they buy brand names they pay for the packaging, for the TV advertising, the prices on offer, the 'vitamins' that are supplemented, etc. Do people honestly need all this? Do people really need fairy characters talking to children out of cereal boxes to induce them to ask their parents to buy that particular product? Surely not. **This is unsustainable commerce**, but is this progress?
In fact, in the last few years there has been an increasing demand for 'no-name products', which usually are made by the same manufacturer that produces brand names, the packing being the only difference, but at a much

lower price. Of course, 'progress' cannot be stopped, but perhaps a mid point between the extremes can be found. Why can society not mix the old with the new, and adopt some good practices from the old grocery system but update that with help-yourself service and the aid of biodegradable plastic bags — which of course did not exist in those years? **This would make for sustainable commerce.**

Society could do this mix, but the will is necessary. If people started rejecting brand names and buying no-name brands instead, if more people shopped in bulk stores, obviously the brand names would have to give way, purchasing costs would go down, the packing and the costly printing would disappear, and the environment would benefit through the saving of hundreds of thousands of trees felled to produce this packing.
Even if the packing were recycled — and there are limits to its recycling — the environment would still be damaged with the de-inking processes, which involve very heavy metals.
Let us consider another example: the automobile industry. Why is it necessary to have a new model each year? Nobody is against progress, but do people really believe that a model of say a Toyota or a Honda car will be substantially different from one year to the next, especially considering that plans for new cars are made years in advance?
Is it sustainable to buy a new model because it has electrically-operated windows, or because the seats are now heated? Does this not sound a bit absurd?

What about magazines? There are hundreds of glossy paper magazines — of course in full color, at the expense of wasting lots of chemical products — entirely devoted to gossip and 'yellow press', and making their publishers very rich indeed. Who needs this garbage?
Naturally, nobody is against freedom of speech, but in a world running out of resources priorities must somehow be established. Do people have no idea of the environmental costs of producing such a fancy junk?
How many trees have to be felled — because, of course, these magazines do not use recycled paper! — how many chemicals are used for ink, how much fuel is wasted in transportation? Where do these items end their brief and useless life?
They are sent to a landfill, because not everything is recycled. Does the reader have an idea of the damage to the environment caused by the heavy chemicals contained in the printed paper if there is — as there usually is — a break in the landfill's plastic lining, since it has a limited life span? **This is unsustainable commerce.**

In North America, at Christmas-time the supermarkets display in their yards rows upon rows of small pine trees cut just to be used for maybe two weeks as a Christmas symbol.
When the season is over, one can see them again sadly piled up on street curbs waiting to be collected for disposal. Is this not a shame? There is a dichotomy here, since on the one hand, children are taught to respect trees, and there are even children programs to plant them; on the other hand, society uses them as a toy to keep an ancient tradition with dubious origins. What about the nutrients that the tree consumed to grow? How does one replace these? This author considers this case to be one of the best examples of humankind's culture of waste.

The same reasoning applies to thousands of products on the market. How is it possible to stop this waste of resources? Simply by not buying the useless ones, by not being influenced by bright and fancy packaging, and by teaching our children to look for quality and price, instead of following the 'advice' of senseless advertising. It is obvious then that the only way to reduce consumption — for instance, in the case of the car — is to keep cars for a longer period, until the cost of maintenance is so high that it makes sense to buy a new one. **This is sustainable commerce,** but, of course, it has to originate in the consumer, since the manufacturer will be happy to sell you a new car every year.

Take another non-sustainable practice: cigarette smoking. Cigarette smoking is carcinogenic, and many measures have been taken to reduce its consumption, including advertising, hikes in taxes to tobacco companies, and other actions. However, consider the increase in taxes. Is it effective? No, it is not, because the tobacco companies just pass on the increase to the consumer. Consumption of tobacco has lessened in North America. Does if affect the tobacco companies located there? No, as their market is now Central and South America. So, what is the solution?

Maybe to very heavily tax tobacco companies with the prohibition to pass on the increase to the consumers, or perhaps to more heavily tax tobacco plantations, or limit their acreage and prohibit imports of tobacco from abroad. The burden that cigarette-smoking puts on society, considering hospital and treatment costs, lost productivity, etc., is immense, and the whole population, smokers and no smokers alike, are paying for that. Therefore, it is unsustainable because there is **no social equity**.
As can be seen in all of these examples, sustainable commerce is linked with society's behaviour, and with the population's good will to put an end to senseless consumption; once again, the solution lies in the hands of the consumer.

## 3.11 Reducing energy consumption

Strategies for reducing waste at the household level have already been discussed in section 2.2.4, so that need not be revisited. This section will instead summarize reducing consumption in both energy and land use. A good procedure to reduce electrical consumption in large office buildings is to have an energy audit made. By far the largest consumption would probably be lighting, using up more than a 40 percent of usage, followed by water pumps and fans, using in the order of 25 percent. The third largest consumption belongs, predictably, to elevators, with about 10 percent. So, only three items account for about 75 percent of energy consumption. Consequently, an energy retrofit project makes sense, possibly saving not only energy but money too. Table 3.3 shows suggested procedures that can be adopted to save energy.

*Table 3.3*      **Suggested procedures to reduce energy consumption**

|  | Procedures | What is thereby reduced? | Additional information in: |
|---|---|---|---|
| **In the household** | | | |
| | Use fans instead of air conditioning | **Energy** consumption | |
| | Switch off lights when a room is not being used | **Energy** consumption | |
| | Install reflecting coats in windows, and have spacers filled with argon gas | Between 30 and 50% **energy lost** through windows | |
| | Adjust the thermostat for the hot water tank | **Energy** from gas or electricity. If from gas, decreases the use of fossil fuels — while reducing one's electric bill | |
| | Lower the temperature settings in winter and at night | **Energy** consumption | |
| | Use the new type of electric bulbs that last longer and utilize less power | **Energy** consumption, because they consume about 5 times less energy that incandescent bulbs | Incandescent light bulbs give off as heat 90 % of the energy that they receive from the grid |
| | Plan the laundry with efficient loads | **Energy** consumption. Also reduces the electric bill | |

| | | | |
|---|---|---|---|
| | Use non-conventional energy sources such as photoelectric panels on the roof | **Use of fossil fuels** and related air pollution | Actual examples in sections 5.3.2.1 and 5.3.3.1 |
| | Insulate walls and roofs with adequate insulators | **Transfer of energy** toward the outdoors in winter, and indoors in summer | Section 4.5 |
| | Install timers in hallways and other parts of the house | **Energy** consumption and longer life for lamps | |
| | Buy appliances complying with EEC (see Glossary) requirements. For washing machines use models that offer different settings in accordance with loads | **Energy** consumption | |
| | When purchasing electronic instruments, such as computers, buy those with a certified logo. Same for appliances | **Energy** consumption | |
| | Use heat pumps | **Energy** consumption, since the equipment works in much more favourable conditions | Actual example in section 5.2.8.1 |
| **Industry** | | | |
| | Research Drastic reductions in energy consumption have been achieved by the iron and steel industry, and large disparities between regions remain | **Coke** consumption Requires high temperatures to produce; in India, for instance, due to research, the energy requirements to produce 1 tonne of crude steel have come down to 18 GJ (GigaJoules), from 40 GJ | |
| | Use cogeneration (CHP) This is the generation of heat and power | **Use of fossil fuels** | Efficiency in these plants can reach very high values of more than 80%, while the average non-CHP plant has an efficiency of about 33 % |
| | Use waste hot water discharged by one | **Energy** consumption | Section 4.1.4.1 |

| | | | |
|---|---|---|---|
| | process into another | | |
| | Use the thermal energy contained in flue gases from boilers, and exhaust from diesel engines | **Energy** consumption | Actual example in section 2.5.1 |
| | Improve insulation in industrial furnaces | **Transfer of heat** and reduce fuel use | |
| | Replace or buy oxygen furnaces instead of open-hearth furnaces | **Energy** used. There could be substantial savings | |
| | Improve technology in high energy based consumers | **Energy (about 90%)**, when smelters use scrap aluminium, instead of raw material. Also, when steel makers utilize scrap metal and electric arc furnaces, it reduces energy consumption by about 1/3, compared to using ore to get pig iron | |
| | Replace carbon-/oil-/gas-fired power plants with fuel-cell power generation units. These devices are themselves integrated energy systems | **A large quantity of fossil fuels.** Conventional power plants have a low efficiency (less than 30%), due to heat losses in flue gases, in condensing water, insulation, etc., and in mechanical work as in the case of diesel engines. Fuel cells are themselves co-generators, since they generate electricity at a high efficiency and also deliver water at a high temperature that can be used to produce more electricity or for heating | Examples of actual installations in sections: 5.3.5.2 for PAFC 5.3.5.3 for PEM 5.3.5.4 for MCFC 5.3.5.5 for SOFC 5.3.5.6 for AFC 5.3.5.7 for DMFC |
| | If an incinerator is used, utilize heat energy in flue gases | **Energy.** This heat can be used to heat water for other processes or uses | Section 2.5 and actual example in section 2.5.1 |
| | Take advantage of hot gases leaving furnaces for hardening and annealing, and from other equipment | Ditto | Actual example of waste heat usage in section 4.1.4.2 |
| | If possible use waste steam or waste hot water from another industrial | **Energy** | Actual example in section 4.1.4.1 |

|  | plants located in the area |  |  |
|---|---|---|---|
|  | Install automatic control devices in industrial burners for boilers, heating, ventilation and air conditioning (HVAC), waste water treatment units, etc. | **Energy** and **emissions** |  |
|  | Replace normal or regular electric motors by those with variable speed units, and adapt their speed to loads | **Energy** and **greenhouse gas emissions** |  |
| **Offices** |  |  |  |
|  | Install low-consumption lamps in buildings | **Energy consumption** Providing the same amount of light, incandescent lamps consume between three and four times more power than compact fluorescent lamps (CFLs). | Reduces heat to the environment, which in turn requires energy for cooling |
|  | Put out office lights at night |  |  |
|  | Install motion sensors connected to employees' desks and equipment (lights, monitor, etc.), so they turn off after a certain time that employees are away from their desk, such as at lunch time | **Energy** |  |
|  | Use more efficient transportation equipment | **Fossil fuels** | Actual example is section 5.5.3 |
|  | Use a more efficient system of street lighting | **Energy** |  |

### *3.11.1 Reducing energy consumption in the urban space*

Urban spaces are large energy consumers, both in electric energy and in fuels. The former usage has to do with street lighting and traffic lights, while the latter goes on the motorized equipment that City Halls need to perform their functions. It also necessary to add the electricity consumption for lighting and operating municipal buildings, warehouses, and different maintenance shops. However, this section deals only with street lighting, since the other forms of consumption have already been commented on in other sections.

For street lighting, the issue is not only saving energy but also, and as its main purpose, to supply good illumination. This means that a project to reduce energy consumption in street lighting systems must look at the whole system, that is, the amounts of lighting and the electric lamp that they accommodate. However, the height and type of the illumination, its degree of tilt, the type of reflector involved, its orientation, the distances between poles, and other such technical aspects are not the concern here. This section is only about two main issues: energy reductions and emissions.

Energy reduction
Regarding energy reduction, normally different types of lamps are used, each one with different characteristics. However, assuming that a small percentage of old incandescent lamps are still around, each averaging about 250W each (for all types), the electricity consumption could be estimated at about 50 kWh/year per inhabitant.
A retrofitting of the street lighting system may involve changing the optics of the light source and, perhaps most important, replacing old high-consumption lamps with lower-consuming, more efficient units. Old incandescent and mercury vapour lamps are inefficient and should be replaced with high-pressure sodium fixtures. For the same light, the new lamps can have a power in the order of 100W each, and with a 10W electronic ballast. Electricity savings in the order of 30% have been achieved with actual retrofitting. Another issue to be considered with street lighting is transmission and distribution losses, which can reach significant values.

Emissions
Most people do not realize that street lighting has also a direct relationship to global warming. Why is this?
Because the electricity consumed could have been generated by carbon-/oil-fired power plants that create greenhouse gases, such as $CO_2$. Thus, when analysing this subject, one has to consider that the savings in energy could be directly related to emissions savings. However, for calculation purposes, the analyst has to take into account the origin of the purchased or generated energy, since the effect will not be the same if the city buys its energy from hydroelectric stations or from coal-/oil-fired power plants.

So, depending on the energy source or origin, the savings in greenhouse gas emissions can have very high values. To get a large-scale and dramatic appreciation of this, it helps to arrive by plane in a large city at night; while the vision is extraordinary, so are the emissions produced to generate that sight.

### 3.11.2 Reducing land use

Table 3.4 lists suggested procedures to reduce land use.

*Table* 3.4   Suggested procedures to reduce land use

| | Action | What is thereby reduced? | Additional information in: |
|---|---|---|---|
| **Municipal** | | | |
| | Establish bylaws related to population density | Land use | Section 3.2.4 |
| | Establish bylaws regarding use of agricultural land around the city | Utilization of green areas | |
| | Encourage construction of high rises | Population spread | Section 3.2.4 |
| | Encourage construction of underground parking lots | Street congestion and land misuse | Section 4.11.1 |
| | Establish good connections in the city | Fuel and time resulting from unnecessary trips | Section 4.8 |
| | Discourage the construction of new developments far from downtown when vacant lots are available | Spread of the city. Need to extend utility services (water, sewer, lights, transportation, roads, etc.) | Section 3.2.4 |
| | Promote the use of car pools | Cars on the streets | |
| | Promote the use of dedicated routes for transit | Congestion | Actual examples in sections 3.3.1 and 4.8.2. Section 4.8 |

## Internet references for Chapter 3

**Source**: U.S. Department of Energy (2004)
Title: *Energy efficiency and renewable energy network (EREN) - Smart community's network*
Comment: Explanation of agreements between communities and industries.

Address:
http://www.sustainable.doe.gov/success/good_neighbor_project.shtml

**Source**: DuPont (2002)
Title: *Business and sustainable development - A global guide.*
*Brief Information on some recycling programs made by Dupont in the U.S.A.*
Comment: DuPont and its Carpet Reclamation Program.
Address:
http://www.bsdglobal.com/viewcasestudy.asp?id=123

**Authors**: H. Bartelings *et al* (2001)
Title: *Economic incentives and the quality of domestic waste: counterproductive effects through 'waste leakage'*
Comment: Study on the introduction of a fee for household waste collection.
Address:
http://weber.ucsd.edu/~carsonvs/papers/589.pdf

**Source**: Thermie (Name of an EU program to reduce energy consumption) (2004)
Title: *EHEN European Housing Ecology Network*
Comment: Paper from the European Housing Ecology Network (EHEN) for identifying and developing good housing practices.
Address:
http://europa.eu.int/comm/energy/en/thermie/ehen.htm

**Source**: Welsh School of Architecture (2002)
Title: *Solid waste pneumatic collection system in the historic center of Leon*
Comment: Excellent information about the installation of a waste pneumatic system in the city of Leon, Spain. The photographs clearly show part of the piping, as well as drop chutes.
Address: http://www.cf.ac.uk/archi/research/cost8/case/waste/leon.html

**Author**: Chandur Bathia
Title: *Air pollution in Agra*
Comment: Measures to control pollution.
Address:
http://fp.thesalmons.org/lynn/india-pollut.html

**Source**: Mail and Guardian (2000)
Title: *Stink over South Africa's foul air*
Comment: Problems for pollution originating in four oil refineries, especially sulphur dioxide.
Address:
http://www.climateark.org/articles/2000/2nd/stinkove.htm

**Source**: The Global Development Research Center (GDRC)
Title: *Planet Earth: Data snapshots of our urban legacy*
Comment: Useful and appealing snapshots of information in the demographic, environmental, social and climatic change scenarios. From the perspective of its spatial impact, for instance, or the country's ecological footprint (section 1.6), this article maintains that The Netherlands needs 15 million of hectares of agricultural land beyond its borders
Address:
http://www.gdrc.org/uem/data-snapshots.html

**Source**: Population Action International
Title: *Sustaining water population and the future of renewable water supplies*
Comment: Comprehensive information about water resources and population. For instance, its Table 2 details water withdrawal related to the percentage of renewable water resources.
Address:
http://www.cnie.org/pop/pai/h2o-toc.html

**Source**: I.C.L.E.I. (2002)
Title: *Local strategies for accelerating sustainability - Case studies of local government success*
Comment: Comprehensive information and case summaries on India, the US, Brazil, Germany, South Africa, Japan, Colombia, Canada and Norway. Some strategies discussed include strengthening local governments, urban planning, poverty issues, developing resiliency (section 1.9), etc.
Address:
http://www3.iclei.org/localstrategies/summary/index.html

**Source**: The World Bank Group (2001)
Title: *Upgrading urban communities – A resource for practitioners*
Comment: A good introduction to this subject with 'before and after' photographs of upgrading. Abundant information on this very important subject. Especially recommended is a section devoted to implementation, where costs, project sustainability, and community involvement are explained. Also worth examining is the alternatives section, which analyses the option of building new houses. As the paper rightly states: *"By tearing down houses you also tear down social networks"*.
A visit to this site is highly recommended.

Address:
http://web.mit.edu/urbanupgrading/index.html

**Author**: George Matovu (2000)
Title: *Upgrading urban low-income settlements in Africa: Constraints, potential and policy options*
Comment: Particular areas of interest are: government responses to low-income settlements, aided self-help, upgrading low-income housing: Past experiences and evaluating community participation in upgrading programmes.
Address:
http://www.worldbank.com/urban/upgrading/afrnd/docs/mdp_low_income_upgrading_paper_2000_final_doc.pdf

**Source:** Stoveland Consult (2002)
Title: *Willingness to pay* (for upgrading slums)
Comment: Short discussion about this willingness to pay is a subject rarely addressed in the relevant literature. These studies are fundamental to determining the payback capacity and ability of slum dwellers to repay the loans. Go to 'Willingness to pay, Lagos 1997' to learn about:

- Methods used to collect data.
- Results.
- What people wanted for the future.
- What people were willing to pay for the different services.
- Summary

The great value of this study is that it refers to actual cases, undertaken in 37 small towns in Nigeria.
Address:
http://www.stoveco.com/wtp.html

**Source**: United Nations – Habitat (2004)
Title: *Best practices briefs*
Comment: This paper has to do with housing, access to financing, land and secure tenure. It describes several projects in Vienna, Sarajevo, The Philippines, East London (South Africa), the Maweni Squatter resettlement scheme in Voi, Kenya, and cases in other countries.
Address:
http://www.bestpractices.org/bpbriefs/housing.html

**Source**: U.S. Environmental Protection Agency (2003)
Title: *Carpet*
Address:

http://www.epa.gov/epr/products/carpet.html

**Source**: American Farmland Trust (2001)
Title: *Carbon Sequestration: A win-win strategy for America's farmers*
Address:
http://www.farmland.org/policy/issues_carbon.htm

**Source:** City of Austin
Title: *Transportation, planning and sustainability (2002-2003)*
Comment: An excellent report packed with information on sustainability in transportation and planning. It includes discussions on: goals, key indicators, and has bar graphs for reporting on different issues, such as gallons of water saved, number of days this region ozone levels exceed standards, reductions in travel time, etc. Visiting this site is recommended.
Address:
http://www.carvermuseum.org/budget/02-03/downloads/ab0203_v1_494.pdf

**Source:** Regional Transportation Improvement Plant – California (1997)
Title: *1997 transportation indicators for Southern California*
Comment: This report compares transportation indicators between 1994 and 1997, with projections to 2020. It portrays impressive figures — such as that the number of hours of delay due to highway congestion, in millions of hours/day, shows an increase of 332 percent. It also shows a healthy decrease of lone driving of 8.4 percent, providing usually hard-to-find information on things such as the increase of people working at home or telecommuting (+ 6.4 percent). Confirming other studies, it shows that the average decrease in vehicle speed in highways was of 30 percent. It is interesting to replicate what the report states about "...*in spite of the billions being spent on the subway and commuter rail, the percentage who travel by trains and buses will actually drop from 6% to 5% (although the absolute number will be greater than presently"*).
Address:
http://www.scced.org/scced/sccedinfo/trans_indic97.html

**Source:** Second Nature – Education for sustainability (2002)
Title: *Education for sustainability*
Address:
http://www.secondnature.org/efs/efs.html

**Source:** Institute of Environmental Studies – The University of New South Wales (1999)
Title: *Education for sustainability – Integrating environmental responsibility into curricula: A guide for UNSW faculty*
Comment: Extensive paper (48 pages) that comprises many disciplines such as the arts, the built environment, engineering, law, commerce and economics,

medicine and science and technology. Argues that instructing students at university is one of the most powerful ways to produce the sustainability leaders that the world needs.
Address:
http://www.ies.unsw.edu.au/Documents/EducationForSustainability.PDF

# CHAPTER 4 - INDUSTRIAL APPROACH TO SUSTAINABILITY

**4.1 Sustainability in industry**

There are three main areas that account for the influence of industry on the environment:

- Contamination through the manufacturing process of the air, water, and soil.
- Waste production during the manufacturing process.
- Resource consumption.

Normally, no clear distinction exists between them, and, more often than not, the three of them blend. Nevertheless, a synthetic description and obviously incomplete list of reduction measures for each one follows:

*4.1.0.1 Reductions in contamination*

Because of its sheer volume, industry is one of the major producers of contaminants, yet it is also where the greatest impact for improving the environment can be achieved. Industry contaminates the environment in its three main dimensions: air, soil, and water, to such an extent that it is able to render many water streams lifeless, many tracts of land sterile, and has atmospheric consequences with effects that are hundreds of kilometres from their source.
Many consequences stem from coal-fired power plants, chemical installations, and oil refineries — probably the industries that pollute the air the most. The oil and chemical industries can also pollute the soil by leakages, and even innocent-looking industrial operations such as pig and chicken farms, along with livestock farming, also contribute to degrading the air with methane and $CO_2$ emissions, while also contaminating both water and soil with the nitrogen and phosphorus in manure.
Many industries in urban areas, such as tanneries, vegetable oil processing plants, sausage manufacturers, dry-cleaners, metal shops, printers, food manufacturers, etc., are also responsible for large-scale contamination.

However, the common denominator in all these cases is that measures can be taken to reduce or even eliminate emissions and contamination — naturally, depending on the type of industry. For example:

- Installing electrostatic filters and scrubbers in power plant stacks, with a very clear policy for disposing of the gathered dust.
- Replacing materials and supplies utilized for certain processes. More environmentally friendly, less dangerous products can often be used: as, for instance, by adopting new refrigeration fluids, since leaks of some refrigerants can have and have had very severe consequences on the atmosphere. Another example relates to a solvent called 'trichloroethylene', a very popular chemical used in industry to degrease metal pieces by means of vapour. This product poses health risks for workers through prolonged inhaling; when evaporated, it remains in the atmosphere. Several alternatives exist, such as:
    - Replacing it, at a higher cost, with other chemical compounds.
    - Improving the performance of trichloroethylene degreasing units. In this regard, manufacturers are reducing emissions by 90 percent, through very small workshop concentrations.
    - Using aqueous-based cleaners at high temperature, coupled with ultrasonic tanks.
- The same holds for the utilization of some agricultural insecticides and pesticides (section 4.3).
- Industrial operations producing noxious fumes, such as paint cabins used in automobile assembly plants, or processes like electroplating, should be furnished with approved air filters, and procedures for the cleaning of filters and treating the sludge that eventually forms.
- Wastewater should be treated and reused in numerous industries (from semi-conductors to the food processing activity), with the goal of recovering metals and chemicals that dissolve, either in suspension form or as a residue.
- Water treatment plants should be incorporated into the building of industrial plants, so as to discharge water in the same condition and quality as when it was taken; or, even better, in order to reuse it. For existing plants, a period should be set for complying with this norm.
- Fumes from furnaces should also be treated to eliminate dangerous gases as well as particulate.

- Hazardous waste should be stored in hermetic steel drums, with a policy of periodic checks for leaks.

A positive step was taken in 1988 with the creation by the chemical industry of the Responsible Care Program (see Glossary), in an effort to decrease the consequences of mismanagement in producing and using chemicals.
This concept is closely related to 'Product stewardship', a centered-product approach to environmental protection (see Internet references for Chapter 4).

*4.1.0.2 Cleaner Production and other approaches*

Cleaner Production has been defined by UNEP (United Nations Environment Program; see Internet references for Chapter 4), as *"...the continuous application of an integrated preventive environmental strategy to processes, products, and services to increase overall efficiency, and reduce risks to humans and the environment. Cleaner Production can be applied to the processes used in any industry, to products themselves and to various services provided in society"*.

This is a preventive approach to environmental management that encompasses all the concepts that have been analysed regarding the environment, and which has been applied in many countries within many industrial activities of various kinds. It can also be measured using an indicator that gauges the ratio of economic efficiency to the cost to the environment.
The numerator of this ratio is computed using Total Cost Assessment, and its denominator employs Life Cycle Assessment (see Appendix, section A.5). In other words, this approach goes **beyond normal accounting** because it incorporates **costs to the environment** into the design of a process or product. (See: Design for the Environment in section 4.1.6.)

Total Cost Assessment (TCA) was developed in the early 1990s by the Tellus Institute in Boston, US. Very briefly, this system works by computing all **costs** involved in a new process, as well as all the **savings** that it produces. A financial analysis then takes place that considers the time factor, and consequently a discount cash flow method (see Appendix section A.2) is used.

*4.1.0.3 Waste reduction*

The tendency to date has to be changed so that industries recover products at the end of their life-cycles so as to revert them, as far as possible, to their initial components. This would mean replacing industries' open, linear processes with sustainable closed cycles that lead to no waste (again, as far as possible). (section 2.8).

Some suggested measures to achieve this include:

- Increase research to reduce the ratio between the weight of a finished product and the weight of all the materials involved in its production (section 2.10).

- Design products to increase their useful life, even if this is at odds with a company's economic interests.

- Use lighter materials, as in the case of cars, airplanes, rolling stock, etc., to decrease the amount of power needed to operate them, since it wastes energy.

- Reduce packing to a minimum, and increase the use of recycled packing materials.

- Some industrial plants should adopt a policy of encouraging customers to send back certain items that can be immediately reused. A current example of one such policy that has been in force for a number of years involves returning toner cartridges used for laser jet printers. This is a 'win-win' situation because customers get substantial reductions in price when they buy refills within a reused cartridge; the factory does not need to order new cartridges, and the environment benefits with this repetitive use.

- It is necessary **to reuse** as much as possible. In the case of water, one automobile plant in Europe was consuming water from a surface source and was ordered by the local government to cut its extraction by 95 percent, and was even prohibited from boring wells to draw water from the aquifer. The plant designed a procedure to indefinitely reuse its industrial water, replenishing only a small part lost through evaporation. It even profited by practically ceasing to pay for the water it used. Another win-win situation.

- More research is needed so each industrial plant can find new uses for the wastes that are particular to its industry. For instance, ashes from waste incinerators can be processed to recover valuable metals.

- Limiting the kind of industrial waste allowed to be deposited in landfills.

- Encouraging the donation of certain products, such as paints, for use by schools, hospitals or low-income people.

- Working closely with a waste broker to find other industries that can use industrial surplus or waste. Table 4.1 outlines the different destinations that an industrial plant can have for its surplus.

*Table 4.1*  **Destination of industrial surplus items**

| Industrial surplus items | Components | Condition | Destination |
|---|---|---|---|
| Sheets, bars, wire, containers | Steel, zinc, aluminium, copper | Brand new or reusable | Another industry through a brokerage system |
| Metal scrap | Steel, zinc, aluminium, copper, lead | | Smelters |
| Chemical products | Any | Either surplus of produced as by products by the plant | Another industry through a brokerage system |
| Packaging material | Wood boxes Steel containers | | Used in the plant. Another industry through a brokerage system |
| | Plastic packing | | Reuse (if possible). If not, once crumbled, utilize as filling for different purposes, including as packing material for the company's products |
| Fluids | Water | | Purification and reuse in plant |
| | Steam | | Sale to other industries Use in secondary processes Use in heating purposes Use in electric power generation |
| | Lathe cooling oils Lube oils | | Recycling or sold as fuel for cement kilns |
| | Hot water | | Sold to other industries. Used for heating |
| Cinder/ashes | | | Sold as filling material or to extract metals |
| Wood, liquid waste, sludge | | | Can be used to generate electricity |

*4.1.0.4 Reducing rejection*

In some areas this is closely related with waste production, since normally the more the waste, the greater the consumption of a raw material. As an example, consider a plant manufacturing blades for razor cartridges. Sometimes, whole batches are discarded because they do not comply with norms of quality control. These blades are made out of steel strips, and their waste is a misuse of a resource (stainless steel), which can be prevented. Certainly, the discarded batch can be re-smelted, but that consumes energy, transportation, labour, etc. Therefore, **improving efficiency** in the production process so as to avoid rejecting such material reduces consumption, and of course saves money.

Another good example of rejection due to mistakes can be found in the automobile industry, which for years has been recalling thousands of vehicles to fix something wrong or dangerous. This is akin to other industries, such as photographic camera manufacturers, who have had to receive and repair a brand new product for fixing and replacement while within the guarantee period.
The paper industry provides another example, as when high-speed machines have to be stopped for some reason and hundreds of metres of paper are lost in the process.

*4.1.0.5 Reengineering*

Reengineering in a sustainable context could be defined as the action of having a fresh look at a certain process and finding ways to improve it that consider two main factors: technological change and commercial and technical competition. It is thought that the meaning of this definition is obvious; nobody can manufacture something and remain competitive in a market with other players while ignoring new materials, using old equipment, or not taking new techniques into account. There are countless examples, such as the assembly of automobiles, where computerized robots, plastics, new alloys and new concepts, superseded previous ways of doing things that of course are very far from the assembly line pioneered by Henry Ford. Perhaps the single best word to condense the meaning of reengineering is **change**.

From the viewpoint of sustainability, this change has fortunately already taken place in many industrial plants where concepts considered natural twenty or thirty years ago are now unthinkable. Not long ago, it was thought a natural part of industrial processes that industries produce contaminated emissions that are spewed by smokestacks, consume water indiscriminately, make poor use of electrical energy, produce and 'store' toxic wastes very loosely, use raw materials very inefficiently, etc.

Today, things have changed, and the environmental movement and the global economy, which among other things make for very strong competition, make it necessary to constantly look for ways of improving processes — in other words to **reengineer** them.

On the other hand, resources are not the only constituents that physically participate in the final artefact. Some inputs are not part of the final product yet are essential to its making. This is the case, for instance, with electricity, water, reactive elements, heat, etc. Water is probably one of the most critical, and from the viewpoint of sustainability, industries should make every effort to reduce consumption of this vital element. Experience shows that there is normally a better alternative regarding its use if research is conducted or processes are changed to reduce water utilization. Some results indicate that:

- Some trades have drastically curtailed the amount of process water and the discharge of useful material that is transported by wastewater. One clear example is the meatpacking industry, which now consumes only a fraction of the water used years ago for animal processing. This has also led to a better recovery of edible products and process blood. One example, in the form of a complete report, is contained in Randolph (see Internet references for Chapter 4).

- The importance of research can be understood by examining other industrial examples that point to very good results. For instance, a highly contaminating plating industry in Tunisia (see Internet references for Chapter 4) reengineered its entire processes using Cleaner Production Principles (4.1.0.2), as they were having serious pollution problems and excessive water consumption levels. The analysis showed, among other things, the following problems:

    o  A waste of organic solvents;
    o  unnecessary waste of some critical materials, such as chrome and nickel; and
    o  excessive use of water.

  After the study, the following results were obtained:

    o  Reductions in toxic emissions for workers and for the environment;

- reductions in consumption of raw materials and water; and
- reductions in the concentration of heavy metals in wastewater.

- Local bylaws should set targets for industries, specifying allocations of water for industrial purposes. A number of communities have taken harsh measures to limit water consumption by industries, golf courses and other large users; some cities no longer issue permits for water-guzzling industries such as paper-makers, chemical plants, rubber manufacturers, etc.

The following example depicts a simple case were reengineering a process can contribute to reducing production costs and wastes.

**The water pump manufacturer example**

A large firm manufactures pumping equipment for different users. One of its products is a type of centrifugal water pump that has diverse applications in harsh environments such as oil platforms, ships, wastewater treatment plants, chemical plants, etc. The pump casing is made out of cast iron, and its impeller of stainless steel alloy, while the driving shaft is made of another stainless steel alloy. Because sales are declining — due in part to their competition asking lower prices — the manufacturer began a study that showed that production costs are higher than expected and the main factor responsible appears to be too much material wasted in the manufacturing of the pump shaft.
This is a piece of precision equipment, using as a raw material round steel bars that come in commercial sizes, and with intensive tooling to bring this commercial size to the design-calculated diameter. Its processing includes adequate machining of both ends to accommodate, in turn, the electric motor and the impeller, along with a further thermal treatment for hardening and annealing. The firm produces 20,000 pumps per year.

The study found areas where larger-than-normal costs led to unnecessary waste in tooling the shaft.
To make the shaft, the process starts with a round steel alloy bar that comes in different diameter sizes; in this case, the closest commercial size for the final diameter design is 5.13 cm (2"). It has to be reduced by tooling the original size to a diameter of 5 cm, thereby wasting in the process a total of 1,740 kg of material (for the 20,000 shafts), which is sent, mixed with cooling oil, back to the steel manufacturer for re-smelting.

The manufacturer was unaware that a new type of steel alloy that is more expensive and stronger than the one he is using had been on the market for two years. His technical department calculates that a shaft made with this new alloy would need only a diameter of 3.7 cm, instead of old one's 5 cm, with the closest commercial size being 3.84 cm (1.5"), and then producing a total waste of material of 953 kg for the 20,000 shafts. Their study also indicated that savings in material would offset the superior price of the new alloy, so the replacement operation would be economically feasible. See Table 4.2.

*Table 4.2*     **Material waste identified in a reengineering process**

|  | Final size needed [cm] | Production of 20,000 shafts ||
|---|---|---|---|
|  |  | Total weight of steel needed [kg] | Material removed by tooling to bring the shaft to specified final size [kg] |
| Old steel alloy | 5 | 1,740 | 92 |
| New steel alloy | 3.7 | 953 | 21 |
| Savings in raw material |  | 787 | 70 |

This reengineering study has therefore already shown advantages in costs and from the technical point of view, since the new alloy has a greater resistance to corrosion than the old one, which is an added sales feature.
Here is a list of advantages that this change will produce for the environment, bearing in mind the 20,000 shafts.

- **Reduction in material wasted** because the use of a bar with a smaller diameter, amounting to 787 kg.

- **Reduction is material wasted** in tooling, taking into account the differences in sizes available and the required final sizes for both alloy bars, amounting to 70 kg.

- **Reduction in fuel** to transport bars from the distributor to the pump manufacturer, due to the difference in weight (787 kg).

- **Reduction in fuel** to transport scrap steel (from the tooling process) from the pump manufacturer to the smelter, due to the difference in weight (70 kg).

- **Savings in tooling oil,** in both cases because there is less material to process.

- Corresponding **electric savings** for the machine shop equipment, since the shaft with the smaller diameter needs less tooling time.

- **Savings in fuel** for handling and transporting equipment (forklift and crane) at the pump manufacturer plant in managing the bars and the metal scrap.

Once the shafts are finished, they are placed in the hardening and annealing furnaces, in trays containing 50 shafts each. Let us assume that, due to some deficient preventive maintenance the temperature in the hardening furnace reached a higher than normal level, say, due to a failure in the natural gas pressure regulator valve, the people in Quality Control declare that those shafts are unfit for assembly, and have to be rejected and sent to the smelter for re-smelting.

The consequences, aside from those pertaining to economics, would be:

- A total **waste of processed material** that has to be sent back to the smelter.

- A waste of **fuel to** transporting the scrap.

Of course, this was an accident and is likely to happen more frequently than one might think. The bottom-line analysis for the whole operation shows that there was wasted processed input (steel bars) and pollution created (fuel and energy spent). Considering that this was only for one single component of the equipment, similar analyses should be made for the whole pump, and for all the other products that the firm produces.
In this very elementary example, it can be seen how the reengineering processes help to obtain savings in materials and energy.

*4.1.0.6 Life Cycle Assessment (LCA)*

This is a much-used methodology that analyses a product from beginning to the end (for more details, see also Appendix section A.5). That is, it considers what goes into the constitution of a product beginning with the extraction, harvesting, or mining of the necessary raw materials, the processing, distribution, use, and any potential reuse, all the way to its final disposal. In so doing, it tries to quantify all the inputs for said product by considering raw materials, water, air, etc., as well as how much waste is generated. However, LCA is not satisfied with the direct input and direct effects and wastes, as it tries to analyse these effects and wastes in a second, third, fourth tier, and onwards.

For instance, in the example of the shaft for the water pump mentioned in section 4.1.0.5, only some effects and wastes were considered starting with its input: a round steel bar. An LCA analysis would also take into account the whole **supply chain** (see Glossary), that is:

- The energy and materials spent to produce the **round steel bar** through hot-rolling a steel ingot.
- The energy and materials spent in producing the **steel ingot** out of pig iron in Bessemer converters (see Glossary).
- The energy and materials needed to get **pig iron** in a blast furnace, out of iron ore and using coke (see Glossary).
- The energy and materials needed to mine **iron ore**.
- The energy and materials needed to obtain **coke** (see Glossary), out of coal, for the blast furnace.
- The energy and equipment needed to mine the **coal** and then to further produce coke.
- Raw materials and energy needed for the manufacture of **mining equipment**, including steel.
- And so on.

It is clear that a **backwards** analysis traces the origin of the steel bar used to make the pump shaft from start to end in the above list (following the words in bold type). LCA analysis considers not only the 'vertical' supply chain but the 'horizontal' one as well. For instance, the energy spent to make the steel to build a Bessemer converter and a blast furnace is also considered.

*4.1.0.7 Input-Output model*

This model, developed by the American economist Vassily Leontieff in 1936, is very important, since it can be used to compute the **direct** and **indirect effects** in an economy for a certain kind of production.

It considers in a matrix the relationship that exists among all industrial activities in a country; for this reason it is sometimes also called an 'inter-industrial relationships matrix', an 'Input-Output matrix', or I/O for short.

The matrix places all commercial and industrial activities in columns; for instance, fishing, steel-making, chemicals, agriculture, mechanical production, textiles, etc. The same industries also appear in rows. For example, an international chemical firm may decide to install in a country a plant for producing several types of plastic textiles and industrial gases and fluids, mainly from hydrocarbon sources, with a total yearly output or production of 10 billon euros. Therefore, the **row** corresponding to the chemical industry is considered.

Taking into account the corresponding raw materials for this industry, there will be a number or coefficient at each intersection of this row with the columns. For the sake of simplification, assume that it is only related to other two sectors: the petrochemical and steel industries. See Table 4.3.

In this case, sales of the chemical industry's production (shadowed row) will be distributed as follows (sales):

- 1 billion euros go as sales to the chemical industry itself (sales to other chemical firms).
- 1 billion euros go as sales to the petrochemical industry.
- 2 billion euros go to the steel industry.
- 6 billion euros in response to public demand for chemical products.

Something similar happens with the steel industry, for instance, that sells:

- 3 billion euros to the chemical industry.
- 3 billion euros to the petrochemical industry.
- 6 billion euros to the steel industry.
- 3 billion euros in response to public demand.

*Table 4.3*   **Relationships between the chemical industry and other industries**

|  | Chemical industry | Petro-chemical industry | Steel industry | Demand | Total production |
|---|---|---|---|---|---|
| Chemical industry | 1 | 1 | 2 | 6 | 10 |
| Petro-chemical industry | 4 | 2 | 1 | 17 | 24 |
| Steel industry | 3 | 3 | 6 | 3 | 15 |

Now, let us consider the first column and its coefficients.

1 billion euros reflects the purchases by the chemical industry of goods from the chemical industry itself (from other chemical firms), which enables the production of a final output of 10 billion euros. Evidently, the relationship between purchase and output will be $1/10 = 0.1$

4 billion euros stands for the purchases by the chemical industry from the petrochemical industry. Here the coefficient will be $4/10 = 0.4$

3 billion euros represent the purchases of the chemical industry from the steel industry. This coefficient will be $3/10 = 0.3$.

An analysis of the other columns under the same concept will produce the matrix depicted in Table 4.4.

*Table 4.4*     **Industries' specific consumption**

|  | Chemical industry | Petro-chemical industry | Steel industry |
|---|---|---|---|
| Chemical industry | 0.1 | 0.0416 | 0.1333 |
| Petro-chemical industry | 0.4 | 0.0833 | 0.0666 |
| Steel industry | 0.3 | 0.1250 | 0.4000 |

Assume now that market studies and projections give a value of the demand for each sector in billions of euros. Combining these demands with the above matrix, and using a computer procedure it is possible to determine how much the output of each sector of the economy should be (remembering that only three sectors are posited in this example), in order to satisfy these demands.
This is the essence of the Leontieff I/O model. Naturally, this brief explanation is just a hint of the whole model. Interested readers have hundreds of publications on this method to turn to for more information. (Leontieff, 1951).
Nowadays most of the countries develop the matrix shown in Table 4.3 for all industries or industrial sectors (the US's Table has 485 sectors).

### 4.1.0.8 Environmental Input/Output model

This is the application of the Leontieff model (section 4.1.0.7) to environmental issues. In this case, the results of the model **give the amount of pollutants, in tons, derived from a certain activity**.

Let us take again the example of making water pump shafts, from section 4.1.0.5. That analysis was just for direct and indirect effects of such manufacture, and in section 4.1.0.6 the remark was also made about how important the application of LCA is to the whole supply chain. Now, using the Leontieff method, is it possible to estimate the **total amount of releases of**

**conventional pollutants into the air** while considering each sector in the supply chain.

For this calculation assume that the total manufacturing cost of a single shaft is US$ 274.49. As a consequence, to produce 20,000 shafts, the total cost would be 20,000 x 274.49 = $5,489,800; this is the value to be used to calculate pollution.

What is it that one wants to compute?
The **total amount of contamination produced in the supply chain** to make 20,000 shafts.
How is the contamination measured?
By computing the total tonnage of the following pollutants:

- Sulphur dioxide ($SO_2$);
- carbon monoxide (CO);
- nitrogen dioxide ($NO_2$);
- volatile organic components (VOC);
- lead particulate emission; and
- particulate matter (PM10) (less than 10 microns in diameter).

To perform this calculation it would help to use the method developed by Carnegie Mellon University's Green Design Initiative (2004). (Economic Input-Output Life Cycle Assessment (EIO-LCA) model [Internet], Available from: <http://www.eiolca.net/> [Accessed 22 May, 2004].)
Accessing the above-cited publication:

1. *Sector name/number:* Write: 'Pumps' and hit 'Search'

2. *Select one of the following commodity sectors for analysis.*
   Choose: 'Pumps and compressors'

3. In the next window:
   Select: 'Conventional pollutants'

4. *Level of increased economic activity (producer price)*
   Write: 5.5 (million)

5. Hit the bar *'Display data for selected sector'*

6. The following result (Table 4.5), will show (while the data in this Table comes from this source, it is not the original Table that resulted from the analysis of this Internet address; it has been modified for this example using the format all other Tables in this book).

*Table 4.5* **Tonnage of pollutants for proposed example (results obtained using Mellon University's Green Design Initiative software)**

| Sector | $SO_2$ [MT] | CO [MT] | $NO_2$ [MT] | VOC [MT] | Lead [MT] | PM10 [MT] |
|---|---|---|---|---|---|---|
| Total for all sectors | 16.858662 | 28.556532 | 13.733691 | 3.765484 | 0.014088 | 2.783687 |
| Electric services (utilities) | 9.999128 | 0.320516 | 4.893757 | 0.039990 | 0.000235 | 0.252680 |
| Blast furnaces and steel mills | 1.886951 | 6.060088 | 0.577759 | 0.220815 | 0.000862 | 0.347101 |
| Paper and paperboard mills | 0.287664 | 0.297202 | 0.175692 | 0.076218 | 0.000005 | 0.045855 |
| Industrial inorganic and organic chemicals | 0.260983 | 0.229547 | 0.215316 | 0.142382 | 0.000156 | 0.025735 |
| Carbon black | 0.194016 | 2.951275 | 0.023635 | 0.111997 | 0.000000 | 0.009311 |
| Petroleum refining | 0.192840 | 0.174129 | 0.120925 | 0.095631 | 0.000001 | 0.013580 |
| Crude petroleum and natural gas | 0.161482 | 0.112461 | 0.271383 | 0.080424 | 0.000000 | 0.003260 |

This analysis covers the complete supply chain for the manufacture of 20,000 steel alloy bars, and, as indicated, it will put (in rounded-out values) into the atmosphere:

- 17 metric tons of $SO_2$, which is the gas blamed for producing acid rain;
- 28 tons of poisonous CO;
- 14 tons of nitrogen dioxide;
- 3.8 tons of VOC, a large contributor to smog and that helps in the formation of ozone ($O_3$), another pollutant;
- 0.014 tons of lead, a very serious pollutant that can have consequences in the kidneys, liver and nervous system;
- 2.8 tons of particulate; smaller particulates are deposited deeper in the lungs and remain there longer than larger particles.

Table 4.5 also shows each sector's contribution to the supply chain. According to the source, this calculation takes into account the US's 485 industrial sectors, yet even with this large breakdown, there is sometimes not enough desegregation. For this reason, instead of computing the absolute value of contaminants production it may be better to compare two different scenarios. For instance, for this case it could be the determination of pollutants using the larger diameter steel alloy against the smaller one. Since there is a difference in weight, there also will be a difference in costs and, as such, a difference in pollutants generated.

### *4.1.1 Industrial ecology (IE)*

This expression was coined by Robert Frosch and Nicolas Gallopoulos (1989) in a paper entitled 'The Industrial Ecosystem View', in a special issue of *Scientific American* magazine called 'Managing Planet Earth'. This seminal paper is also known as 'Strategies for manufacturing'. (See also Tibbs in Internet references for Chapter 4.)
Industrial ecology can be defined as the study and analysis of the potential interactions between industrial activity and the environment. That is, until a few decades ago, industry operated with complete disregard for the environment, and this discipline tries to modify this approach by integrating and complementing both in a harmonized way.
It analyses material consumption and process reengineering, and one of its main principles is that instead of treating wastes, it is better not to create them. For instance, the steel industry can very efficiently treat the wastewater it generates in its processes, yet it would be preferable not to use so much. But this is only a part of the concept; another is to use resources in a closed loop, mimicking nature.

If manufacturing worked according to this concept, products would also be designed taking into account their destination at the end of their lives, when considering that the **components of a product should as far as possible return to their original condition**. The industry should understand that it is just 'borrowing' from the environment to manufacture something, but once the product is no longer in use, its components should be returned to the environment, thereby repaying the 'loan'.

This image tries to convey the idea that somehow industry has to close the loop with the raw materials it uses. Figure 4.1 shows an example of this statement applied to the production of paper bags for garden refuse. It should be noted that in this particular example recycling does take place, but with a mix of recycled paper and virgin fibre, which must be added for mechanical resistance.

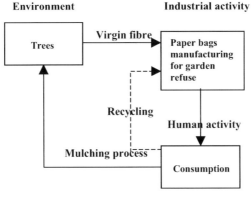

*Figure 4.1*     Closing the loop

This illustrates the need for joint action between the government and the industrial and commercial sectors to establish policies for persuading people to use recyclable paper bags, reusable, rigid plastic receptacles, or any other continuous-use container.

In some countries, this is actually taking place at a national level, where the government, assisted by commerce and industry, is strongly promoting the use of paper bags for shopping as well as other bags made from woven materials. The Plastic Bag Environmental Levy in Ireland (see Internet references for Chapter 4) is clearly an example to be followed.

In this case, supermarkets charge $0.25 per plastic bag, such that since 2001 the number of bags used by supermarket customers has dropped by about 90 percent. The enormous damage that this plastic 'curse' is causing to animals in rivers and in the marine environment, in harbours, littering streets and vacant spaces, very strongly suggests that measures like this have to be adopted. A harsher example exists in South Africa, where retailers face hefty fines if they sell plastic bags.

Another approach that may be worth considering is what this book calls the 'reverse component process'. This means reusing in a process components that have been restored to their **original condition**. Take the PET (polyethylene terephthalate) bottles, for instance, which are used for soft drinks: they are currently recycled to a lower grade product, 'fibrefill', which is then utilized as a filling material, for instance, in pillows and cushions — which, of course, is putting it to good use.

However, a new method developed by a US company is to reduce the scrap to its original components to manufacture again a bottle-grade resin that is the original raw materials for bottles. This is even better, since the industry will not need to use more hydrocarbons — a non-renewable resource — to manufacture these bottles.

Auto body parts, such as bumpers, use another type of plastic called thermosets, which cannot be reduced to its original components, and then use them again. It is in these cases where research should be aimed at getting a reverse component process for plastic of such types.

However, not everything is negative. Many programs and increased research are finding new uses for plastics that cannot be reduced to their original components. Two interesting new applications are as plastic lumber and railways ties. The important thing here is that not only does the plastic waste find its way into useful things, instead of occupying space in landfills, but mainly in replacing wood lumber, thereby saving a lot of trees.

*4.1.1.1 Eco-efficiency*

This is methodology, developed by the BASF German chemical company (see Internet references for Chapter 4), consists essentially of analyzing a product through its lifecycle (see Appendix, section A.5). It thereby studies a certain product from the stage of obtaining the raw material until its final disposal at the end of its life — or, alternatively, when it becomes a component of another product, such as a pigment, for instance. The eco-efficiency method would analyse all the impacts and consequences of this pigment, considering its social, environmental, and economic dimensions. Thus, it develops several options or alternatives that would take into account:

- Materials consumption;
- energy consumption;
- emissions to air, soil, and water;
- risk potential for misuse; and
- toxicity potential.

The result of this analysis for each alternative or option is plotted in a system of coordinate axes, with environmental impact in the **y**-axis and costs in the **x**-axis. From there, it becomes easy to determine the best alternative, option, or process. The process has been enhanced with the addition of socio-efficiency of products and processes.

### *4.1.2 Industrial metabolism*

Metabolism was defined in section 2.10. Industrial metabolism — pioneered by Robert Ayres (Ayres, 2001) — is an attempt to mirror **biological metabolisms**. Industrial metabolism can be broadly defined as the way the flow of energy and materials go through the whole economic process, that is, starting with the initial extraction of natural resources, then through the industrial process plus labour, and then to consumer use, and finally to disposing of wastes. More specifically, it refers to the process followed for the manufacture of a single product, just as 'biological metabolism' refers to the process of a living organism.

In biological metabolism, waste is not an unwanted by-product of some process but a component of a process; in other words, it is a part of a larger process. For instance, in a marine environment (see Figure 4.2), phytoplankton, which are very small plants drifting in the upper layers of the sea, receive nutrients from the bottom of the sea and uses sunlight to photosynthesize them into food. Zooplanktons, which are very small ocean animals, feed on the phytoplankton, and are in turn eaten by other larger organisms and fish. Both of these produce detritus — which is waste from the food chain, from dead fish and tissue, and from decayed plants — and that is decomposed by bacteria into nutrients, which are again absorbed by the phytoplankton, thereby closing the cycle.

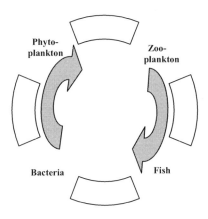

*Figure 4.2*    **The circular flow in the marine ecosystem**

Unfortunately, the human world does not work this way In many processes, waste is just 'rubbish', and as such is discharged into the environment; so instead of having a closed cycle, as in nature, man has it open. Therefore, this pattern looks more like the branches of a tree, as in Figure 4.3, which uses water as an example.

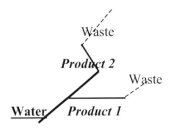

*Figure 4.3*          **The open or linear path of wastes**

This has a triple effect, because it pollutes the environment, depletes natural resources, and increases costs. Industrial metabolism tries to mirror nature in the use of wastes, as depicted in Figure 4.4.

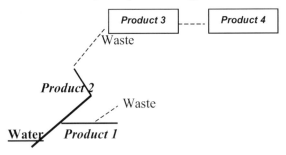

*Figure 4.4*          **Utilization of wastes in an open system**

The cradle-to-grave concept mentioned in section 2.12.6 is wise in that, instead of thousands of people disposing of their waste all around, it is safer to have the manufacturer doing it since the wastes will then be handled just by companies, and properly treated.

It also makes economic sense, since manufacturers will be forced to design their products with easily disassembled components and to recycle them. Naturally, most of the time the consumer is paying for any disposal, as is evident when one buys new tires and is charged a disposal fee, as with changing the oil in one's car.

### 4.1.3 Materials flow analysis

The Gross Domestic Product (GDP) is an internationally used indicator that shows the 'wealth' of a country. This indicator measures all the economic contributions made from the economical point of view, including value added, and is used as a comparative parameter to measure its economic health, to analyse trends, and to establish comparisons. It is the best-known economic indicator; however, it is not adequate for determining the state of a region, since it refers only to its economic aspects. A sustainable indicator would incorporate all of the above but would also *deduct* the hidden costs to the environment and the society of the region.

Consequently, this is an imperfect tool as it does not take into account these social and environmental consequences, or the variations in the natural capital (section 1.12). For instance, a country exporting its crops includes in its GDP the benefits accrued from their sale domestically and internationally. However, many areas do not receive enough rainfall and this is a phenomenon that cannot be predicted with certainty. Such crops will therefore need irrigation, which is undertaken with water extracted from aquifers or from rivers. Water in aquifers carries dissolved salts, and when a farmer uses it to irrigate crops or when the sun evaporates it, salt is left on the soil. Is this problem significant?

This problem is especially serious in arid regions, where there is normally not enough rainfall to carry the salt from the soil; for that reason, the degree of soil salinity is a valuable indicator of soil quality. Soil salinity diminishes the capacity of the roots to take water, thereby altering the soil's carrying capacity (section 6.12) to produce enough food. Some countries have prepared maps of their territory that show in percentages the salt levels in crops areas for different zones.

Therefore, to return to the example of the crop-exporting country, it is probably losing some of its natural capital due to soil salinity. Besides, when the aquifer contains dissolved sodium, this is also responsible for changing soil properties, reducing infiltration from rainfall water and favouring erosion. However, none of these effects show in the country's GDP, and this is the reason why researchers have been looking for some other indicator that is able to measure the extent to which a country is using (or depleting) its own resources (and those of other countries). One of the techniques used is Materials Flow Analysis.
This tool investigates the amount of materials that flow through the economic system of the country within its region. A leading institution in this area is the Wuppertal Institute in Germany.

Stephan Bringezu and Helmut Schütz of the Wuppertal Institute (see Internet references for Chapter 4), have prepared in 2001 a comprehensive report entitled *Material use indicators for the European Union, 1980-1997*, as Eurostat Working Papers. In this publication, they define some indicators for materials flow analysis and extract some conclusions. Table 4.6 draws information and definitions from this paper.

*Table 4.6*        **Material Flow Analysis**

| Indicator | Meaning | Formula |
|---|---|---|
| **Input indicators** | | |
| Direct Material Input (DMI) | Measures the input of used materials in the economy, including imports | DMI = Domestic + Imports |
| Unused extraction (UE) | Material extracted but that has no economic value. The rock extracted while excavating a road tunnel belongs to this category, unless it is crushed and use to make concrete | |
| Total Material Input (TMI) | | DMI + UE (local) |
| Total Material Requirement (TMR) | Includes unused materials from other countries | DMI + UE (from other countries) |
| Domestic Process Output (DPO) | All materials used in the domestic economy before flowing into the environment | - Air emissions in manufacturing processes<br>- Industrial wastes |
| **Output indicators** | | |
| Domestic Processed Output (DPO) | Total mass of materials used in the domestic economy before discharges into the environment. Includes:<br>. Emissions into air<br>. Discharges into water sinks<br>. Industrial wastes<br>. Domestic wastes<br>. Incinerator discharge<br>Recycling is NOT included | |
| Direct Material Output (DMO) | The total amount of materials output leaving the economy after use in the economy and/or in others' economies | DPO + Exports |
| Total Domestic Output (TDP) | Total amount of material outputs to the environment released in the domestic territory from economic activities | DPO + UE (local) |
| Total material Output (TMO) | Includes exports and indicates the total of material that leaves the economy | TDO + Exports |
| **Consumption indicators** | | |
| Domestic Material Consumption (DMC) | Total amount of material used in the economy, without unused extractions | DMC = DMI - Exports |
| Total Material Consumption (TMC) | Primary material requirements associated with domestic production and consumption activities | TMC= TMR – UE(From other countries) |
| Net Additions to Stock (NAS) | Measures the physical growth rate of an economy | Added materials and/or products – discarded materials or products |
| Physical Trade Balance (PTB) | The trade surplus or deficit of an economy | |

| Efficiency indicators | Services provided or economic performance are related with input, output or consumption indicators | Example: GDP/TDO measures economic performance related to material losses to the environment |
|---|---|---|

The above-mentioned publication at the Internet address indicated regularly has excellent information on these indicators for European countries, Japan and the US, and very clear explanations and examples of each case.

### 4.1.4  *Industrial integration*

This section is devoted to commenting on the Kalundborg industrial complex in Denmark, a vertical and horizontal industrial integration that is most probably unique. What is important here is that it began a new idea that may have profound consequences, namely, that of the **inter-industrial materials flow management**. This is usually done within a company, but was never undertaken before involving many industrial plants. The main gains in this project are reflected in these impressive figures:

- Water consumption has been reduced by 25 percent;
- oil consumption decreased by about 20,000 tons/year;
- 80,000 tons of ash are used in the construction and cement industries, instead of going to landfills;
- 200,000 tons of gypsum a year are produced, substituting the extraction of the natural product;
- a large amount of commercial fertilizer is saved by using a complex by-product; and
- many other products are recycled, such as newspaper, cardboard, iron, and other metals, glass, etc.

Fortunately, the Kalundborg experience gained international recognition, and other locations have tried to emulate it. One of the followers of this approach was the region of Styria, in southeast Austria, which has applied a recycling program between companies within the region. The results, as indicated by quantitative values as well as the number of recycled materials, have been just short of spectacular, and include 17 different kinds of wastes. According to the publication in Internet references for Chapter 4, this Styria undertaking has been more diversified and complex than the Kalundborg case.

For a brief analysis or this industrial complex, see section 4.1.4.1, especially Figure 4.5, regarding relationships between and the circuits created among the different products. It is interesting to see how some wastes now produce valuable products, such as sulphur, gypsum and yeast.

Section 4.1.4.2 comments on a metallurgical operation where sulphuric acid ($H_2SO_4$), mercury (Hg), and other metals are produced as industrial by-products.

Section 4.1.4.3 illustrates the utilization of bagasse from sugar cane operations to generate electric energy in Belize, in a waste-to-energy process. In this case, there is a double benefit since the burning of the bagasse both avoids dumping this waste in a landfill while producing electricity that would otherwise have to be imported.

---

### 4.1.4.1 Case study: Industrial integration, Kalundborg, Denmark

*The city of Kalundborg in Denmark is one of the few examples — if not the only one in the world — of industrial integration on a large scale. It also involves different types of industries and users, since eight different companies have developed what they called an 'Industrial Symbiosis', whose main feature is that discarded items from one company constitute some part of the inputs of another.*

*This cycle starts with the Asnæs power plant, which generates electricity and supplies water to the population. Low-pressure steam and hot water are delivered to about 4,500 households, replacing the need for oil-fired stoves. There is also steam taken by an oil refinery and two chemical plants. The fly ashes produced by the power plant are input for a cement plant, as well as for construction undertakings. Cooling water from condensers in the power plant is piped to a fish farm, which produces trout and salmon.*

*Wastes from households and other users are utilized to generate biogas and compost for soil amelioration. More waste from the power plant finds its way into the manufacture of construction boards, and slurry from one of the chemical plants is utilized as fodder.*

*Figure 4.5 schematically shows the different interconnections for this complex*

> *Naturally, it is not often that one can find a set of complementary industries. As specified by the City of Kalundborg, geographical distance between plants is of paramount importance, not only because costs are linked with distances, but also because heated water in this case, as well as low pressure steam, lose their thermodynamic properties because of transmission losses.*

### 4.1.4.2 Case study: A metallurgical process using the thermal content of flue gas for heating and to extract commercial products

*A company located in South America mined an open pit mine to extract gold-bearing minerals. However, as the open pit progressed in depth, the ore came highly contaminated with sulphur. This is inconvenient because sulphur interacts with the floatation process (see Glossary) used to obtain metallic gold. The solution was to remove the sulphur from the ore before it becomes part of this process.*

*The procedure is rather simple. The ore from the open pit is ground into manageable chunks and then fed to a rotary kiln, which at the same time acts as a fine grinder through the action of free-falling steel balls. This pulverization and the heat make the sulphur evaporate from the ore, which can then be safely fed to the floatation process.*

*The resulting gases can just be vented to the atmosphere after a cyclone treatment to remove particles, preventing air pollution, but the flue gas also has a high content of sulphur and other pollutants. The solution was both environmentally beneficial and economically feasible. As the flue gas leaves the kiln, some heat is removed from it and transferred to the cool air entering the kiln, thus recovering energy. The gas is then washed and chemically treated to extract the sulphur, which is used to produce highly valuable sulphuric acid.*

*The residual gas is further processed to get, amongst other metals, mercury, which not only has commercial value but is also a very dangerous contaminant for human health and the environment. The residual sludge can be treated to recover water and then disposed of as a tail in a tailings pond.*

### 4.1.4.3 Case study: Generation of electricity by using residues from other industries. The case of Belize
http://www.ji.org/projects/006/exec_summary.htm

*The economic foundation of this small county located in Central America, with shores on the Caribbean Sea, is sugar production and citrus juices, fruits, lumber and tourism.*

*As in many other countries, it has problems with electricity supply. A hydro power plant and a fossil fuel-fired power plant currently generate electricity, which adds to energy imported from Mexico at a high cost (US$0.21 kWh). The hydro plant is considered unreliable because of low rainfall in the dry season and flooding in the rainy one. A scheme is under study to build another hydro power plant (at Chalillo), but it has aroused a lot of resistance as it is thought that it will not solve the problem, and since it will flood a rain forest and archaeological land. Another option is to install a diesel power plant that uses fuel oil.*

*An alternative is to build a co-generation (that is heat and energy), waste-to-energy biomass plant, fed with bagasse from a sugar cane industry that produces about 400,000 tons of it per year. The energy from this plant, which will be located close to sugar mills, will be used for in-house purposes, the surplus being fed to the national grid, thereby alleviating the economic burden of importing energy. According to certain sources, the bagasse-fired plant will produce 18 MW, so twice the energy expected from the hydro power plant, and at a cost of about US$ 0.06 to US$ 0.07 per kWh. As a secondary fuel, the plant could use citrus waste, mainly orange processing wastes, as well as wood wastes from sawmills and other sources. An additional benefit relates to land use, since in this way the orange wastes will not be sent to a landfill.*

*This book is not the place to discuss the advantages and disadvantages of each option. However, it appears that the waste-to-energy scheme has more advantages from not only the reliability point of view, but also considering the impacts on the environment posed by each option, as well as the burning of fossil fuels in the case of the diesel power plant. Nevertheless, if it is true that the utilization of wastes will reduce land use and the potential*

> *danger of leaching from the landfill, it is also certain that the plant will require water from a river for condensing and other processes, and that the discharged water to the river, although uncontaminated, will be at a high temperature. Another concern relates to the economic sustainability of the undertaking, considering international demand and prices for sugar.*
>
> *The purpose of this brief analysis is to illustrate that waste-to-energy plants can be serious competitors with other energy sources, and they therefore should be seriously considered.*

*4.1.5 Dematerialization*

This involves designing products with a lesser content of raw materials. A good example is provided by the automobile industry, which in the past produced very heavy units with a lot of steel, chrome and other heavy metals. Nowadays, there is a use of plastics, aluminium, and special alloys for the engine, no metal bumpers, etc.; this has produced much lighter vehicles with several benefits over the old ones. The advantages are lower consumption of raw materials, less energy needed to operate the vehicle, less inertia, which means higher acceleration and also more rapid deceleration, the impact of which is a very concrete lessening of brake wear and of the dust they produce, the possibility of using lighter batteries, etc.

Another dramatic example of dematerialization is found in the railway industry, which has sharply reduced the weight of the rolling stock, which means needing less energy for the engine to pull them, less stress on the tracks, a lesser use of raw materials, etc.

As seen, dematerialization also means less consumption of energy to produce something and to operate it, and the use of energy only when necessary. This is an everyday experience when, for instance, someone enters a building and steps into a hallway, and electric sensors switch on and illuminate the site.

176  Introduction to Sustainability: Road to a Better Future

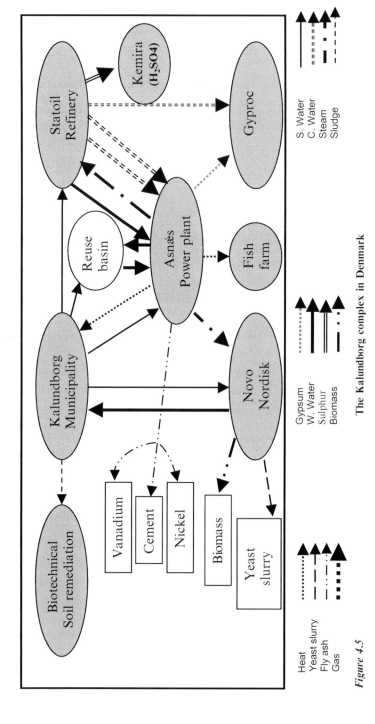

Figure 4.5   The Kalundborg complex in Denmark

Source: Center for Industrial Symbiosis, 2004
Based on information kindly provided by Noel Brings Jacobsen

## 4.1.6 Design for the environment (DfE)

This title is self-explanatory, as it literally means that industry should design its manufacturing with the environment in mind. The components in a normal industrial process are product design, production volume, manufacturing costs, working capital, profit, etc.; however, the environment component is nowhere to be seen. The environment should be considered just as another component or input, or as another raw material of sorts, as is the fuel used for combustion in the boilers, water for processing, resources utilized, etc.

This subject links up with Industrial Ecology (section 4.1.1), where the necessity to involve the environment in the design process is put under scrutiny. How does this work?

For instance, a computer manufacturer designs its equipment in such a way that it can be upgraded during its useful life, leaving room for additional memory cards, video cards, the installation of a CD burner, etc. At the same time, the computer should be designed for easy disassembly at the end of its life.

The following example, whose concepts and material were taken from the publication *Design for the Environment* (see Internet references for Chapter 4) also illustrates how DfE can help.

> **Fabric for office furniture**
> Fabric used for office furniture is usually woven with plastic fibres, which in turn come from hydrocarbons. Colours are obtained through dyes from different origins; for instance, acid dyes contain organic acid groups, and others, like basic dyes, are derived from hydrocarbons, many of them toxic. By entering into an office without enough ventilation on a summer day, it is possible to smell some faint chemical odours that could come from these fabrics, due to weak gas emissions.
> In this example, a manufacturer not only changed the manufacturing process but also the content for its fabric. He uses renewable natural resources such as wool and *ramie fibre* (see Glossary). Dyes to color the fabric are non-toxic, non-carcinogenic, and contain no heavy metals. The fresh water used in the process is treated and ends up in the same state as it started, as pure water. Fabric scrap from the manufacturing process is not waste, as it is made to produce a compost to be used in gardens and crops, where it gradually decomposes.

## 4.1.7 Indicators

Indicators are treated in section 6.1, and the reason that they are included here is their utilization in industry. Indicators provide very useful, albeit sometimes over-condensed, information about the current situation in a particular industry — such as, for instance, the relationship between water utilization per unit of product. Once measures are taken to correct or improve something, the need arises for information about how the process is performing regarding a particular issue. Probably the most common and known indicators in industry are related to Quality Control, where certain characteristics of a product, such as width, hardness or flexibility, are fundamental for its good performance in the hands of its user. Certain characteristics of a product can be nearly impossible to measure with an indicator, and then it becomes necessary to have a qualitative input, as for instance, about the satisfaction or degree of comfort of a customer using a product. One actual example is when some industries, such as companies that make shaving razors, routinely request some of their employees to shave at work in order to get a test result that no instrument nor indicator can supply.

Indicators plotted against time are a valuable tool to determine the trend of a certain characteristic, and to check if values stay within specified limits. A timely and judicious use of this tool can alert its makers that something is happening, permitting them to cease production before too much material has to be rejected.
The most internationally-used indicator, the Gross Domestic Product (GDP) measures the wealth of a country and is useful in establishing comparisons, although it is criticized on the grounds that does not consider externalities (section 1.11).

### 4.1.8 Waste exchanges

This is another concept pursued in many countries. Many industries have surpluses of steel bars and sheets, aluminium profiles, Portland cement, pipes, boards, sidings, etc., than may be needed by or useful to another industry or industries.
The system uses a brokerage organization where all the available material is listed in a computer database. At the same time, other industries' needs are posted in the same manner, so it becomes simple to match surpluses and needs, and put the companies involved in contact. In some ways, this system is sensitive to geographical distances, with items such as steam, for instance, or when transportation costs can offset the economic advantages in price and availability. The following Internet addresses provide good information about waste exchange firms:
Exeter City Council
http://www.exeter.gov.uk/business/b2b_services/trade_refuse/waste_interchange.xml

Cuyahoga County
http://www.cuyahogaswd.org/business/exchange.shtml

City of Cape Town – Integrated waste management

http://www.capetown.gov.za/iwex/iwe4.doc

King County
http://www.govlink.org/hazwaste/business/imex/
Austria
http://www.wk.or.at/ooe/Abfallboerse/Geschichte_Waste_Exchange.htm

### 4.1.9 Comparison of methodologies

A summary of the methodologies analysed is shown in Table 4.7.

Table 4.7    Summary of actions and methodologies for industrial sustainability

| Methodology | Section | Characteristics |
|---|---|---|
| Contamination reduction | 4.1.0.1 | Can be very effective when appropriate legislation is enacted and strict controls are in place. This reduction can also be extended to agriculture, to decrease the content of nitrogen and phosphorous in the soil if there is a willingness to pay for crops with less fertilizers and herbicides |
| Cleaner production | 4.1.0.2 | Preventive environmental management that has proved effective |
| Waste reduction | 4.1.0.3 | Fundamental action that involves a series of different measures |
| Rejection reduction | 4.1.0.4 | Calls for an increase in quality control to avoid unnecessary scrap |
| Reengineering | 4.1.0.5 | By analysing processes by considering new tools and methods, as well as better information, this can reduce consumption of some inputs, such as water, raw materials and fuel |
| LCA | 4.1.0.6 | Very useful technique to analyse the amount of goods and services provided for manufacturing products, in what is called the supply chain |
| Input-Output model (Leontieff, 1936) | 4.1.0.7 | Mathematical tool developed to analyse inter-industrial relationships |
| Environmental Input-Output model | 4.1.0.8 | Very useful technique. Really an application of LCA, but intended to determine the amount of contaminants produced for manufacturing a product. Very valuable to analyse from this point of view the difference between two different processes or alternatives. Software is freely available on Internet for these calculations, with a user-friendly mechanism. Founded in well-established principles and data from US government organizations |

| | | |
|---|---|---|
| Industrial ecology (IE) | 4.1.1 | Important concept that analyses interactions between industry and the environment. Can yield very good results when trying to find better uses for wastes |
| Eco-efficiency | 4.1.1.1 | Concept developed by a chemical company using the LCA approach. However, this method also analyses, besides environmental, social and economic aspects, and employs a good visualization technique |
| Industrial metabolism (IM) | 4.1.2 | Fundamental notion that studies the flow of materials and energy through the economic process |
| Materials Flow Analysis (MFA) | 4.1.3 (Table 4.2) | Analyses the flow of materials and energy in the economy in a system for different industrial sectors |
| Total Materials Requirements (TMR) | 4.1.3 (Table 4.2) | Analyses the flow of materials in the economy in a system for different industrial sectors, but also takes into account the environmental costs of extraction, disposal, as well as foreign components |
| Industrial integration | 4.1.4 4.1.4.1 4.1.4.2 4.1.4.3 | Related to the idea of building eco-industrial parks, in that it agglomerates different industries in a geographic area that can share energy, sewage, as well as having wastes/inputs links |
| Dematerializing | 4.1.5 | Reduces the weight and variety of components in manufactured products |
| Design for the environment (DfE) | 4.1.6 | Tries to include the environment in the process of designing a product; in other words, the environment is considered at the same level as the input(s) for the product |
| Indicators | 4.1.7 | Supply information on some impact produced by a project, a process, or a product. These also allow comparing these values with international and local standards. Much used for monitoring and control purposes |
| GDP | | The oldest and most common measure of economic development utilized worldwide |
| Waste exchange | 4.1.8 | Brokerage between surplus in one company and need from another. A very good alternative for industries trying to get rid of surplus or no-longer-needed raw materials and products |

## *4.1.10 Recycling in industry*

It is unnecessary to emphasize the importance that recycling has for the environment, and also from the economic point of view. Consider, for instance, that more than 14 MW/hr are required to produce one ton of aluminium, while aluminium recycling requires only 5 percent of this amount — to say nothing about the production of contaminated releases to the atmosphere to convert alumina into aluminium. One must also take into account that in aluminium manufacturing many producers have to import the raw material, bauxite, which means a disbursement of hard currency for the country, and the necessity of building an adequate dedicated harbour infrastructure to receive bauxite from overseas. Richard L. Ottinger & Mindy Jayne (see Internet references for Chapter 2) cite an impressive fact when they say, *"The U.S. throws away enough aluminium to rebuild the country's commercial aircraft fleet every three months."* Indeed, no further comment is needed!

If almost everything could be recycled, then society could solve its problems of waste, but unfortunately, it is not so easy. Consider the following very important components of waste: paper and plastics.

**Paper**

Paper consumption in the US is about 340 kg/per person-year. One might as what stops industry from recycling it all, thereby saving millions of trees, millions of cubic meters in landfill space, as well as considering that recycling paper uses about 60 percent less energy than the energy to manufacture it from virgin wood pulp.

The answer is simple: because not every kind of paper can be recycled.

Paper and boards contaminated with food, such as pizza boxes, are not recyclable because it is difficult to eliminate the food residues that usually contaminate this type of packing. The same applies to special papers like those covered with chemical substances, such as wax paper, wax milk containers, and juice boxes.

There is another technical problem related to the number of times that paper can be recycled. Paper is made out of wood fibres, and when fibres are subject to recycling, they break down and become too short to be used for paper-making. Besides, something else to think about paper recycling is that there must be a market to buy products made out of recycled paper.

Many stationary stores offer both types of paper: made with virgin and with recyclable fibre. Not surprisingly, paper made out of recycled paper is a little more expensive than paper made with virgin wood fibre. What is important to remember is that by buying paper made with recycled fibre one helps the environment, and hurts it when buying paper made with virgin wood. This should be the most important aspect to consider, especially when there is only a small difference in cost between these two types of paper.

Fortunately, many big consumers of office paper follow this concept and use mostly recycled paper. Canada is one of the largest paper producers in the world, although directives in the federal government mean that all their own office paper is recycled.

**Plastics**

A greater problem is posed by plastic recycling, and understanding it requires considering the different kinds of plastics available in the market. Although there are many kinds, researchers in the plastic industry have agreed on six basic resin categories — as depicted in Table 4.8. Plastics are essentially polymers, consisting of large molecules of some monomer (a chemical compound such as ethylene, an oil derivative) obtained through a process called polymerization. Each kind of resin is especially suited to a certain purpose, such as to give flexibility, breakability, hardness, floatability in water, ability to be squeezed, etc., that meets the special needs of the product manufactured with it. This is an essential requirement considering the very large variety of uses for plastic products, from simple plastic bags to complex car parts with very different characteristics.

However, this very diversity conspires with the recycling of the different resins, because different plastic products cannot be melted together due to their different melting points, which calls for sorting before the plastic waste can be processed. How is plastic sorted?
In many countries, it is mandatory that each piece of plastic product, especially household items, be identified with a logo that has number inside it. Figure 4.6 depicts the familiar logo used in plastic products.

*Figure 4.6*  **Logo and number for plastic recycling**

The only purpose of this identifying number, which runs from 1 to 7, is to facilitate sorting the type of plastic used in manufacturing a product. The recycling process is simple because after melting the waste plastic, impurities are removed and it comes out as large strands of resin. These strands are washed and ground into pellets, which constitute the raw material for other plastic production procedures.
Unfortunately, not all seven categories of plastics are converted again into plastics. By far only the two first categories, numbers 1 and 2, have widespread recycling use and account for more than 90 percent of plastic that is recycled. Commercial and marketability reasons account for this, since it is cheaper to use virgin raw material than to utilize the pellets from plastics with higher numbers, and also because there is not enough of a market for products that can be obtained with them. This is why many municipalities accept for recycling only number 1 and 2 plastics. This unfortunately means that the balance has to be buried in landfills, using up a lot of space. This is why it may be better to incinerate those plastics, with the savings in land and with additional advantage of producing energy in the process (due to their hydrocarbon content). However, many people argue that this is dangerous because such incineration can produce dioxins and furans (section 2.5).

In sum, it is necessary to consider that recycling, as good as it may be, is, for various reasons, the least desirable alternative among the 'four R's', which are: reduce, reuse, recover and recycle. One of these is that some materials, as just seen, cannot be recycled. Therefore, it is better to follow the hierarchy, which is to first reduce the amount of materials and raw materials, and then to reuse any wastes as far as possible.

Table 4.8 shows the different categories of plastics, their uses, and their recycling capabilities.

Table 4.8 Different categories of plastics, their main uses, and potential for recycling

| Number | Resin | Found in: | Recycled in: |
|---|---|---|---|
| 1 | Poly-ethylene Terephthalate [PET] or [PETE] | Transparent bottles for: soft drinks; water; cooking oil, honey, grocery bags, cleaning products. | This is the most recyclable plastic. It can be converted into: Food containers, polyester fibrefill, sports clothing, sleeping bags, carpeting, shower curtains, bathtubs, tennis balls, car parts, boats, furniture, urethane foam, extruded products, and many other plastic products |
| 2 | High density polyethylene [HDPE] | Non-transparent bottles and containers for: juices, cider and water jugs, hair products, powder detergents, ice cream containers, produce bags food wrap, hard plastic for file containers | Plastic lumber, soft drink base cups, drums, curb-side containers for recycling (blue boxes), toys, medicine containers, detergent, and oil bottles |
| 3 | Polyvinyl Chloride [PVC] or [V] | Credit cards, clear plastic containers, pharmaceutical bottles, pipes, salad dressing bottles, vegetable oil bottles | Packaging, binders, panelling, speed bumps, sewer pipes and telephone cables |
| 4 | Low density polyethylene [LDPE] | Containers for: Food storage, talcum powder Toys Bread. Plastic baskets, garbage bags | Plastic shopping bags, mailing envelopes, garbage cans, floor tiles, playground equipment |
| 5 | Polypropylene [PP] | Plastic fibres for upholstery and luggage, food containers, bottle caps | Battery components, ice scrappers, pallets, storage bins |
| 6 | Polystyrene [PS] | Hard plastics for VCR cassettes, TV and computer frames and casings, CD containers, plastic silverware, carry-out containers, meat trays, packing foam | Office supplies, egg containers, foam packaging |
| 7 | Other resins Made using layers of | Milk cartons, leach bottles, snack bags, microwaveable containers, squeezable | Mainly plastic lumber, roadside posts. Not a great deal of recycling |

|  | different components | bottles |  | potential, because the different constituting layers are difficult to separate |
| --- | --- | --- | --- | --- |

### *4.1.11 Conclusions on industry*

While there is no doubt that the industrial approaches to sustainability are a subject that has just begun to be enunciated, and is far from being discussed at any depth, it is probably within industry where the largest contributions to sustainability will be made. From the Industrial Revolution to the Computer Revolution, industry has shaped society's life in many different ways. How one lives and learns and travels and spends one's free time, etc. — all of these and more things about our lives have been affected by the influence of industry.

Most things people enjoy — such as air conditioning, air travel, access to remote learning, health care, education, easy, fast and cheap phone communications, entertainment, advanced medicine, etc. — depend on industrial goods and services. They have created individuals who are completely different from human beings who lived at the beginning of the $20^{th}$ century, touching on areas that include even our needs. Unfortunately, it has also begun a frantic and insatiable demand for goods that industry tries to fulfil, in the process using large quantities of natural resources — both renewable and non-renewable — and producing increasing contamination of the air, soil and water. The price society is paying for this is too high, and so severe as to change the atmospheric conditions of the planet.

Yet there appears to be an awakening about the problems caused by industry. It looks as if people are worrying about the future, and about their children's children future. Perhaps there is still time for the industrial community to take pause and reconsider the whole scenario, and to make a commitment. The good news is that many, many people, very many industries, and most governments are beginning to take some action, although not yet in a coordinated effort.

Naturally, it is easy to blame the industry for everything, yet that would be unfair. Industry would not exist if there were no demand for its products, and this points to society. Obviously, people are influenced in our buying through advertisements, prizes, better performance, more power, miraculous effects, etc. Yet the problem is that a person **should buy only what one believes one needs, not what advertisements make us believe is needed**; from this perspective, what the human race needs is to be educated.

Just as children are taught to brush their teeth before going to bed, or not to throw candy wrappers in the street, people should be taught to care for the environment. People should be convinced to ask for more durable products, for less packaging, for more efficiency in cars and appliances, etc., and, of course, taught how to save energy, decrease the amount of wastes generated, and use less water.

Last, but not least, it is the responsibility of governments to enact bylaws to change some habits. This could be interpreted as dictatorial, although sometimes harsh measures are necessary to preserve the rights of the majority. Many cities around the world have banned cigarette smoking indoors, including in offices, theatres, restaurants, shopping centres, etc. Some of these measures have been in force for a couple of years. What is the result?
Everybody in those cities can now enjoy a meal without contending with smoke; the number of pulmonary diseases due to second-hand smoking is decreasing, and, surely, more disposable income is available as people are smoking less and no longer buying cigarettes.
As the main producer of manufacturing goods, industry should take steps on its own, as a contribution to a sustainable society. These steps could be:

- Substantially increase efficiency in industrial plants in water and energy consumption. To this end, water should be used and reused in a continuous, repeating cycle. As for energy, when using fossil fuels efforts should be made to employ co-generation. Industries ought to take advantage of renewable energy sources, perhaps not for their production needs but for air conditioning and heating, for hot water and another uses.

- Treating their own sewage and using that treated water for toilets and to water their lawns. It should be mandatory for new plants to have double water piping systems: one for drinking, cooking, and sanitation, and the other for toilets. A time limit should be established for old plants to follow this practice.

- Smokestack emissions should be strictly controlled, including the projected destination(s) of any dust collected from any of its filtration. Processes that control the spewing of harmful emissions should be improved, perhaps by changing a process or equipment, or by stopping them.

- With the cradle-to-grave policy, industrial plants must reengineer their products for a complete reuse at the end of their lives. **Recycling them is not enough**. Industrial plants should strive to find methods, procedures, and/or solutions to recover from wasted products the components that were used to manufacture it. This is currently done for almost all metals and glass, and should be expanded to other items. In some countries, governments have put a deadline for manufacturers that establish that by 2010 85 percent of a car should be recycled. Again, even as good as this sounds, it is not satisfactory to require that 85 percent of a car should become original products. As we saw in section 2.12.6, GM plans to manufacture a 100 percent recyclable aluminium car that would completely revert to its original raw materials.

- To get a final single product with the minimum amount of material. This was discussed in a basic example, where it was assumed that it was possible to make a lighter shaft for a pump using aluminium, instead of steel. It is necessary to consider the relation between the initial diameter of the aluminium bar and the final dimension of the shaft. The bottom line is the need for profound research in each industrial plant about improving current production methods, the use of new materials, the acceptance of new products in the market, the analysis of new technologies, etc.; in other words, to perform reengineering analysis of each production process, including of course water and energy usage.

- **Integration** is a key word. Perhaps the most convenient solution in the future is constraining the location of industrial plants to industrial parks, where the surplus and/or waste of one industry could be the input for another. This horizontal integration can have multiple benefits regarding transportation, general expenses, safety, etc.

- Restrictions should be enacted regarding packaging. Many people will argue that this, in essence, is a manufacturer's prerogative; and so it is. However, when one considers that packaging occupies a large percentage of each landfill, then it becomes everyone's problem, and the government must intervene to defend a very scarce resource: land. Table 4.9 shows, by volume, the content of package in landfills in selected countries.

*Table 4.9*  **Amount of packaging in landfills**

| Country | Packaging |
|---|---|
| Canada | 33 % |
| Florida (USA) | 28 % |
| New Zealand | 28 % |
| Germany | 50 % (in 1991) |

See also 'Packaging' in Internet references for Chapter 4.

## 4.2 Sustainability in transportation

Transportation is important as it affects every aspect of human life, and because energy spent on transportation accounts for more than 20 percent of the world's total primary energy use (Riley, Internet references for Chapter 4). This is a very complex issue with its own definition for sustainability, given that **the main purpose of an urban transportation system is to provide access to people's destinations such as those for education, health, shopping, work, entertainment, institutional and banking activities, etc.** Therefore, transportation must fulfil the three pillars of sustainability as follows:

- Provide frequent, fast, and reliable communication for people and freight.
- Not be an economic burden for the society it serves, as a subsidized service.
- Be available to every geographic sector of a city;
- Provide a viable alternative to the use of cars;
- Fares charged should be affordable to the poorest sectors of the population;
- Offer a safe journey;
- Interact with the environment by having the least impact on it.

These indeed amount to a tall order, because some of these goals are in various ways incompatible. For instance, to offer service that is frequent and fast calls for more units on the streets, involving increased expenses that translate into increased fares. It also means much more pollution and congestion, militating against the goal of having the least impact on the environment.

Many choices are now available for people's transportation. In an urban scenario, these options have developed as a result of urban dynamics: people moving to the outskirts of cities, population growth, affordable cars and gas prices, seeking greater comfort, etc. However, these have also produced many side-effects, such as longer travel times, excessive pollution, lack of public transit, accidents, traffic congestion, etc. Table 4.10 shows a comparison between eleven different modes of transportation available today, considering thirteen different criteria to gauge them. The Table is self-explanatory, although some explanation of the modes and especially of these criteria is in order.

**Modes of transportation**
Busways:
These do not refer to 'dedicated lanes', that is, lanes in urban streets only used by buses, usually during designated hours alone. These involve

'dedicated roads' for buses' exclusive use, so are therefore free of any other kind of traffic, at all times.

Electric car sharing:
   This is a new concept in that a pool of electric cars is available — through a monthly fee — to transport people from their homes to their destinations (see below under 'Economics').

**Criteria**
Different modes of transportation are evaluated according to their compliance with certain criteria. In this example, Table 4.10 shows these criteria and comments on qualitative measures about each mode.

*Economics*
This has to do with operating a system profitably. Many public transit systems are currently government subsidized, because money collected in the fare box is insufficient to pay for the associated investment, maintenance, fuel, wages, etc. In many cases, it is not a matter of ridership, because even high volumes of passengers will often fail to collect the necessary funds for a system's operation.
The economic dimension of car ownership is paramount, since the expenses to own, maintain, and operate a car are usually greater that the fixed cost of riding the public transit system. In fact, the analysis will mostly turn on the number of hours a car is driven every day. It makes no sense to own a car and have it parked for at least eight hours a day, while the owner is at the office, and then for about another ten hours, parked at night in a garage.

Meantime, owners are paying for their investment and the corresponding depreciation, insurance, parking charges, etc., just to have their car lie idle over long periods. Perhaps an analysis should be done to compare the degree of satisfaction caused by car dependency compared with all these economic costs. This is why the new notion of the 'car-share' is beginning to take hold, and gain converts.
This idea tries to maximize the benefits of car ownership by many different people sharing its use over the whole day. In addition, being electric and used for short rides, it can have its batteries recharged overnight, which makes environmental common sense. The concept is indeed appealing, since it not only maximizes the use of a resource but does so without polluting the environment while freeing up parking space. Its practical application is another matter, and its success remains to be seen.

*Environment*
In this column, some appraisal will be made of each mode of transport. Obviously, any mode of transport utilizing fossil fuel damages the

environment and destroys property. Thus, modes that use electricity and the ways of generating it will be considered. After all, environmental concerns are prompting a technical revolution in search of friendly alternatives to car use, not the overuse and decline in fossil fuels.

While many different solutions are being put forth, ranging from solar energy to hybrid engines, given very stringent bylaws regarding air pollution it would appear that the fuel cell, whose only wastes are $CO_2$ and water, which is the preferred solution. The electric car, a very old idea indeed, dates from the early 1900s and has not yet found general approval due to its relatively short range and the need to recharge its batteries.

## *Social*
Social criteria analyse transportation modes in accordance with their impact on society. In many cities, because of the extensive use of the car, some of the following inconveniences arise from this social perspective:

- Not everyone owns or can drive a car, so they must use public transit. However, when it exists, this service –may have low ridership, be infrequent, and involve lengthy travel times.

- Highways take up a lot of valuable land, which could be better used for housing, parks, or commerce.

- Aside from these inconveniences, citizens may suffer due to delays in public transit because of jams, especially in the CBD (Central Business District), and from polluted air generated by cars' emissions.

Table 4.10  Transportation comparison

| Transportation Mode | CRITERIA | | | | | |
|---|---|---|---|---|---|---|
| | Economics | Environment | Social | Infrastructure | Land use | Accidents |
| Walking | | | | Need sidewalks, bridges, traffic lights | | Yes |
| Bike | Subject to ridership | | | | Bikeways needed | Frequent |
| Streetcar | Subject to ridership | Good | Good | Reasonable | Sometimes dedicated right of way needed | Unusual |
| Light Rapid Transit (LRT) | Subject to ridership | | | Expensive. Capital intensive | Can be a blend of surface, underground and aerial routes | Unusual |
| Subway | Subject to ridership | | | Very expensive. Capital intensive | | Unusual |
| Train | Subject to ridership | Damaging | | Extensive. Capital intensive | Dedicated corridors needed | Unusual |
| Busways | Subject to ridership | Damaging | | | Dedicated corridors needed | Unusual |
| Bus | Subject to ridership | | | | Street space | Possible |
| Electric Car-share | | | | | Street space | High chances |
| Private car | Expensive to maintain | Produces harmful emissions, which intensify during idling periods | Promotes social inequity, because of infrastructure needed for car owners is paid for by everyone | Extensive and capital-intensive to build and maintain roads, bridges | Street space, and also needs space for parking lots | High chances |
| Electric car | | | Promotes social inequity, because of infrastructure needed for car owners is paid for by everyone | Extensive and capital-intensive to build and maintain roads, bridges | Street space, and also needs space for parking lots | High chances |

| Transportation Mode | CRITERIA | | | | | | | |
|---|---|---|---|---|---|---|---|---|
| | Relationship weight and passengers | Health | Contribution to traffic congestion | Accessibility | Travel time | Delays | Stress |
| Walking | | Excellent benefits | | Limited | Long | | |
| Biking | | Excellent benefits | Yes | Limited | Long | | |
| Streetcar | High | | Yes | Large | Short | Unusual | |
| Light Rapid Transit (LRT) | Because of high volume, its is moderate | | | | Short | Unusual | |
| Subway | High, because they are usually trains | | | Limited by closeness to lines | Short | | High with some people suffering claustrophobia |
| Train | Very high | | | | Usually short | | |
| Busways | High | Generates air pollution | | Limited by closeness to lines | Medium because no traffic and stopping in bus stations | Feasible but not probable | |
| Bus | High | Generates air pollution | Yes | | | Feasible | |
| Car-share | Medium | | Yes | Good in CBD areas | | Feasible | High, due to driving, delays and traffic jams |
| Private car | Very high | Need to keep a body position for long periods. | Very large | Large | Extensive because of average speed, traffic jams, road repairs, accidents, etc | Very frequent because of traffic conditions. Also provokes delays for people riding the public transit | High, due to driving, delays and traffic jams |
| Electric car | High | | Yes | Limited by battery | Extensive | Very frequent | High |

In many cities, City Hall has important and especially visible undertakings, such as drilling tunnels to connect two parts of the city separated by a hill, building urban intersections, or improving car traffic in certain thoroughfares. There is no doubt that these public works will enhance connectivity in the city, facilitating traffic, and probably helping with avoiding road accidents; so, from the infrastructural perspective of these are necessary undertakings. But what about the social component?

If questioned, City Hall will probably defend the idea on the grounds that these undertakings will benefit the population. Yes, but what population?

These undertakings only benefit people driving cars, so what about those who do not have the means or the desire to own a car? Since it appears that these works will benefit a minority, where then is the social equity? On the contrary, these works create a huge inequity from the social point of view, as patrons of the public transit system have to finance something that they do not use.

Naturally, nobody can be against progress in a city, and no doubt these works will indeed bring progress. And yet public monies come from all segments of the population, so it would be fairer to make a *selection of projects* to be executed with those funds that is considering not only transportation projects, but also projects related with other areas.

It is also essential to *reach a consensus* about projects to be financed using public monies, and this is not always easy. For example, assume that two main problems are evident in different sectors of a large city. The low part is in a valley and runs along a river, and has a history of street and house floods during heavy rains due to a lack of appropriate storms drains and ramparts in the riverbanks. The other part of the city is more commercial and at a higher elevation, and has serious traffic-related problems, causing traffic jams due to the few lanes on a main avenue and some very busy intersections.

Each sector believes that its problem is most important, and will offer many reasons to support its claim. People from the high part of the city who lack the threat of floods probably think that the delays produced by traffic jams, the disproportionate travel time, the accidents, the fuel consumption, time lost, etc., are more important than sporadic flooding in the lower part of town. In turn, those in the lower part of town may consider that flooded houses, floating furniture and homeless people are by far more significant that traffic congestion.

It becomes necessary to *prioritize different projects* to consider the social, economic, and environmental implications of each side, considering that usually there are not enough funds to develop them all. Chapter 7 addresses this issue.

From the social point of view in relation to transportation, it is sometimes sad to witness how short-sighted 'planning' and economic interests can destroy, instead of improve, something that works and is benefiting everyone. Consider the case of the city of Los Angeles, for instance, which until the mid-50s had an excellent streetcar network that extended 60 kilometres from downtown, covering not only that core but also 42 incorporated cities, including far away settlements such as San Bernardino, and doing all this without any government subsidy.

It appears that economic interests — which many people attribute to a giant bus manufacturing company, and a large tire producer – led to the system's collapse, replacing it with smelly, air-polluting buses. The city of Los Angeles has now built two LRT lines and one train line, but what is ironic is that the first line very closely follows the route used by the streetcar more than 50 years ago. Some analysts try to explain the change by saying that ridership was at that time declining because of the city's phenomenal expansion, and its more intensive use of cars, but what probably should have been done was to expand the rail system and not to build more and more highways.

Fortunately, hundreds of cities around the world, especially in Central and Eastern Europe, have kept their streetcars as an efficient, clean, comfortable, and very civilized mode of urban transportation, even improving its intermodal connections to subways, suburban trains, and peripheral buses. Vienna is an excellent example of this very rational urban transportation system in a large city, and with a fare structure that is fair to its citizens. The same may be said about cities such as Amsterdam, Den Haag, Budapest, Prague, Munich, Brussels, etc., just to name a few. Figure 4.7.

Figure 4.8 shows social concern in the design of modern streetcars when they offer a low-floor access, adequate space for wheelchairs and prams, restraints to avoid movement, electronic indication of stops and transfers, etc. In North America, the city of Vancouver, Canada, also has a very efficient multimodal transportation system, including an LRT, trains, buses, trolleybuses, and the Seabus, composed of two ferries that link the two shores straddled by the city of Vancouver, through the Burrard Inlet. Here too, the system was developed with a social interest in mind, offering good coverage of wealthy and poor neighbourhoods, and adequate intermodal connection to avoid unnecessary walking when transferring

### *Infrastructure*
This criterion calls for conducting an analysis of the works needed to operate each mode of transportation. This is mainly an economic issue, but it is also land-related as it has to do with dedicated rights of way. Its analysis involves the calculations on the net present value of each alternative, and their rates of return. However, an analysis of these undertakings calls for considering their

194  *Introduction to Sustainability: Road to a Better Future*

**social rates of return** (see Glossary), rather than an economic rate — that is, considering the production factors at their actual price and not at their market price. In each case, a fundamental factor common to all is ridership, or the volume of people using the system. This makes it essential to have information about the human density of the areas traversed by the future line.

*Figure 4.7*     **Modern articulated streetcar in Vienna**

*Figure 4.8*     **Interior of a modern streetcar
A sustainable mode of urban travel**

## Land use

There is no doubt that there exists a direct relationship between the different undertakings and the use of the available land, and this is not a straightforward problem. The urban use of land is directly linked with human density and has consequences for infrastructure issues; therefore, it appears that a long-term policy has to be developed in order to assure that there will be enough patrons for a transit line, especially a rail one, and in a certain corridor. To this end, the policy followed in Curitiba (section 3.3.1) is very illuminating as one that has yielded impressive results. As a counterpart, two cities in North America have developed, for **political** rather than for reasons of **rationality**, subways that run in areas of low density, which have correspondingly low revenues.

The use of land is not such a cut and dry issue. In most cities, and in downtown areas where there is probably not enough space to build a rail link, the lines usually go underground. Many cities have already had these problems in the past, when rail tracks went through tunnels into the core of the city. It is usually a good idea to study the possibility of using these old tunnels for new urban transportation projects. Underground work is very costly, and considering that large cities often have an already built network of tunnels for water, sewer, power, communications, etc., or because of geological soil conditions, it is often better to consider elevated rail alternatives, as was very recently done in Bangkok and Kuala Lumpur.

Because of the density of downtown areas, stations are often a short distance from one another — perhaps no more than 400 meters apart. When the line reaches less populated areas of the city, it is quite possible to build a line on the surface, or even one that is elevated on concrete columns, which greatly reduces costs. In these cases, the distance between stations is increased, also allowing the trains to increase their speed and reduce travel times.

## Accidents

This criterion relates with events leading to injuries or the loss of human lives, as well as any material damages associated with the interaction of this mode of transportation with others. Two modes are most relevant in this context: bicycles and cars. The first are vulnerable since they provide no defence for their riders, especially when bikes share the road with other vehicles. Even dedicated lane in roads do not protect riders from being hit by vehicles travelling in the same direction, or from injury due to slippery conditions caused by rain or ice. It would be more logical and a better choice to have dedicated bike space on sidewalks, but of course, this calls for wider sidewalks. These mean risks too, because pedestrians inadvertently stepping into bike lanes can sometimes be seriously hurt.

Regarding car accidents, these are often much more serious since, even if the driver is protected, the high speed and collisions can cause great injury even when the driver is not to blame for the mishap. For instance, one need only observe what happens on highways in car pile-ups resulting from some initial accident. The cars approaching from behind have no time to reduce their speed and stop, especially if weather conditions such as rain, snow or ice greatly impair safety by preventing quick stops.

*Relationship between weight and passengers transported*
This indicator analyses the relationship between the amount of material needed to transport people and their weight. In a walking person, this ratio is one, since each person transports their own weight. This increases a little on a bike, and it increases several times over when another mode of transportation is involved. Just imagine how much a medium-sized car weighs, how much energy it consumes, and establish its relationship to the weight of a sole driver. Since it seems that vehicles with only one person make up most car traffic, society is wasting resources, energy, physical space, and money under such conditions.

Something has been done in this respect, especially as compared to car weights two decades ago. Without a doubt, cars are better now, as they use aluminium, plastics, no ornaments, no metal bumpers, etc. This, however, is not enough, accounting for the current trend to develop mini cars that fit one driver and a passenger. This still has a large ratio, yet it takes up less space on roads and in parking lots, while consuming less energy. All the same, from the vantage of safety, these vehicles are more fragile than conventional cars, and they are riskier.

*Health*
This is an indicator of the utmost importance. In selecting the mode of transportation, it is obvious that the car is not the favoured one from this point of view because of its noxious discharges; as everyone knows, this is one of the main reasons — although not the only one — why local authorities are enacting and enforcing more stringent bylaws regarding emissions. Of course, every municipality aims to have zero emissions from vehicles in the city, and research has advanced in this regard since some large car manufacturers are already planning to produce fuel-cell operated vehicles in the future (section 5.3.5.1). However, this is not easy, as the technology is not quite ready to deliver this; in addition, from an economic standpoint not everyone will be in a position to get rid of old cars to buy new, more costly ones.

## Contribution to traffic congestion
Understandably, in light of this, LRT (Light Rapid Transit), subways and trains are the preferred solution, provided that the LRT or trains do not intercept urban streets at the same level, such as to make for long queues and to be a source of often-fatal accidents. Many cities do involve roads with multi-modal heavy, two-way traffic — that includes cars, cabs, streetcars, trolleybuses, buses, and trucks — and often without very wide thoroughfares. Some avenues in Hong Kong are good examples of this.

Obviously, then, a sustainable policy requires adopting some measures to alleviate this, bearing in mind the environmental element of the fumes spewed. One such measure is to restrict use by private cars of the CBD area during certain hours, and keeping this area open only to buses and cabs. Other measures could include the use of electric car-shared vehicles, or even declaring the whole area exclusive to pedestrians. Some old cities even curtail passage by heavy trucks and buses not only to the area but also in their neighbourhood, to prevent damage of its architectural heritage through vibrations. The pedestrian area is then greatly increased, and as an additional advantage, it also boosts commerce because of the greater number of people walking in the area.

Another aspect related to transportation is related to how businesses are being run nowadays. Those who once commuted to work every day now work from home, and the Internet has also modified many people's habits with abundant information at the tips of their fingers; consequently, few need to visit libraries for information that is on people's screens at home. E-commerce is yet another aspect leading to fewer trips: the entertainment industry, through DVDs, keeps people at home, and courses also facilitate distant learning. This new pattern evidently reduces the number of people driving cars, thereby relieving traffic congestion.

## Accessibility
It is not surprising that the private car is the most favoured vehicle for accessibility, since drivers do not rely on fixed transit lines or schedules, and can go will to areas the transit system does not reach. Naturally, this comes at a price, in the shape of economic and environmental costs. Accessibility is also bound to the geography of urban areas. If a city is cut in two by a river, with a single bridge to join it, then nothing is gained by using a car that must use the same route as public transit. This makes it essential to analyse the city's accessibility and connectivity. In many old cities, as well as in those that developed on a rough topography (contending with deep creeks, rivers, hills, forests, large cemeteries, etc.), it was always necessary to find ways to enable connectivity, avoiding the need to drive downtown in order to get from one area to another, as often happens.

*Travel time*
This criterion is common to all surface modes of transport that share the road with pedestrians, cars, bikes, streetcars, etc. The economic significance must be factored into the consideration of travel time, as it can cause hundreds of millions of dollars in losses due to lack of productivity. This was the main reason given to build highways. However, since highways feeds themselves, attracting more and more vehicles than they were designed for, average speeds on urban highways are decreasing, conflicting with the very purpose of their initial construction. Many highways se aside a special lane for cars that have more than the driver; this is a good measure as it speeds up travel times, while at the same time making for a more rational use for vehicles.

*Delays*
This criterion relates with others, including the number of vehicles, the width of the road, ancillary construction such as bridges, the size of cars, etc. In this context however, it refers to delays produced by unforeseen circumstances, such as accidents, heavier traffic than expected, etc. Again, public transit is more advantageous that car use, and less susceptible to delays, when it has dedicated roads, considering the amount of people transported, and given the obligation of cars to yield to buses when there is no dedicated road..

*Stress*
Far from being the least important, this last criterion is quite the opposite. Car driving on urban roads or on highways produces a lot of stress since attention has to be paid not only to one's driving, but also to others'. Cars can come too close, on the side or just centimetres from one's rear bumper, people jaywalk, the car just ahead can come to a sudden stop, traffic lights unexpectedly change, etc. Accidents can easily happen, especially if the driver is talking over a cellular phone, at the same time, etc.. None of this happens on the public transit system, where patrons can even enjoy the journey reading a good book or dozing off.

### 4.2.1 Case study: The Transmilenio bus system in Bogotá, Colombia

*Information on this case study is taken from Diaz (2003; see Internet references for Chapter 4).*
**Background information**
*The city of Bogotá, Colombia, has 7 million people and has built a rapid transit network based on dedicated lanes for articulated buses. The system is brand new since the first 41 km extension was in service in 2002.*
*This case is important not only because has begun to solve the*

*problem of Bogotá's chaotic traffic, but as it also shows how economies that cannot afford to build subways (at a cost of more that US$70 million/km) have instead found a very sustainable solution costing about US$5 million/km.*

*The first bus line has two dedicated lanes, two running in each direction, along the middle of an avenue that has no links to the other three traffic lanes of normal traffic that run in each direction. Stations are built in islands and serve both directions. Ticketing is similar to those for LRT and subway systems, since people can access the boarding platform after buying a ticket in a booth and going through a turnstile; the system also utilizes smart cards. Passengers access these boarding platforms from both sides, using bridges over the ordinary traffic lanes.*

*Economics*

*It is important that the system moves in a working day about 792,000 passengers, or about 35,000 people per hour. The system is private and not subsidized, and operating expenses are met from the proceeds of the US$ 0.36 fare, which includes transfers. Monies collected from its operation cover the capital investment, operating costs, maintenance and earn a profit.*

*A 25 percent tax on gasoline is being levied to support further expansions, which are scheduled until the year 2016. At that stage the city is projected to have population of about 8.5 million, and every citizen is expected to live within 500 meters from a Transmilenio line.*

*Operations*

*This first line uses 470 articulated buses and 235 smaller feeder buses operated through control stations that can pinpoint the location of each bus, since they are all equipped with a Global Position System (GPS). Larger buses, which have a seating capacity of 160 passengers, are diesel- or natural gas-powered, and run every 2 minutes at peak hours. Feeder lines connect other parts of the city with Transmilenio stations, using smaller buses that have a capacity of 80 passengers.*

*Social*

*The advantages the system has brought for Bogotá's citizens are considerable. This author was in Bogotá in early 1998, and can bear witness to the extremely poor conditions of the transportation system at that time. It was run on derelict buses with small capacities, afflicted by fumes seeping into the passenger cabin, dim*

> *lighting, and extremely long travel times because of traffic, as well as many accidents and breakdowns. According to Diaz, travel time was reduced by 32 percent, translating for users into a gain of more than 300 hours/year for them.*
> *There has been a gain in social equity since the implementation of the Transmilenio system, when private cars used 95 percent of available road space.*
>
> *Health*
> *Air pollution is blamed for the death of 1,200 Colombians per year from pneumonia associated with air pollution, and with 52,764 accidents resulting in 1,174 deaths.*
>
> *Final comments*
> *There is no doubt that this project, like the Curitiba project (section 3.3.1) that inspired it, complies with every aspect of the definition of sustainable transportation. Both epitomize the meaning of the word 'sustainability', through transportation systems that are reliable, fast, frequent, and moderately priced, not only fulfilling their aim of providing access, but improving the health of its riders by greatly enhancing the environment.*

## 4.3 Sustainability in agriculture

A sustainable agricultural system is one that can indefinitely meet the requirements for food and fibre at socially acceptable economic and environmental costs. (Crossen (1992; see University of Reading in Internet references for Chapter 4)
While one might think that agriculture and farming are clean industries, unfortunately, they are not because:

- *Fertilizers* containing nitrogen, potassium and phosphorus introduce large amount of minerals into the soil. When nitrogen migrates to drinking water sources such as groundwater, it can cause severe illness. Conventional wastewater treatment plants do not eliminate nitrogen and need to use sophisticated and expensive technologies, such as **reverse osmosis** (see Glossary). Phosphorus content, when discharged into rivers and lakes favours algae growth, which reduces the amount of available oxygen in the water that fish and aquatic organisms need. Potassium is considered a non-toxic metal.

- *Pesticides* also contaminate soil and, after harvesting, winds can disperse their chemicals over a large region. One example is the area along the Amu Darya River, in Kazakhstan, which was extensively cultivated with cotton many decades ago. The abandonment of that operation produced a barren area contaminated with chemicals that went airborne, dispersing over a large area when there are strong winds.

- *Water overuse* can endanger the very existence of water sources, as well, sometimes, as the sinks where they discharge. Again, this happened with the diversion of the Amu Darya River, which led to the shrinking of the Aral Sea, which has lost more than 60 percent of its water; its salt concentration has thereby increased, as a result devastating the natural habitat for sea creatures. There has been also a climate change with more dusty and hotter summers and harsher winters. The economic, ecological, and social consequences of this have been catastrophic.

The Ogallala aquifer of the United States extends 1,300 kilometres from north to south, and is about 600 km wide from east to west, lying beneath from southern South Dakota to northwest Texas; it also has large extensions in Nebraska and Kansas. As the aquifer's **rate of extraction has exceeded its rate of replenishment,** in some parts water supplies for irrigation have been exhausted.

The two examples above show what happens when an action favours economic interests with complete disregard to its environmental consequences. These both stand as clear examples of **unsustainable management** of a precious and scarce resource.

- Farming is also a large pollutant of aquifers and rivers, especially with nitrite contamination, due to improper management of animal wastes, mainly from poultry and swine facilities, and their ammonia emissions.

- *Air pollution* is also produced by the exhausts of mechanical, driven farming equipment.

The application of sustainable measures to farming operations is clearly essential. Very briefly, these consist in using organic instead of conventional farming; this involves utilizing natural products for pest and weed control and fertilization, as well as a clear determination of consumer willingness to pay for products with reduced fertilizer and herbicide contents. This issue goes beyond the scope of this publication; the reader may consult hundreds of publications and Internet papers that deal with all aspects of agricultural sustainability.

## 4.4 Forestry sustainability

Whoever has ever flown over land where forests have been logged will probably agree that it is one of the saddest views on earth. Walking through such a landscape will only aggravates this impression — since there will be tree stumps but no vegetation, and bare lands that are undergoing erosion by water runoff. Without doubt, logging is necessary, but what is not is indiscriminate logging and unsustainable forestry practices when logging rates often far exceed sustainable recovery levels.

In addition to commercial logging, two other factors are responsible for forest degradation: forest fires and insects. *Canada's Forests at the Crossroads* (see Internet references for Chapter 4) reports that, cumulatively, between 1980 and 1996, insects have affected 69 million hectares of forests, 43 million have been damaged by fire, and 16 million hectares by harvesting. As for the types of forest logged, the harvesting is of second and even third rotation forests, although 90 percent of logging in Canada is being done in primary forests, which is disturbing from the sustainability point of view, since old forests have a high biodiversity and are also valuable because they are wilderness.

Many people believe that planted forests can replace natural forests. While this is probably true regarding the harvesting of timber, it is not exactly accurate in relation to the ecosystem as it alters the habitat of many different natural species. The planting of 'factory forests', for example, where old growth forests once stood, can dramatically reduce species diversity (Gamlin, 1988). Clearly, more is needed than the convenience of single-species or monoculture forests if entire ecosystems are to be sustainably exploited on an industrial scale (Tibbs, 1991). There are reports that in tropical forests valuable species are logged without regard for neighbouring uncommercial yet valuable species — such lianas and vines — which are vital parts of biodiversity.

Perhaps one concept that can be used to illustrate forests' disappearance is their carrying capacity (section 6.12). According to *Canada's Forests at the Crossroads*, the carrying capacity of a forest in a zone of the province of British Columbia (western Canada) is of 2.4 million $m^3$ per year. Unfortunately, this carrying capacity has been breached by logging 3.8 million of $m^3$ per year. Obviously, **this is not sustainable.**

From an environmental point of view, it is clear that logging produces extensive damage not only to the soil but also to the wildlife, by destroying its habitat. Something similar happens with human settlements, usually primitive, very fragile cultures that are pushed out of a territory where they have lived for hundreds of years — as is happening mainly in Brazil, but in other parts of the world as well. In short, the forests' social component is also important.

The source cited above proposes a series of indicators for every domain. From the perspective of sustainability, only two of them will be considered: 'Sustaining long-term production', and 'Regeneration'.

The first one takes into account:

- A comparison between actual harvest rates and rates fixed by the government.
- Regeneration trends.
- Productivity limitations.
- Insects and outbreaks of fire.

The second has a view on the sustained productivity of ecosystems: forests can be left to regenerate naturally or by planting, and this indicator shows their ability to regenerate.

The Scottish Forestry Commission (see Internet references for Chapter 4) commissioned a study to determine the contribution of Scottish forestry to its economy, using the Input/Output analysis approach (see Glossary). This study considers the inputs to the industry such as labour, equipment, fuel, etc., and its outputs, including lumber and wood products. It determines the three types of multiplier effects (section 6.9), which are **output, employment** and **income**, and draws an interesting conclusion about backward and forward effects (section 6.5).

## 4.5 Sustainability in the construction industry

Miller, *et al.* (see Internet references for Chapter 4) report that energy estimates spent on buildings, including their construction, operation and disposal, account for over 40 percent of total energy consumption. This total energy consumption estimate comes from a Life Cycle Assessment (LCA) (see Glossary and Appendix, section A.5), and involves the energy spent for the extraction, processing, and manufacturing of all the elements participating in a building, as well as for its demolition and disposal.

The measure of this energy consumption is an indicator under the name of 'Embodied energy of building materials', which can be associated with the total energy needed to acquire building materials, from the time of their first extraction, to their manufacture and installation. Sam Mukhtar (see Internet references for Chapter 4) has produced values, calculated in MJ/kg, for some common materials, at least for New Zealand. Tracy Mumma (see Internet references for Chapter 4) also provides extensive information about embodied energy for housing.

Tall buildings are also large consumers of electric energy and water. Energy can be saved in many ways in tall buildings:

- Using high-efficiency appliances and boilers (water heaters);
- Using natural gas instead of oil or electricity in kitchen appliances;
- Installing sensors in hallways to switch on lights when needed;
- Employing central air conditioning units operated with natural gas;
- Utilizing photovoltaic panels (section 5.3.2) on the roof and in walls exposed to the sun's rays to generate electricity from the sun.

Measures to save water in modern buildings may include a purification plant in the basement where wastewater can be treated and reused for flushing toilets and for garden irrigation, as well as tanks for collecting and storing storm-water for later use.

The EU has a program called European Housing Ecology Network (EHEN; see Internet references for Chapter 3) that promotes cooperation among housing associations, municipalities' housing technical divisions, and researchers working in energy conservation, with a view to targeting the construction of low-energy dwellings. The Network's aim is to identify and develop good practices in all aspects of ecology involved in or affecting the provision, management and servicing of housing. The program has construction plans for locations in Denmark, Ireland, Italy, Portugal, Spain, the Netherlands, and the UK.

Results from the EHEN are very promising. In the case of apartments built in Skotteparken, Ballerup (Denmark; see Internet references for Chapter 4), they emphasized hot tap water and water heating systems. This housing development for 100 families claims an energy demand for the above-mentioned issues of almost 50 percent of traditional methods, using only an increase in capital of 8 percent. This reduction was obtained with solar panels (section 5.3.3) of 6 $m^2$ per apartment. Hot water is kept in 5 $m^3$ storage tanks, controlled by an Energy Management System that does not keep the water constantly running between the tanks and the solar panels, but which uses a pulse or intermittent system; this decreases heat losses.

According to the source, the area has an average of 1,750 hours of sunlight per year. Naturally, other measures taken to save energy are:

- Extra insulation, mainly in the ceilings.
- Double-glazed windows.
- Ventilation systems with heat recovery.

In the US, the National Institute for Standards and Technology (NIST), in collaboration with the US Environmental Protection Agency (EPA), has developed a software program called BEES (Building for Environmental and Economic Sustainability; see Internet references for Chapter 4) that selects cost-effective, environmentally-preferable building products, and it involves nearly 200 building products. The software measures these products' environmental performance using LCA, in agreement with ISO 14040 international standards (see Glossary).

### 4.5.1 Comparison between singles dwellings and multi-family buildings

It is likely that most people, especially those with young children, prefer to live in a house with a garden, a driveway for the car (in addition to a garage for one or two cars), a yard, and maybe some room for a swimming pool, etc. While this may be a natural wish, it is unfortunately not efficient from the point of view of sustainability nor from an economic one. Many reasons support this statement, including:

- Land use
  Even a small house needs its own plot of land, requiring the addition of sidewalks and driveways.
  Single family units such a this, usually with an entrance on one street and the back yard facing a parallel alley or street — which are common at least in North America — take up a lot of space while they may be occupied by only 4 to 6 people. The result is that a small number of people use a lot of space.
  This in turn, causes the sprawl of populated areas, generally at the expense of agricultural land or forestry. This has a ripple effect because to reach these new settlements it usually becomes necessary to expand or create new transportation routes, to construct new roads (that use up more land), as well as to extend utilities many kilometres for the water, sewage, electricity, cable and telephone networks.

- Soil
  More dwellings means that more soil will be covered over, and such construction decreases the soil's capacity to absorb rainwater, in turn causing flooding, run-offs and erosion.

- Energy
  A single detached dwelling leaves five sides exposed to the elements. This means that heat enters the house in summer by transmission through four walls and the roof. The same happens in winter, when heat from within the house escapes to the outside.
  Heating a single dwelling in winter requires energy to drive the fan, which pushes heated air through it, plus gas to warm up that air; energy is also spent to drive the compressor that cools down the house in summer. Of course, this equipment also occupies physical space, besides which there is a need for an electric or gas water heater, and for a holding tank for the hot water.
  Street lighting needs to be expanded to cover the new development.

- Water
  Each dwelling consumes water according to the number of people and their habits. As mentioned above, this involves extending the water mains and connections to each dwelling, the installation of water meters and the need for their monthly reading, which is a labour-intensive task. Each dwelling may also waste a lot of water — which is made fit for human consumption at a considerable cost — to irrigate the lawn, wash the sidewalk and the car, fill the swimming pool, water the trees, etc. Extended street sweeping is necessary to cover the new settlement, with high equipment and labour costs.

- Wastewater
  Wastewater from the kitchen, toilets, dishwasher, and the washing machine discharges into an underground sewer. This wastewater, which comes from several dwellings instead of a single one, greatly increases the length of the sewer network for domestic collection, and the length of the main trunks to transport it to the wastewater treatment plant, aside from increasing the number of manholes that need to be inspected and maintained.

- Solid waste
  Every single dwelling is required to place non-recyclable garbage in a plastic bag (which is seldom biodegradable; see Glossary), for collection at great expense by hauling trucks, to be transported to a landfill and buried.

- Inefficient use of land
  To shed light on this subject it may help to use an analogy with the daily use of a car.

Everybody agrees that a car to transport only one driver is a waste of energy considering that what people are doing is just taking the car's components — its body, seats, transmission and plastic — for a free ride! A car is intended for transporting people, not to shift around the car!

The widespread acceptance of this fact has led to the adoption of various measures: dedicated lanes on highways for cars with more than two people, the carpool scheme, the park-and-ride system where cars are left in a parking lot while their drivers board a bus, a train, or a subway, etc. Everybody should make an honest appraisal of their driving habits and calculate how many times a year their car carries its full complement of people: that is, the driver plus three or four people.

Why is this comparison raised here? Because something similar happens in single houses. For instance, in many parts of North America, Europe and Asia, gardens and yards are covered by snow from November to April, so about five-and-a-half months, or about 46 percent of the time, preventing any practical enjoyment for the house's inhabitants. So the question arises as to whether it is sustainable to have that land unused for almost half a year? As in the case of the car, owners are paying for something they hardly use, or that is not used at all, in the process preventing the land from being used in some more rational way by others.

How to solve this problem? The solution was found many years ago with the construction of high-rises. Buildings that are 20 or 30 stories tall offer many advantages over single dwellings. They can usually be built even in the middle forests. A building or a set of buildings can have a communal swimming pool, recreational amenities for children, space for barbecuing, even a small commercial centre, among other facilities.

It is thereby not necessary to add anything else to point out the advantages of high rises over single houses from the point of view of land use. This type of housing can provide comfortable shelter for thousands of people within a space that, if used by single houses, might only provide for the dwellings of less than 100 people.

Such buildings have the following advantages:

- They usually have underground parking lots, making it unnecessary to use valuable land as single houses do. Land use is also greatly reduced as there is only need for one sidewalk and one road.

- Soil
  There is by far more uncovered soil, which allows for the percolation of rainwater into the ground.

- Water
  Water consumption can be dramatically reduced, as compared to equivalent single dwellings, since wastewater can be indefinitely utilized in a closed circuit. To this end, wastewater from kitchens, toilets, and utilities is sent to a purification plant in the building's basement, where it is then re-circulated through dedicated piping for flushing the toilets. Another independent system that has no physical connection to this one is used for drinking water.
  Considering that toilet flushing consumes may be 4.5 litres of water per flush, a calculation of the number of units in the building and the number of daily flushings makes for a very high figure in water savings.

  Any excess treated water can then be used for the garden, and the surplus is discharged into the sewer. However, an additional advantage is that with this method the amount of water reaching the water treatment plant is considerably reduced, translating into smaller facilities and in more time for the water to be treated in the digesters (see Glossary) before its final discharge.

- Energy
  Energy requirements for heating and cooling are greatly reduced because each apartment has only one or in some cases, two walls that are not shared with the other units. Of course, no losses occur through the roof, since even the roof on the highest floor can be covered with a garden. Naturally, the elevators make for a considerable consumption of electrical energy, as do the need to illuminate the stairs and hallways, and the required water pumps. However, the roof and part of the walls can be covered with solar panels (section 5.3.3), which can collect heat to provide warm air or water and help to reduce the need for electricity.

  If the building is supplied with heat pumps (section 5.2.8.1), it is then possible to dig several very deep wells (possibly to a depth of 90 metres) to be used for heating and air conditioning in connection with heat pumps operation, thus reducing energy consumption.
  When high-rises are built, the apartments' different appliances are purchased in large quantities, greatly reducing costs, and the builder can then choose the most energy-efficient appliances.

- Solid waste
  High-rises allow people to sort their waste into recyclable items, such as bottles, paper, and cans, for depositing in large containers, usually at the back of their building. This system's economic advantage is evident: only one truck is needed to collect the garbage, and another for the recycled items. Compare this with the number of trucks needed for a similar population living in single homes.

- The large concentration of people increases density, enhancing the operation and the economy of the public transportation system.

- Economics
  Real estate taxes sharply decrease for each apartment, compared to single dwellings, and the same holds for insurance. Maintenance costs are also drastically reduced, given for instance that replacing shingles on the roof of a medium-sized house can amount to about US$2,500.

- Social
  These buildings usually allow for more social interaction than single dwellings. Another factor worth considering is safety, since apartment break-ins, although still possible, are much less frequent than of houses.

- Wood
  The construction industry is a large consumer of wood products for frames, beams, trusses, etc. Naturally, this has a direct impact on forest exploitation, and is one of the reasons for their disappearance or shrinkage. High-rises do not use wood but Portland cement, steel and aluminium, drastically curtailing the need to log for the construction of dwellings.

### 4.5.2 Case study: Sustainability in paradise – The Maho Bay resort complex, Virgin Islands, USA

*The mountainous US Virgin Islands lie southeast Puerto Rico, at about 18° latitude north. The three most important islands are St. Croix, St. Thomas, and St. John.*
*The islands are so poor in fresh water resources that a large portion of drinking water in the first two islands is produced by a desalination process from seawater; by law, all residences and hotels are required to collect rainwater. Rainfall is highly variable, and most of it returns to the atmosphere by evaporation*

*and by plants' transpiration. Aquifers are scarce, the only important one being in St. Croix.*
*On St. John Island, there is a set of resorts that, after 18 years of operation, has attracted one million guest-days. How is this possible? Because sustainability was borne in mind during the construction and operation of these resorts, where everything counts, where everything must be recycled as much as possible. Most of the following data were taken from a paper published by Stanley Selengut (see Internet references for Chapter 4), an entrepreneur who helped to urbanize St. John.*

*Mr. Selengut also developed the Maho Campground, which has 114 tent-cottages. Besides its privileged climate and natural beauty, this area in the middle of the Caribbean Sea offers many sports like sailing, scuba, night snorkelling, park events, fishing, windsurfing, kayaking, tours, and, as he says, an intimacy with the great outdoors.*
*The cottages are equipped with one bedroom, a living room, kitchen, and an open sun deck. It is a paradise, and despite the scarcity of fresh water, thanks to ingenuity and wise use of scarce resources, the resort draws thousands of visitors each year. The reason this resort is mentioned in this book is to showcase entrepreneurial abilities that can deal with a very fragile environment, co-existing with economic growth, within a polluting industry: tourism.*

*Without judging the environmental advantages or disadvantages of some measures that were taken in this case, let us analyse some of the main characteristics of this unique place:*

*The wood-floored bungalows are made of recycled materials, are elevated on platforms, and do not sit on land. They do blend with the foliage, so there is no visual contamination.*
- *All bungalows are connected by elevated walkways to protect plants and avoid erosion, while all the utilities are installed below them, disturbing very little of the ground.*
- *For personal hygiene, there are 'wash houses' with water-saving toilets, spring-loaded faucets and chain-operated cold-water showers.*
- *Mosquitoes and other bugs are dealt by natural measures, such as by reintroducing lizards to the island.*
- *A sister resort, Harmony, is also designed with sustainable principles to the extent that it is entirely solar- and wind-powered;*

> even the construction equipment was powered in this way!!! It was also built with recycled materials; for instance, the floor decking is 100-percent recycled newspaper, and the sidings are made of a composite of cement and recycled cardboard, like the roof shingles.
> - Ceramic floor tiles use post-industrial glass waste from a local manufacturer, while bathroom tiles and furniture tops have 73 per cent post-consumer glass bottle content.
> - Photovoltaic panels installed on the roofs provide energy for lights and appliances, and sensors shut off the lights when a guest leaves the place.
> - Walkways are built with plastic lumber mixed with sawdust.
> - These sustainability efforts are not restricted to construction, since there is even a store that purchases in bulk and then sells products in a minimum package.
> - The establishment has a restaurant whose wastes are composted and then used to grow vegetables; it also utilizes biodegradable products for cleaning.
> - Conservation efforts include baking bread and other food with a solar oven that can reach a temperature of $220°C$.
> - Wastewater is treated by using natural bacteria that break down and separate the solid matter, after which the water is used to flush toilets and for an organic orchard and gardens.
> - Paper is shredded and then used for packing, and metal cans are compacted and accumulated until enough aluminium in the container makes shipping both 'wastes' cost-effective.
> - Collected rainwater amounts to 1.5 million litres/year, and it is used for laundry, housekeeping, and bathhouses.
> - Laundry services use 100 percent biodegradable detergents and reuse wastewater.
>
> This case study is an excellent example of how economic growth, through the construction of an eco-resort, is compatible with sustainable development. This is expressed best by repeating Mr. Selengut's own words: "Why choose to build an eco-resort rather than a more traditional hospitality facility? Because it is much more profitable! What makes sense from an environmental and conservation point of view also saves money"

This is not a minor matter, since all the $CO_2$ that is produced these days requires more forests than ever to absorb it. The U.S. Department of Energy offers some very good information on this subject (see Internet references for Chapter 4).

**Internet references for Chapter 4**

**Author:** Peter N. Pembleton (2002)
Title: *The role of industry: Specifying sustainable development- Indicators for sustainable indicators in industry*
Address:
http://www.unido.org/en/doc/3559

**Source**: The University of Reading (2000) - Crossen (1992)
Title: *Sustainable agriculture*
Comment: Provides useful information and definitions of important sustainable indicators for the economy, transportation, leisure and tourism, overseas trade, energy, land use, water resources, forestry, fish resources, climate change, ozone layer depletion, acid deposition, air and fresh water quality, marine, wildlife and habitat, land cover and landscape, soil, mineral extraction, waste, radioactivity,
Address:
http://www.ecifm.rdg.ac.uk/sustainable_agriculture.htm

**Source:** Randolph Packing Co.
Title: *Reducing in waste load from a meet processing plant - Beef*
**Address:**
http://www.p2pays.org/ref/01/00466.pdf

**Source**: Indigo Development (2003)
Title: *Creating systems solutions for sustainable development through industrial ecology*
Comment: Information on Industrial Metabolism, Urban Footprint, Dynamic I/O models (with an example), Life Cycle Assessment, Design for the Environment, and other techniques.
Address:
http://www.indigodev.com/Tools.html#Urban%20Footprint

**Author:** J.D. Miller, *et al.* (1996)
Title: *Transportation energy embodied in construction materials*
Comment: This paper presents an actual case example of this computation for the construction of a new residence for students at Sussex University, built on the seafront in Brighton, U.K. The author recognizes that the results are site-specific as are other factors such as location; however, as the author states it appears that the study helped to identify the factors that significantly affect the energy consumed in the transportation of building materials.

Address:
http://alcor.concordia.ca/~raojw/crd/reference/reference001500.html

**Author**: Sam Mukhtar (1998)
Title: *Embodied energy in building materials?*
Comment: Very interesting analysis.
Address:
http://www.strategicdata.co.nz/mukhtar/sam2.htm

Title: *Measures of sustainability*
Comment: This publication provides very valuable information about embodied energy in buildings, with a comprehensive table with the values of main materials in MJ/kg and MJ/m$^3$. Other very useful concepts are also discussed, such as externalities, the ecological footprint, and Life Cycle Assessment.
http://www.cdnarchitect.com/asf/perspectives_sustainibility/measures_of_sustainablity/measures_of_sustainablity_embodied.htm#top

**Author**: E. Diaz (2003)
Title: *Bus rapid transit and other bus service innovations – Bogotá's Bus Rapid Transit System: Transmilenio*
Comment: Comprehensive description of the project with pictures showing aspects of facilities. There is also information about investments and the type of organizations that were formed to build and operate the system, as well as comparisons of pollution emitted by the system regarding an initial baseline.
Address:
http://banking.senate.gov/03_06hrg/062403/diaz.pdf

**Author**: Todd Litman – Victoria Transport Policy Institute (1999)
Title: *Transportation cost analysis for sustainability*
Comment: Analyses transportation costs including non-market environmental and social costs, and provides recommendations to incorporate total cost analysis in studies regarding transportation. This paper provides a great deal of valuable information — such as detailed in Table 1, regarding transportation costs categories, internal and external costs per vehicle-mile, and a wealth of related information. Reading of this paper is recommended.
Address:
http://www.stfx.ca/people/x99/x99fhk/drivevswalk.pdf

**Author**: Todd Litman – Victoria Transport Policy Institute (2003)
Title: *Sustainable transportation indicators*
Comment: Very good explanation of the difference between goals and objectives related to indicators. See Table 1 for Sustainable Transportation Performance Indicators (STPI). There is also a Table showing Sustainable

Mobility Indicators, as well as a Table 3 that presents a list of Sustainable Transportation Indicators.
Address:
http://www.vtpi.org/sus-indx.pdf

**Authors**: Martin J. Bernard III and Nancy E. Collins (2001)
Title: *Shared, small, battery-powered electric cars as a component of transportation system sustainability*
Comment: According to the authors, they explore six trends in the US that may change the way people use the transportation system. These six trends are demographics, questions about the future of the transportation infrastructure, growing interest in alternative means of transportation, the growth of the Internet and electronic communications, individual preferences moving toward access to goods and services instead of ownership, and renewed interest in urban living.
Reading of this paper is recommended.
Address: http://www.stncar.com/shared/ssbecs.html

**Author**: Northeast Sustainable Energy Association (NESEA) (2001)
Title: *Information about sustainable transportation*
Comment: Offers three strategies for reducing the environmental impact of transportation, as well as a wealth of information about different alternatives, such as electric and hybrid electric vehicles in the market. A second section provides the Web addresses to seven related papers.
Reading of this paper is recommended.
Address:
http://www.nesea.org/transportation/greenindex.html

**Author**: Heather A. Cheslek
Title: *Water reuse and conservation the United States Virgin Islands*
Address:
http://ceeserver3.mit.edu/cheslek.pdf

**Authors**: G. Pastore and M. Giampietro - Instituto Nazionale della Nutrizione, Rome, Italy (1998)
Title: *Ecological approach to agricultural production and ecosystem theory: The amoeba approach*
Address:
http://www.ilri.cgiar.org/InfoServ/Webpub/Fulldocs/Aesh/Ecologi.htm

**Author**: J.R. Betthany - Saskatchewan Interactive Agriculture Management (2002)
Title: *Soil salinity*

Comment: The cause of soil salinity is shown in a diagram. This paper also provides pictures illustrating different soil conditions and a chart indicating the field, forage, and vegetable crops' tolerance to salinity.
Address:
http://interactive.usask.ca/ski/agriculture/soils/soilman/soilman_sal.html

**Source**: Energie-cités in co-operation with the municipality of Ballerup and Cenergia Energy Consultants
Title: *Solar district heating*
Address:
http://www.agores.org/Publications/CityRES/English/Ballerup-DK-english.pdf

**Author:** Robert Q. Riley (1996)
Title: *Energy Consumption and the Environment.*
*Impacts and Options for Personal Transportation*
Comment: Excellent introduction to this subject with links to economic implications, global warming, and new technology.
Recommended reading.
Address:
http://www.rqriley.com/energy.html

**Authors**: Helmut Schütz and Stefan Bringezu (2002)
Title: *Aggregated material balance and derived indicators for the European Union - Method guide and statistical data*
Comment: The authors refer to a study conducted with two major objectives: a framework and method description for drawing up aggregate materials balance. A methodological guideline was developed on economy-wide materials flow accounts and balances with derived indicators.
Address:
http://www.wupperinst.org/Sites/Projects/material-flow-analysis/u33.html

**Author**: Hardin B.C. Tibbs (1992)
Title: *Industrial Ecology: An environmental agenda for industry - Managing for the global environment - A complex challenge*
Comment: Extensive and enlightening approach regarding industry, linear and other flows of wastes. Author rightly points out that: *"The 'extract and dump' pattern is at the root of our current environmental difficulties. The natural environment works very differently. From its early non-cyclic origins, it has evolved into a truly cyclic system, endlessly circulating and transforming materials, and managing to run almost entirely on ambient solar energy"*. An excellent description of the Kalundborg process is also presented (section 4.1.4.1).
Highly recommended reading.

Address:
http://www.sustainable.doe.gov/articles/indecol.shtml

**Source**: Department of the Environment and Local Government (1999)
Title: *Plastic bag environmental levy in Ireland*
Comment: Presents information on this important initiative
Address:
http://www.mindfully.org/Plastic/Laws/Plastic-Bag-Levy-Ireland4mar02.htm

**Source**: U.S. Department of Energy
Title: *Building technologies program*
Comment: Offers energy solutions to reduce utility costs in multifamily buildings
Address:
http://www.eere.energy.gov/buildings/multifamily/index.cfm?flash=yes

**Source:** US Environmental Protection Agency (2004)
Title: *What is product stewardship?*
Comment:
As they rightly state *"Product stewardship recognizes that product manufacturers can and must take on new responsibilities to reduce the environmental footprint of their products"*
Address:
http://www.epa.gov/epr/about/index.html

**Source:** U.S. Environmental Protection Agency (2004)
Title: *Packaging – Industry initiatives*
Comment: Details illustrative examples of how some large industries such as Coca Cola, Dupont, Texas Instruments, Procter and Gamble, just to mention a few, are considering and some of the implementing this issue.
Reading of this paper is recommended.
Address:
http://www.epa.gov/epr/products/pindust.html

**Source**: BASF Actiengesellschaft (2003)
Title: *BASF's eco-efficiency analysis*
Address:
http://berichte.basf.de/en/2003/datenundfakten/sd/oekoanalyse/?id=V00-f*3n44n37bir0wi

**Source**: Oikos – Stiftung für Ökonomie und Ökologie (2003)
Title: *Managing socio-efficiency of products and processes – Further development of the BASF Eco-Efficiency Analysis by the social sustainability dimension*

Address:
http://www.oikos-foundation.unisg.ch/academy2003/paper_schmidt.pdf

**Author:** Stanley Selengut (1999)
Title: *Maho Bay, Harmony, Estate Concordia, and the Concordia Eco-Tents, St. John, U.S. Virgin Islands*
Comment: Excellent paper detailing the particulars of this eco-tourism undertaking, written by the entrepreneur who developed it.
Address:
http://www.yale.edu/environment/publications/bulletin/099pdfs/99selengut.pdf

**Source**: Canada's Forests at a Crossroads
Title: *Section 3 – The forest industry*
Address:
http://pdf.wri.org/gfw_canada_industry.pdf

**Source**: The Scottish Forestry Commission
Title: *Scottish forestry: An Input-Output analysis executive summary*
Comment: Forests and their economic impact in Scotland. Reading of this paper is recommended.
Address:
http://www.forestry.gov.uk/Website/OldSite.nsf/LUPrintDocsByKey/hcou-4u4jmj

Title: *The recycling-system Styria - An example of Industrial Ecology in Europe – The concept*
Address:
http://www.kfunigraz.ac.at/inmwww/eng.html

**Author:** Tracy Mumma
Title: *Reducing the Embodied Energy of Buildings*
Address:
http://hem.dis.anl.gov/eehem/95/950109.html

**Source:** National Institute of Standards and Technology (NIST) (2002)
**Author**: Barbara C. Lippiatt
Title: *BEES 3.0 – Building for environmental and economic sustainability – Technical manual and users guide*
Address:
http://www.bfrl.nist.gov/oae/publications/nistirs/6916.pdf

**Source:** This case study was submitted to UNEP IE by: ECOMED, Agency for Sustainable Development of the Mediterranean, Via di Porta Lavernale, 26, 00153 Roma, Italy. It was reviewed by UNEP IE, in November, 1995.
Title: *Cleaner production for reducing water consumption at a metal plating industry* (1993)
Address:
http://www.emcentre.com/unepweb/tec_case/basicm_27/house/h1.htm

**Source:** United Nations Environment Program – Production and Consumption Branch
Title: *Cleaner production (CP)*
Address:
http://www.unepie.org/pc/cp/understanding_cp/home.htm#definition

**Source:** Business Assistance - Design for the Environment - DfE Examples (2002)
Title: *"Re-imagine" the manufacturing process*
Comment: Very interesting actual examples of DfE
Address:
http://www.moea.state.mn.us/berc/dfe-examples.cfm

# CHAPTER 5 - ENERGY SUSTAINABILITY

*"Right now more energy passes through the windows of buildings in the U.S. than flows through the Alaska pipeline."* K. Bidwell and P. A. Quinby (see Internet references for Chapter 1)

## 5.1 Introduction

This chapter deals with one of the most important problems that the world faces today. Conventional sources of energy are diminishing, and some researchers have already hinted at a time horizon for their depletion. It certainly means that the oil and carbon reserves that remain for future generations will not suffice for their needs. It is possible that those people might find better uses for these energy sources as raw materials — rather than contaminating the atmosphere by burning them, as humankind now does.
The technology already exists — and, naturally, it can be improved — for the sustainable use of **non-conventional energy or renewable energy sources (RES),** with the aim of slowing down our extraction and consumption rate of fossil fuels, while at the same time doing a favour to the planet's battered environment.

Consequently, this chapter briefly analyses the different possibilities humankind has at the present for developing and utilizing these non-conventional sources. Naturally, these descriptions can only glimpse at each alternative, since it is not the purpose of this book to analyse each one at depth. The aim is simply to make the reader aware of the different options, their main characteristics, and some actual applications or cases. Examples of operating stations are provided to show that these schemes are not only possible, but are really in operation and already contributing to decreasing the amounts of fossil fuels now employed.

Of paramount importance is the cost per kWh. In general, the cost per kWh of non-conventional alternatives has been declining because of the use of new materials, higher efficiencies and increasing demand. However, with the possible exception of wind energy, the cost per kWh generated for non-conventional alternatives still remains higher than electricity generated by conventional sources; this is one of the reasons why progress has not been faster. In other words, the utilization of these promising techniques is greatly affected by the influence of economics. Besides, consider that conventional utility plants have a price per kWh that relates to the size of the undertaking that is large plants — as is usually the case — offer the benefits of **economies of scale** (see Glossary).

However, to directly compare costs between these two different modes of generation is incorrect, being much like the proverbial comparison between apples and oranges. A proper comparative assessment of conventional and non-conventional energy sources requires calculating the production costs as well as the 'hidden' costs that is the **environmental and social costs** associated with the extraction of resources, their processing, and their further utilization in a power plant. Thus, for to calculate the real cost per kWh from a coal-fired power plant requires also taking into account:

- The Life Cycle Assessment (see Appendix, section A.5) of coal extraction, transportation and utilization. That is, how much energy — which translates mainly into $CO_2$ contamination — is spent to mine coal, and how much is used to transport it to where it is consumed?

- The cost of the pollution produced by smokestacks from power plants, measured in $CO_2$ and other sulphurous gases, which leads to global warming and acid rain.

- The amount of residues from smoke filters that will be dumped into the soil, and the cost to clean it up.

- The number of people affected by pulmonary diseases due to smokestack gases. If in any doubt about this effect, we need only consider the problems generated in Krakow and Katowice, in Poland, due to the coal exploitation and industrial utilization in that area;

- The energy and pollution caused by making boilers, turbines, condensers, electrical equipment, etc., should also be considered.

Similar things apply to oil-fired power generation plants.

- What about nuclear plants? After decades of operation, it is said that society gets good value for its money and this with 'clean' energy production. Consider that it is deemed clean only because no noticeable radiation emanates from the plants or smokestacks, but it is important to also bear in mind that nobody has yet figured out what to do with nuclear wastes, which for centuries may contaminate soil and water.

Finally, if someone discovers a process to treat radioactive waste or to render it harmless, or if a place is found where it can be safely buried, how much will that cost? Even in such a case, this generation is merely 'passing the buck' for coming generations to get rid of radioactive wastes that we produce. It appears that this generation is extending its footprint (section 1.6) not only in space but also in time, to the future.

When we calculate all of these costs, what is the actual price per kWh produced by coal, oil or nuclear plants, as compared to by RES? Obviously, to undertake an evaluation requires performing the same analysis for non-conventional sources, since they also generate pollution, thus increasing the cost per kWh generated. What might these costs be?

- For wind energy (section 5.3.1) it is again necessary to apply the Life Cycle Assessment to learn the amount of pollution caused by extracting the raw materials used for the construction of blades and gear boxes for wind turbines, as well as for the cement or steel towers supporting them.

- Any associated energy and contamination costs must also be considered. These arise in the process of extracting, refining and purifying metals to manufacture the silicon wafers employed in the production of photovoltaic cells (section 5.3.2). The land space that solar panels would eventually use must also be regarded as an associated cost.

- In the case of biomass utilization (section 5.3.4), environmental costs will show through Life Cycle Assessment applied to the manufacturing of components for methanol (section 5.3.4.1) and ethanol (section 5.3.4.2) production, and to make boilers, turbines, generators, etc.

This calculation has probably already been made. If so, it will show that the cost of electricity produced by non-conventional sources is more than competitive when compared to that produced by conventional sources. However, this is not the point. From the standpoint of sustainability it is necessary to take into account the economics of a venture, although it is probably more important — especially at the present time — to put more emphasis on the environmental and social aspects of the business.

This means that **even if the cost of electricity generated by non-conventional sources were higher than that generated by conventional sources, society should go ahead at full speed and start phasing out the**

**latter**. In other words, policies should be implemented for the replacement of conventional sources by non-conventional ones **even if the cost per kWh is higher than in conventional plants.** This way, the conservation of the environment can take precedence over the economy, and people can still grow but through better development.

This is not news, and can be done. In the US some states are implementing the zero tolerance policy for car emissions. Their bylaws do not consider that this policy will mean a higher cost for purchasing a car. In other words, local authorities — quite rightly — deem that the environment and people's health are more important than economics. The same applies to energy generation.

Conventional energy plants have been working for more than a century with the purpose of bringing comfort and benefits to people, but mainly to make money; this is fine, since it is at the heart of the capitalist system, yet no attention has been paid to the environment, with results that are now here for everyone to see. Some progress has been made indeed, in the last decades of the $20^{th}$ century, but unfortunately in some ways it looks like a case of being too little, too late.

Existing energy-generating plants that use coal and oil are deemed not sustainable not only because they use non-renewable fossil fuels, but because their emissions contribute to global warming — with its known effects on the world climate and the melting of the glaciers. Eventually, coal and oil reserves will be depleted and the world will have to find other energy sources.

Nuclear energy is also non-sustainable not only because it uses non-renewable and thereby finite resources, but mainly because of the already-discussed problem with radioactive wastes. One proposed solution is to encapsulate and bury radioactive wastes in deep caves, but it is only a temporary solution since there does not yet in existence a metallic enclosure that will prevent leaks over the long term, and because geological phenomena such as earthquakes can liberate such highly dangerous materials. One plan has even thought of burying such containers deep in the sea, but again, science has not yet produced any material that can resist corrosion for a thousand years. Perhaps energy generated by fusion provides some hope, but until now, after of decades of research, no solution is in sight.

Of course, there remains hydro energy, although the world is running out of adequate sites for these undertakings. Besides, large hydroelectric dams, which generate the cleanest source of conventional energy, usually provoke the loss of very large extensions of agricultural land, as well as erosion, climate change and social problems. They are also capital intensive and normally alter an area's ecology. The Aswan Dam is one example of this last: while producing considerable economic benefits for the Nile Valley, it deprives of nutrients the lands that lie downstream, requiring a larger use of fertilizers. Meantime, the ecology and geography of the Nile Delta are changing. For instance, near the Rosetta Mouth of the river, there are no more of the sardines that once lived in the mud carried by the Nile. Severe coastal erosion has also resulted: the beach is being invaded by the sea, as no more sediments are carried by the river to resupply the seashore with sand.

From the social point of view, a large hydroelectric undertaking can produce disruption in the way of life of many people, as has happened at The Three Gorges Dam in China — which will displace more than one million people and flood countless villages and agricultural land. Naturally, huge economic benefits will be derived from the availability of cheap electricity in a poor area of China.

## 5.2 Brief technical information on energy conversion equipment

Production of electricity is usually a procedure that requires several steps taken in sequence, as follows:

### 5.2.1 Coal-fired, gas-fired or oil-fired power plants

Production of electricity is usually a procedure that requires several steps taken in sequence, as follows:

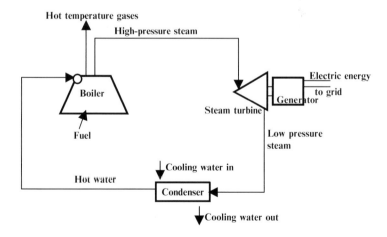

*Figure 5.1* **Sketch of a coal-fired power plant**

## 5.2.2 Nuclear power plants

They work in a similar fashion:

- First stage: atoms split in a **reactor**, generating heat in the process; this is the equivalent of burning fuel;
- Second stage: Heat is transferred to an **intermediate fluid;**
- Third stage: This fluid transfers its caloric content to water in a **heat exchanger**, producing steam, just as in a boiler;
- Fourth stage: Steam drives a prime mover, such as a **steam turbine**, which in turn drives an **electromagnetic rotor;**
- Fifth stage: The rotation of the rotor, which is part on an electric generator and within an electromagnetic field, generates electricity, which is then distributed through a grid.
- Fourth stage: Same as in the closed cycle described above.

## 5.2.3 Gas turbines

These work according to a different principle, since the combustion takes place within the machine, and not in another place, as occurs with boilers. The gas engine includes a gas turbine (the difference between a gas turbine and a steam turbine is that in the first, hot gases make it turn, whereas in the second case, steam makes the turbine rotate). When the gas turbine rotates, so does a compressor attached to the turbine shaft.
The process is as follows:

- Air is sucked out of the atmosphere and compressed in the **compressor**.
- The compressed air is sent to **combustion chambers** where fuel is injected, producing combustion and hot gases.
- The gas stream is directed against the blades of the turbine, making it rotate.
- A turbine-driven generator produces electric energy.

This cycle is open in the sense that, once burnt, the working fluid (hot gases) is discharged into the atmosphere.

## 5.2.4 Wind turbines

These use the wind's kinetic energy (which is a function of the wind's velocity). They usually consist of three large blades attached to a rotor that is connected to an electric generator through a gearbox. Wind impinges on the blades, thereby making them rotate and driving the electric generator.

## 5.2.5 Diesel engines

These engines utilize the stored chemical energy of diesel or gas. A diesel engine uses pistons that are displaced within cylinders in an alternating, lineal motion. A piece of equipment called a crankshaft, which is connected to an electric generator, converts the alternating motion into a rotatory one.

The engine sucks air from the atmosphere, compresses it within the cylinders by means of the pistons, and then injects pulverized fuel into the compression chamber, where the mix of air and fuel is ignited. This produces gases whose expansion result in the pistons' movement, as well as in motion by the crankshaft and the electric generator, which latter produces electricity. This is also an open-cycled process whose efficiency is better than that of steam turbines, while still being low: a large part of the fuel's energy is spent in mechanical friction, in heat losses in cooling the engine and in the exhaust gases.

## 5.2.6 Hydropower plants

Both small and large installations of these plants use hydro turbines (different types will suit a given site's characteristics). The turbine is attached to an electric generator, whose 'fuel' is the **potential energy** of the water — that is, the energy yielded by the height of the reservoir (head) where the water comes from — and the rate of the water's flow. Micro-hydro plants have outputs of up to 100 kW, while mini-hydro installations produce up to 1,000 kW.

## 5.2.7 Biomass

Energy from coal, gas, oil and wind comes from the sun, but we cannot utilize it directly, but must use an intermediate. The same is true about the energy one needs for living: it originates in the sun, but people cannot utilize it directly, but only through an intermediate — in this case, **plants.** The human body makes use of this energy when inhaling the oxygen they produce, and by ingesting the nutrients they contain.

The sun generates its own energy through a process called **nuclear fusion,** a process that scientists are trying to reproduce in laboratories, albeit without very much success to date.

By using energy from sunlight, water, and carbon dioxide ($CO_2$), a plant's chlorophyll enables it — through its green leaves — to convert the sun's energy into sugar and oxygen ($O_2$). $O_2$ is then a 'waste' or a by-product of this process, which is called **photosynthesis**. The plant itself uses the balance of energy in order to grow, and it stores the surplus energy. See Figure 5.2.

Photosynthesis involves a reduction process — that is, it takes oxygen from the $CO_2$ molecule, and in so doing it absorbs energy from sunlight. When wood burns in the presence of oxygen, the process is reversed, since it is a process of oxidation that releases the energy absorbed in photosynthesis, as well as $CO_2$. As biomass consists of carbon and hydrogen, these components can be processed to give up a fuel known as biogas.

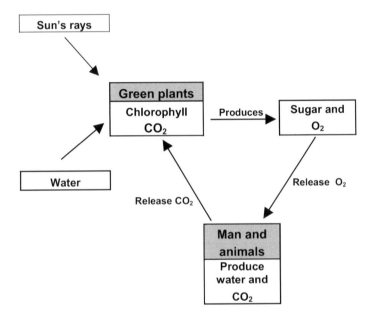

*Figure 5.2*     The photosynthesis process

One can see that in all these cases of energy production, some intermediate is necessary. Devices or prime movers are needed, including boilers, steam turbines, gas turbines, diesel engines, wind turbines, or hydro-turbines. But these devices merely extract the energy contained in various intermediaries: a fossil fuel, the wind or water (all of them of solar origin), to then generate electrical energy by driving a generator. This raises an obvious question: Is there any way to directly convert directly solar energy into electricity?
Yes, there is.
The only way one can directly convert solar energy into electricity is by employing the **photovoltaic effect** (section 5.3.2).

*Nolberto Munier*

## *5.2.8 Geothermal*

This section deals with both **ground-source heat pumps (GSHP)**, for cooling or heating purposes, and with **geothermal energy**. The utilization of the former can drastically reduce electrical consumption, while the latter can be used for heating purposes and to generate electricity. Therefore, the concern about sustainability raises interest in their utilization.

### *5.2.8.1 Heat pumps*

In order to understand how a heat pump works it is necessary to have an idea of how an air conditioning unit or fridge works, that is, of the refrigeration cycle. Both systems have the same elements: a compressor, a condenser, an evaporator, a valve called an expansion valve, and pipes connecting these elements (Figure 5.3). The expansion valve separates the two sides of the compressor: that with the high pressure (discharge, output) and the one at low pressure (intake, input)

A working fluid is also part of the system. This exists alternatively in a liquid and in a gaseous state (like the water and steam in the water steam cycle of Figure 5.1), and the entire cooling principle is based on this alternation. The fluid has a very low boiling point, i.e., it necessitates very little heat to evaporate, or to pass from the liquid state to the gas state (unlike water, which needs to be heated to 100°C to evaporate).
A liquid will evaporate by taking or absorbing heat **from** somewhere/something else.
For a gas to condensate requires giving or transferring heat **to** somewhere/something else.
Now it becomes clear just how the system works:

- The compressor compresses the refrigerant — which is in a gaseous state — at a relatively high pressure (1).
- The hot gas is piped to the condenser (2) where it condenses, i.e. becomes a liquid, delivering heat in the process. A fan takes air from the atmosphere, pumps it through the condenser and in so doing absorbs the heat from the refrigerant and discharges it into the atmosphere.
- After leaving the condenser, the liquid refrigerant reaches the expansion valve (3). Bear in mind that the expansion valve separates the two parts of the circuit. The high pressure is where the liquid is now. A small orifice in the expansion valve allows some of the liquid to enter the evaporator (4), which is held at a low pressure because of the suction produced by the compressor.

- Because the liquid is now at a low pressure it will quickly evaporate, thereby **taking heat from the area surrounding the evaporator** (which could be the interior of a fridge or of a car, or air that passes through the evaporator coming from a room).
- The refrigerant leaves the evaporator at a low temperature, and low-pressure gas is fed again into the compressor, to repeat the same cycle.

**Note:**

A simple experiment can be performed to check how low pressure makes for fast evaporation: Moisten the back of one hand, immediately raise to the mouth that damped area and inhale with the mouth. A cooling of the damped area will be felt immediately, which results from the vacuum or low pressure created by the suction that evaporates the humidity. This required taking some heat from somewhere, in this case the damp spot on the hand, which is thereby cooled.

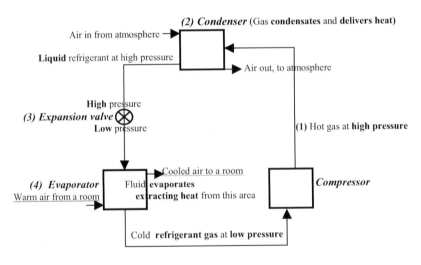

*Figure 5.3*      Air conditioning cycle

Heat pumps work through a simple concept that uses the system explained above, but one that can work in reverse: that is, it can be used for heating, instead of to cool.

This type of equipment usually involves placing the condenser and the compressor outdoors. The evaporator is situated indoors, along with the ducts that also provide heated air to a house. These are usually called **air-source heat pumps (ASHP),** because they use air as a medium for condensation (in summer) or for evaporation (in winter). In summer, the gas condenses using the air as a cooling system (even it is hot outside), whereas in winter the liquid refrigerant evaporates, thereby taking heat from the air (even if it is cold outside).

This system (in both its versions) works well and has been installed in millions of houses around the world. Yet it is necessary to realize that it works in very unfavourable conditions, since it absorbs heat from a very cold environment (in winter), and discharges heat (in summer) to a very hot environment. The fact that this equipment has to work with variable outdoor conditions means:

- The Energy Efficiency Ratio (EEF), that is, the relationship between input and output, is affected by outdoor temperatures. In summer, the higher the outdoor temperature, the lower the efficiency. In winter, the lower it is, the lower the efficiency. This translates into more electrical consumption for the same output.

- The cooling effect is also affected by outdoor temperatures, since the higher the outdoor temperature in summer, the lower the cooling, and the reverse is true in winter.

- The efficiency of the system is also affected by debris or dirt in the condenser/evaporator.

- The noise produced by the condenser's fan can be annoying.

Because of these drawbacks, which translate in more energy consumption and in turn more $CO_2$ emissions, the impact on the environment can improved through the ground-source heat pumps, which is the object of this section.

This concept is again simple, as it involves using the ground as a transferring medium, instead of the air in the outdoor condenser/evaporator. In other words, the ground can be used to discharge heat in summer, or to provide heat in winter. This is because when solar energy hits the earth, although some of it is reflected energy still is absorbed and stored in the ground and in water (about 45 percent), which thereby act as heat accumulators. These heat sources can be tapped by an arrangement known as ground-source heat pumps.

To take advantage of the ground's thermal conditions, one need only bear in mind that its temperature is rather constant (between 12 and 16 degrees Celsius) at a depth of about 3 metres below the surface, excepting at very high latitudes.

**Some brief technical information about how GSHP systems operate**

Ground-source heat pumps (GSHP) consist in adding to the above mentioned ASHP a pipe, usually made out of plastic, which is buried in the ground, and in a form of a loop that is several meters long (perhaps 50 or 60) in a horizontal arrangement. This circuit is integrated with a water pump to circulate water within the coil and the condenser/evaporator. In the GSHP system, the compressor is kept indoors. The system has a reverse capacity in that the summertime condenser becomes an evaporator in the winter, while the summertime the evaporator (becomes a heater in winter.

Summer operation:
Water cooled by the ground's temperature in the looped pipe is pumped into the condenser, turning the refrigerant into a liquid. The air circulating through the evaporator is chilled once the liquid refrigerant evaporates, and is then sent through ducts to cool the indoor environment.

Winter operation:
Water in the looped pipe is heated by the ground's temperature and pumped to the evaporator (which in summer acts as a condenser), turning the refrigerant into a gas. The air circulating through the heater (which in summer acts as an evaporator) is warmed by this process, and then sent through ducts to heat the indoor environment.

In some applications, the loop is placed vertically through holes drilled to a depth of maybe 90 meters or more. This means that the system can be used effectively for large buildings within a congested downtown area by drilling a battery of wells that have enough depth.
The system described here uses water as cooling fluid, possibly with antifreeze and in a closed circuit — meaning that the water is constantly re-circulated. Other systems use open circuits, whereby water is extracted from a river, for instance, its thermal properties used, and then discharged back into the river.

Probably the largest office building in the world using this system is the Waterfront Office Building in Louisville, Kentucky, USA, completed in 1994. It has four 40-metre-deep wells, and uses an open circuit that discharges the water into the Ohio River. The Geothermal Heat Pump Consortium (see Internet references for Chapter 4) claims that the system's cost was US$1,500 per ton of refrigeration. By comparison, conventional systems with centrifugal chillers, cooling towers, and ancillary equipment, costs from US$2,000 to US$3,000/ton of refrigeration (see Glossary).

**Economics**

In summer, a conventional ASHP system needs to condensate the refrigerant using air at a temperature of, say, 32°C. Clearly, this compares poorly with a similar scenario using water at 12°C.

In winter, the evaporator must extract heat from outdoor air that could be at, say, 2°C or less, yet using groundwater it can extract heat from water at temperatures as high as about 11°C.

Some researchers (see azcentral.com, in Internet references for Chapter 5), say that a house with about 180 m$^2$ of floor space can be heated or cooled for as little as US$1/day. While this figure may appear to be low, there can be no doubt about the savings and their magnitude. The same source says that the initial investment of several thousand dollars can be recovered in 2 to 3 years, claiming that a 20- to 40-percent savings can be achieved against conventional units.

*Sustainable sources* (see Internet references for Chapter 5), claims that studies by the US's Environmental Protection Agency (EPA) indicate that these systems have the following increases in efficiency: See Table 5.1.

*Table 5.1*  Comparison of efficiencies of GSHP against other sources

| Equipment type | Greater efficiency reached by GSHP |
|---|---|
| Gas furnaces | 48 % |
| Oil furnaces | 75 % |
| Air-source heat pumps | 40 % |

*5.2.8.2  Geothermal energy*

Another very powerful and inexhaustible source of heat is the thermal activity in the Earth's core. This is not a new discovery; the Lardarello geothermal plant in Italy — the first of its kind in the world for electricity generation — was commissioned exactly 100 years ago, in 1904, and is still generating.

How does it work?

Water and steam are sometimes trapped in some sort of reservoirs in the earth's crust due to geological formations. A well bored to that reservoir will bring very hot water and/or high-pressure steam surging to the surface. This water and/or steam can be used for heating purposes or for the generation of electricity through a steam turbine and electric generator — or it can be used for both purposes. Many geothermal plants are in operation around the world, such as in the US, Italy, France and Iceland. In this last, the capital Reykjavík probably has the largest application of district heating using geothermal energy. Many thermal stations around the world, such as in Bath, England, or Budapest, Hungary, use geothermal water for leisure and curative purposes.

Different types of geothermal reservoirs exist in the earth's crust. Some produce pure steam, others produce only hot water, and yet others give rise to combinations of the two. If the water temperature is hot enough and at enough pressure, it will also generate steam when it is released to the surface. It does so because in the reservoir the water is under a greater pressure than that found on the earth's surface; as a result, when it releases at the much lower pressure, it evaporates into steam in a process called 'flashing'.

Advantages of geothermal plants:

- An inexhaustible source of energy and, as a consequence, sustainable;
- Produce heat and energy without polluting;
- Need little space to be developed, so are probably the most land-use effective option;
- Unaffected by rain or wind regimes, and operate 24 hours a day, 365 days per year.

Disadvantages of geothermal plants:

Although only distilled water is used in boilers, turbines and condensers, these are notorious for problems associated with water impurities. Thus, immediate concerns are raised regarding the dangers to the generation equipment under discussion, since they utilize geothermal steam and water that is rich with minerals, gases and impurities.
Sinclair Knight Merz Pty Ltd. (see Internet references for Chapter 5) indicates that maintenance costs for this type of equipment amount to about twice the usual cost for plants using fossil fuels, mainly because of these problems with corrosion, deposition and erosion.

**Economics**
It is difficult to establish values for geothermal plants because of the many factors involved, for instance, the need to drill new wells, when the yield of an exploited well is declining. 'Economy of geothermoelectric generation' (see Internet references for Chapter 5) provides very detailed calculations on this issue. See also Ian A. Thain (see Internet references for Chapter 5), for a brief but good description of problems with the Wairakei geothermal plant in New Zealand.

**Society's options**
Since energy sources are diminishing, common sense says that humankind should contemplate two different but complementary options:

- Reduce fossil fuel consumption in order to extend the life of existing reserves, not only by lessening the dependency in fossil fuels, but also by using them more efficiently.

- Switch to non-conventional sources.

The first option has been in force for decades, and now there are 'fuel-efficient' (at least by present standards) aircrafts and car engines, power plants and manufacturing processes. Even so, it appears that these advantages are offset by population growth, with its demand on services, and the consequent creation of more consumption.

Therefore, it is reasonable to think that society's options point to the second alternative, at least for the time being, to alleviate the load and demand on fossil fuels. Fortunately, serious attempts have been made by governments to switch partially to non-conventional sources of energy: countries such as the UK, Germany, Denmark, Spain, Austria, Canada, the Netherlands, Belgium, the US and India, to mention only the most advanced, have taken great strides toward using non-conventional sources of energy.

## 5.3 Non-conventional sources for energy generation

Several alternatives of non-conventional energy sources or renewable energy sources (RES) are available. What follows will include brief descriptions of each one, although readers are cautioned not to expect exhaustive accounts on each system: that would take up hundreds of pages while distorting the purpose of this book. This is not a technical manual, so this information will only be illustrative. This book's main purpose is to let readers know about both the existence and the potential of different non-conventional sources of energy generation. Information has been added, when it is available, to comment on the advantages and disadvantages of each system, their potential, some technical limitations, a brief word on the economics involved, and a few actual installations will illustrate the subject.

Although all the mentioned points are important, this last is especially significant to the aim of fostering sustainability, since it shows that these plants 'are not castles in the air' or far-fetched technologies that will take shape some time in the future. These technologies exist today, they are generating energy around the world, and they will probably be the key electricity sources of the future.

This book aims to give readers — researchers, policy makers, stakeholders or those implementing some new energy source within a community — an idea of the most suitable plant for their local need. To that end, of course, this author would also encourage consulting an advanced bibliography, some manufacturers and energy experts regarding their use. That is, the aim here is to furnish a starting point for such projects, and to help with a discussion about the options.

Two main reasons urge replacing non-renewable energy sources with renewable ones. One is that the world will eventually run out of coal, oil and gas, since of course reserves are limited. But even if new and richer oil and gas fields and carbon deposits were found, or if new techniques can improve extraction and combustion, the main problem is atmospheric pollution, especially with the production of $CO_2$ as a flue gas release, since it makes for climate change by increasing the greenhouse effect.

This is why the utilization of renewable sources is so important. To simply give a broad idea of the order of magnitude involved, bear in mind that calculations show that, to produce one GWh (see Glossary) of electricity, renewable energy plants generate an average of 6 metric tons of $CO_2$. Meantime, coal, gas and oil power stations produce **120 times more** than this much $CO_2$ per GWh.

## 5.3.1 Wind energy

### Technical aspects

**Something to think about**
Only 2 percent of the sun's heat reaching the earth creates winds, building up kinetic energy. By using wind turbines, humanity converts it into electrical energy.
Enormous progress has been made in the area of harnessing this wind energy, which is centuries-old endeavour with the famous examples of Dutch and Spanish windmills, and water pumps operated by windmills on farms in many countries.
Modern technology utilizes the same principles yet through a completely different approach, namely, by using turbines with vertical and horizontal axes, and computer-assisted adjustments of these blades — which are as aerodynamically shaped as airplane wings. Many of these wind turbines dot the landscape in Germany, Denmark, The Netherlands, Austria, India, and elsewhere, as more than 20,000 wind turbines are now operating around the world.

Statistics show that Germany has the largest amount of installed wind power, followed by the US, Spain, Denmark and India — with a total output, as in the case of Germany, of thousands of megawatts. The second largest wind farm in the world is in King Mountain, Texas (see Internet references for Chapter 5); according to "*Globe Business*", July 2001, there will be 214 wind turbines with an installed generating capacity of 278 MW around the globe — which thus prevent the release of 20 million tons of $CO_2$ by substituting for fossil-fuel fired power plants.

The largest offshore wind park is located in the shallow waters off Middlegrunden, about 2 kilometres outside of Copenhagen. It involves 22 wind turbines placed in an arch, with their columns anchored at a depth of 3 to 5 meters. Each turbine is 63 meters tall, and has rotors of a 75-metre diameter. Wind turbines are usually mounted on conical steel or concrete towers of different heights (of an average of 50 metres), depending on the area and the output. Smaller wind turbines use a steel pole supported by guy wires.

The blades are attached to a shaft that drives an electric generator — usually through a gearbox — to increase the generator's rotatory speed.

The relationship between wind velocity and electrical output is **not linear** but **cubic**; that is, the energy produced increases proportionally to the cube of the wind's speed — which also means that reductions in wind speed also have a considerable effect on energy output. These devices are subject to the variable intensity of winds; consequently, electricity production is intermittent. However, this does not appear to be a problem because electricity from small units can be stored in lead-acid batteries, as DC (direct current), and since modern units can level off power produced at their rated capacity.

Wind turbines are usually assumed to have 20 years of life.

Wind energy depends of three factors: **wind speed, air density,** and **rotor area.**

The wind speed is a natural characteristic of a zone, so not all locations will suit this factor. Air density is directly linked to output: the denser the air, the better. Thus, care should be taken when considering locations at a high altitude, due to the thinner air. Rotor area depends, of course, on the length of the blades and the height of the tower to accommodate them.

## *Safety*

The rotor ensemble has safety systems to brake or stop the unit when strong winds may imperil the structure, since the main safety problems have to do with excess speeds. For instance, if the generator for whatever reason disconnects from the grid, the absence of a load will tend to speed up the rotor to dangerous speeds. Hence, the necessity of installing automatic braking systems. One of the systems simply changes the pitch of the blades, thereby stopping the rotation.

### Economics of wind turbines

From the economic point of view, wind turbines are competitive with the producing energy from coal-fired power plants. Yet this comparison does not factor in the hidden costs of the latter — such as the depletion of fossil fuels, pollution, global warming (from the emission of $CO_2$), etc. As a result, it appears that wind turbines are economically more advantageous that even modern coal-fired power plants, once the environmental costs are considered. Wind turbines can generate electricity in some places at a cost as low as US$0.03 per kWh.

Different sizes of wind turbines are available on the market to suit different needs and conditions. In wind farms, it is customary to install large units producing as much as 1.8 MW to generate electricity to feed the grid.
Regarding investment costs, they vary widely depending on the size, winds, etc. For instance, offshore wind turbines can call for an investment of about €1,700 per kW, and large wind turbines, due to economies of scale, generate energy at a lower cost than smaller ones. Small turbines, rated from 1 kW to perhaps 100 kW, are adequate for household use but do not offer the benefits of economies of scale. Nevertheless, an economic calculation will probably show a little savings — say over a 20-year period — compared to using power from the grid, after considering all the expenses of purchasing the unit, its installation, more taxes paid because of a property's increased value, yet less taxes paid because of incentives, maintenance, insurance, etc. However, the most important part is the savings in fossil fuels, and in the air pollution produced if the electric utility supplied the household, instead of the wind turbine.

A wind turbine installed on a farm will probably generate an excess of energy. In such a case, it is possible to inject that excess into the grid, selling it to the power company. Many states in the US have provisions for such arrangements, and normal selling prices for that excess electricity produced on such farms is the same as the price for electricity bought from the electric company. The importance of wind turbines becomes evident when one considers that about 100 turbines, depending on an area's characteristics, can supply more than 75,000 homes. Because they work in a 'third dimension' — since they occupy no physical space other than for the tower and the service roads — the land where they stand can be used for other purposes as well.

### Environmental aspects

Without doubt, these units bring considerable benefits in saving fossil fuels, and in the cost of their extraction, refining and transportation. They do not pollute, even if some people claim that they contaminate a landscape visually. They are elegant structures, grouped in so-called **wind farms**, and if they alter the landscape, the same can be said of a bridge or a road. Other than this, they do not generate any type of pollution while offering opportunities for thousands of people to engage in their construction, erection, and maintenance.

*Sustainable aspects*
Wind turbines are a sustainable way of generating energy because they:

- Use an inexhaustible 'fuel': the wind.
- Are not polluting in any way, except during the manufacture of their parts.
- Substitute for fossil fuel consumption.
- Allow for electricity to exist in remote areas, thereby providing a better standard of living.
- Create jobs for thousands of people engaged in their manufacture, sale, installation and maintenance. It is said that the wind turbine industry creates 22 direct and indirect construction and manufacturing jobs for each MW of installed capacity, and wind projects create one operation and maintenance job for every MW of installed capacity (see *Renewable energy* — Internet references for Chapter 3).

Naturally, some noise is produced by the blades' movement, normally as a function of their speed, which could provoke complaints from neighbours. Countries such as the Netherlands, the UK, Sweden and Denmark have wind turbines installed offshore in shallow waters, eliminating the problem generated by noise.

Nevertheless, there is widespread agreement that the noise should not be more than 30 dB (see Glossary) at a distance of 350 meters; this is the level of noise of a quiet bedroom. To assess the intensity of discomfort that this noise could produce, as a comparison the noise level in an office is in the order of 55 dB.

In this connection, Wind Flow Technology Ltd. (see Internet references for Chapter 5) cites a survey done in Europe in the mid-90s on the annoyance factor of wind turbines in sixteen sites within three countries. Its main finding was that the number of people expressing annoyance by wind turbine was small. Complaints have also arisen, especially in California, about the thousands of birds killed by the rotating blades.

From the perspective of industrial ecology, it has been argued that wind turbines do not recover the energy that was employed to manufacture their constituent parts, for their erection and decommissioning. However, a Life Cycle Assessment (Appendix, section A.5) has shown that the energy spent is recovered within a couple of months.

As a bottom line, it appears that wind turbines constitute an efficient way to get renewable energy, and one hopes that the natural logic of ongoing technical progress will boost their efficiency even more.

### 5.3.2 Photovoltaics (PV)

**Something to think about:**
*The amount of solar energy that hits the surface of the Earth every minute is greater than the total amount of energy that the world's human population consumes in a year.* (US National Renewable Energy Laboratory)

Edmond Becquerel discovered the photovoltaic effect in 1839, when he found that some substances exposed to light generate electricity. This phenomenon can now be seen in pocket calculators that are not battery operated.

*Technical aspects*
As mentioned previously, this is the only way to directly convert solar energy into electricity. Its foundation lies in Quantum Theory. Many are surprised to learn that Albert Einstein was awarded the Nobel Prize for Physics in 1921 "*...for his services to Theoretical Physics, and especially for his discovery of the law of the photovoltaic effect...*" — and not for his work on the Theory of Relativity.

In order to explain this complex issue in a simple and obviously very plain way, consider that the sun emits electromagnetic radiation, which Quantum Theory postulates as photons, which are understood as both waves and particles. When these photons impinge on photovoltaic material, they transfer energy to it and thereby produce the movement of electrons; this is electrical energy. The human body feels such transfers of energy in the form of heat, such that when we are exposed to the sun's rays, we get warm, while our unexposed parts feel no such warmth.

In the industrial production of electricity through solar energy, this is generated in cells that are arranged in modules, which in turn form arrays and are connected to the grid through a converter, whose name derives from the fact that it converts direct current (DC) produced by the photovoltaic (PV) array into alternate current (AC) in the grid.

Each cell is formed by a negative phosphorous-doped silicon layer and by a positive boron-doped silicon layer. However, this process demands large surfaces to produce sizable amounts of electricity, putting a toll on land use. It is estimated that about 1,400 square kilometres (that is, a square with about 37 km per side) are necessary to generate 1,000 MW.

Nevertheless, PV is extremely important and is a fundamental component of spatial crafts, probes and satellites, as it is the only source of electricity available in outer space. At present (2004), the efficiency of PV is quite low, about 14 percent. But reliable estimates affirm that by 2020 it will reach 20 percent. Direct 12 volts current generated in a PV array uses a DC-to-AC inverter for utilization in AC devices, and rising the 12 volts to standard 110 or 220 volts.

The unit of measure for a PV module's output is in Wp, which means 'peak power' — that is, the power that a module produces in watts when it is exposed to a radiation of about 1,000 watts per square metre, or the amount of energy delivered by the sun in summertime at about midday.

From the point of view of their construction and installation, PV installations have these characteristics:

- Panels can be installed on roofs in the form of flat roof tiles or shingles, as well as in other places where they do not use up land space.
- They can be used very effectively as a means of transportable energy, with panels mounted on trucks. Of course, the best examples of this transportability are the PVs installed on the satellites and probes sent to outer space.
- They are modular.
- No maintenance is needed since there are no moving parts; due to this, these systems are robust and have a long life, and are practically maintenance-free.

To increase the efficiency of PV installations, sunlight is concentrated by the use of parabolic optical devices equipped with a mechanical contraption to track the sun along its journey through the day. However, this can also build up excessive heat in the panels, and sometimes it is necessary to provide a cooling medium. The same can happen in hot regions when panels are installed on roofs, so there is a need for ventilation to keep them cool.

At present, there is an estimated total installed capacity of PV of about 3,000 MW.

### Economics of photovoltaic

The price remains high, at around US$6/kW, but the forecast is for reductions to about US$2.5/kW, in constant dollars, by 2020. Besides, in many countries the use of PV leads to tax savings.

PVs are ideal to power lights and small appliances in rural homesteads, however, large photovoltaic installations exist around the world, in countries like Greece, the US, etc. Germany has installations in Hemau, Bavaria, which, since 2003, is the largest PV plant in the world. It produces enough electricity to meet the needs of about 4,500 people. See relevant information on this plant in *World's largest solar power plant* (see Internet references for Chapter 5). This report affirms that the operator is promising a **very positive** return on investment (ROI) of about 7 percent, on an investment of €18.4 million.

**Environmental aspects**

PV does not produce any negative environmental effects by way of pollution to the air, land or water. It makes no noise, and does not consume any non-renewable resources. Its 'fuel' is an inexhaustible source: the sun.

**Sustainable aspects**

- PV probably offers the best deal for the environment among non-conventional sources, perhaps being challenged only by mini hydro power plants.
- It provides thousands of jobs for the construction and installation of solar panels.
- PV constitutes an ideal alternative for remote locations or as back-up systems when problems develop in the grid. For instance, when electrical service is interrupted by storms, falling poles, cable failures, sabotage, etc.
- In India, the Punjab Energy Development Agency (PEDA) has installed an array of solar panels to generate enough electricity to operate 50 water pumps. They extract water from a depth of 6 to 7 meters, supplying 140,000 litres of water per day: enough to irrigate between 2 and 3.2 hectares for most of the crops (PEDA, see Internet references for Chapter 5).

**Disadvantages**

- For large installations, its greatest drawback is probably related to land use since, as mentioned, they cover large spaces. However, they can also be installed in uninhabited, desert areas.
- As the system works with sunlight, on cloudy days the production is low, and it is nil at night.
- Electricity must be stored in bulky batteries.
- Since PV produces DC, AC appliances need inverters.

Section 5.3.2.1 discusses a large-scale housing project built between 1997 and 1999 in The Netherlands, that uses photovoltaics to generate 1 MW of electricity. It is highly recommended that the reader consult the website for this undertaking. It contains abundant technical information along with full colour photographs that illustrate various aspects of the project.

Section 5.3.2.2 shows a similar arrangement but for a commercial use in a supermarket in Finland.

---

### 5.3.2.1 Case study: 1 MW decentralized and building integrated PV system in a new housing area, Amersfoort, the Netherlands

*Data for this case study was taken from the publication "1 MW decentralized and building integrated PV system in a new housing area of Amersfoort. Case studies: Netherlands" (see Internet references for Chapter 5).*

*This case shows that large-scale housing projects integrating a PV system at the district level are feasible, and relates with the generation of 1 MW PV system involving 500 houses.*

*Readers are encouraged to consult the mentioned Web site to see with very clear pictures how PV panels were placed to form eaves that link many houses, and their placement in other parts of the buildings.*

*The arrangement not only fit in with a very pleasant and beautiful environment, but it used no land space. The total project cost amounted to €9,227,621, or €18,455 per house. The mentioned paper states that the cost of electricity is of €1.15 kWh.*

*This paper makes an important point by asserting that the project has shown that PV is a building component.*

---

### 5.3.2.2 Case study: Solar modules made integral to hypermarket roof - Tampere, Finland

*Data for this case study is taken from "Solar modules integrated into roof of hypermarket".*

*This renovation project of the Lielahti Citymarket in Tampere, in south Finland, involved installing solar panels on the building's roof, totalling 330 $m^2$, in order to produce about 39 kW of energy; this is sufficient to fulfil the needs of the market's shops.*

*This paper argues that a big advantage of the system is that in the*

> *summertime, when the demand for cooling and air conditioning is at its peak (around midday), the electricity produced by the panel also peaks. In the summer, it will cover about 4 percent of total energy needs.*

### 5.3.3 Solar collectors

Some of the energy received from the sun can be harnessed through photovoltaic devices (section 5.3.2) and by using heat collectors that heat water for different uses.
A solar collector is a simple gadget with a black metal plate that works as a solar heat absorber. Embedded in this plate is a metallic coil, and the whole ensemble is enclosed in a double glass cover. Let us explore the components of this collector:

The black metal serves a twofold purpose: as an absorber of sun energy and to support the metallic coils. Black surfaces catch the sun's energy well, as they do not reflect it. Quantum Mechanics theory explains this behaviour in terms of an interaction of electromagnetic waves from the sun with matter. This principle is used every day by people who wear dark clothing in winter, in order to absorb the sun's rays, and light apparel in the summer to reflect them.

The metallic coil conducts water, which is fed from its lower part, and discharges hot water through thermal convection (see Glossary) at its upper end. The heat collected by the black plate transfers to the water by conduction.
The glass enclosure has two purposes: First, it captures the infrared radiation (non-visible light) that enters the enclosures but cannot escape. This is the well-known greenhouse effect, which is described in any book on optics. It is also why the interior of a car that has shut windows and is left in a parking lot on a summer day is so hot that even its steering wheel cannot be handled.
The second purpose of the glass enclosure is to prevent the convection heat from dissipating into the atmosphere. This very principle is used to heat air, when it is forced to the bottom of the device and made to circulate upwards between the black plate and the enclosure. Heat is transferred by conduction from the black plate to the rising air.

For large industrial applications, the system works differently, since it employs mirrors. A project in the US called Solar II (see 'Solar thermal electricity', in Internet references for Chapter 5) involves 1,800 parabolic trough mirrors around a tall tower. The mirrors track the sun's movement and reflect the energy received to the top of a central tower. This tower contains a molten salt solution that absorbs the energy and heats the water, generating steam; this steam drives a steam turbine, which in turn runs an electric generator. The molten salt stores the heat and keeps a high temperature of about 565°C; therefore, even after sunset steam can still be generated for some time using the molten salt's caloric content. As with other systems (see photovoltaic systems in section 5.3.2), peak production is at noon, coinciding with peak demand.

This report maintains that the solar plant will generate electricity for more than 350,000 homes, which is indeed a considerable use of solar energy. Some countries, such as India, are very well suited for solar energy, given the number of sunny days per year and the amount of energy in $kWh/m^2$ that is received annually due to their nearness to the Equator.

**Economics**
Solar heating in large installations and with concentrators can be very competitive.

*5.3.4 Biomass*

Biomass is organic matter produced by plants and animal wastes. As described in section 5.2.7, photosynthesis converts the sun's energy into sugar and $O_2$. Plants, in turn, transform this sugar into energy and store it as starch. When people or animals eat plants and breathe $O_2$, this energy is transferred, used by the organisms, and any balance is eliminated as waste and as $CO_2$.
By the same token, when firewood burns the energy stored in that wood is released, together with $CO_2$. However, plants absorb that $CO_2$ again, repeating the cycle, so there is no release of additional $CO_2$ into the atmosphere. This is the 'carbon cycle' defined in section 2.7.

Not all biomass is in a solid state, since some residues, such as the black liquors from the paper industry, sewage sludge and sludge from other industrial processes, are liquids. By 1990, the use of biomass worldwide accounted for 13 percent of total generated energy, although this figure conceals a great discrepancy between developed countries, which use only 3 percent, and developing countries, with 33 percent.

*Types of biomass*
Different types of biomass are wood, forest waste, crop residues, municipal waste, some industrial wastes, some grains, etc. It is also possible to utilize sawdust, peanut shells, bagasse, rice hulls, walnut shells, etc. Biomass is utilized to produce electricity by means of these devices:

*Use by direct combustion*
This involves the direct combustion of biomass in boilers to generate steam for industrial purposes, or to propel a turbine for electric generation. One example is the burning of bagasse — a sugarcane waste — in sugar mills.

*Use by gasification*
As we are about to see, it is possible to obtain the fuel **methanol** — which is used for cars or even in a blend with gasoline — by means of crop wastes, wood and wood wastes, animal waste, and many other 'wastes', such as from food processing.

*5.3.4.1 Methanol*

This is obtained from feedstock — a raw material containing carbon and hydrogen — through a process called **gasification** and through **steam reforming.** It involves subjecting the raw material to a high temperature and pressure, and purifying it to obtain a synthetic gas called 'syngas', a mixture of carbon monoxide and oxygen. Because its source is a renewable resource, this is a **sustainable fuel**. A ton of feedstock can produce about 700 litres of methanol. Bagasse, for instance, is an excellent feedstock for methanol, because of which countries like Brazil and Cuba are able to drastically curtail their dependence on fossil fuels.

After eliminating undesirable gases and impurities from syngas, a catalytic process leads to obtaining methanol. The advantage of producing methanol is that it permits the use of feedstock of many kinds, such as waste from industries, forestry, municipal wastes, etc.
Methanol is an alcohol, a biofuel, and its utilization does not add any $CO_2$ to the atmosphere, since the process returns the $CO_2$ that was taken earlier during the chlorophyll transformation in green plants, so it closes the energy chain.

However, methanol synthesis generates high amounts of $CO_2$, discharging it into the atmosphere. This implies a low efficiency in methanol production; nevertheless, it is possible to achieve high methanol production rates by the injection of hydrogen, but this requires high investment costs. This is why an integrated process with pulp-making makes for higher efficiency.

*5.3.4.2 Ethanol*

This is another fuel made by a process of fermentation of starch or sugar from sugarcane, and it is also obtainable from wood. The biological process produces a gas, which includes methane, carbon dioxide ($CO_2$), and water vapour. However, and because it is in competition with the production of food products, its price is not competitive enough to be used in cars. Besides, ethanol derived from corn cannot be considered sustainable because it uses limited resources.

Besides their uses as fuels, Ethanol and Methanol are attractive alternatives for vehicles operated by PEM (Proton Exchange Membrane) fuel cells (section 5.3.5). In a PEM cell, hydrogen and oxygen are mixed, and the result is electricity and water. These cells utilize hydrogen, but as this is a very flammable product to store it is better to have it in some carrier such as ethanol and methanol.

*5.3.4.3 Biodiesel*

Another fuel made from oils and fats that can substitute for or be blended with diesel oil is biodiesel. Brazil is one of the countries promoting its use as a fuel: recently, in December 2003, it planned to start using biodiesel for railway locomotives, in a proportion of 20 percent biodiesel to 80 percent diesel fuel. Biodiesel can be made from any vegetable oil, although in this case it will be made from soya beans since Brazil is one of the world's largest producers of this crop.

*5.3.4.4 Methane*

This is a flammable gas obtained from landfills due to the anaerobic (see Glossary) digestion of waste. Landfills account for more than 1/3 of methane emissions, making this very important especially since this gas produces the greenhouse effect (global warming) far more that the other gas that is responsible for the greenhouse effect: $CO_2$. Landfill gas is seldom pure methane, but it is made up of about 50 percent methane ($CH_4$) and 45 percent carbon dioxide ($CO_2$)

Besides, being a flammable gas, methane can generate electricity when burnt in a gas engine to drive an electric generator. This is clearly a **sustainable resource** since it comes from an inexhaustible source: waste. The benefits of this kind of utilization are as follows:

- It reduces the emission of methane into the atmosphere.

- Generates electricity.
- Replaces the use of fossil fuels.
- Burns cleaner than fossil fuels, even if not as efficiently as natural gas.
- Wind energy, photovoltaics and solar energy (as shown elsewhere in this chapter) are unavailable through the 24-hour day, due to intermittent winds and as solar devices only work in the daytime. This is not a problem with landfill methane, which is always available.
- It is the only non-conventional energy resource whose elimination leads to a benefit, as mentioned.
- Its utilization also reduces the danger of landfill fires.

Gyungae Ha, a Korean corporation (see Internet references for Chapter 5), extracts methane from a landfill and utilizes it to fire a boiler. The firm calculates that the use of this free fuel produces a savings of US$3,400,000, considering the market value of methane, and that it will greatly improve the air quality by curtailing the landfill's methane emissions, as well as by reducing 84,000 tons/year of $CO_2$.

Farm animals generate a considerable quantity of wastes. For instance, a pig generates 2.5 times more waste than humans, and producing methane. Besides, the wastes that materialize in cow, pig and poultry manure contain large quantities of nutrients that rains transport to rivers, which favours the growth of algae — depriving fish of needed oxygen. This is a strong reason to process manure. Overall, manure can have a dual application, as fertilizer or utilized to produce methane.

The decomposition of manure process methane. Dung or manure from farm animals, such as pigs, is usually collected in trays below the pigpens' grates, where it is stored in hermetic tanks called digesters. There, an anaerobic (or airless) decomposition takes place, producing biogas, which is piped to storage tanks for further use as fuel in boilers to generate steam. In turn, this steam drives a steam turbine electric generator.

In Canada, a plant to treat manure and obtain biogas has been built in Vegreville, Alberta (see EnviroZine, in Internet references for Chapter 5): the processed manure from 7,500 head of cattle will be used to produce electricity as of June, 2004. This manure is expected to generate 1 MW of electricity, and the plan is to increase this to 3 MW: enough to supply power to a town of 5,000. There is also an integral use of the waste since the liquid resulting from the process will be treated to get rid of ammonia, and, after adjusting its pH (see Glossary), it will be utilized for crop irrigation. Without a doubt, this is a very sustainable project. It is worth citing what the report says about the 'waste': *"This new **technology treats cattle manure as a resource as opposed***

***to a waste****. It is a new and very cost-effective approach that addresses social, economic, and environmental issues associated with manure management".*

According to David Hall (see Internet references for Chapter 5), in Norfolk, UK, the largest biomass plant in Europe is already in operation, producing enough energy to power 70,000 houses. It burns poultry litter — a blend of straw, wood chippings and poultry droppings — and, together with another two smaller plants owned by the same company, these the only plants in the world that operate with this 'fuel'. They have the added advantage of also producing as a residue a fertilizer that is rich in phosphates and potassium.

*5.3.4.5 Pyrolisis*

Pyrolisis occurs when waste is heated without the presence of oxygen. This process yields gas, liquid fuel and char. These residues can all be processed, refined, and used as fuels. Engines or boilers can use the resulting gas and oil, and the char can be gasified. One example of pyrolisis on a large scale occurred in the tire fire mentioned in section 2.12.5, where thousands of tires burnt without oxygen, producing oil.

### 5.3.5 Fuel cells

It appears that the future for fuel cells is brilliant. The ceramics industry (see Internet references for Chapter 5) forecasts a market of US$95 billion for fuel cell technology.
These devices can operate many appliances, from laptops and cellular phones to large generating plants. Naturally, it is not possible to operate such a broad range of applications with a single type of fuel cell. Table 5.2 gives a glimpse of types, applications and other characteristics of fuel cells. Whatever the type, the operating principle is common to them all. Many different sources such as methanol, natural gas, methane, etc., are used to feed fuel cells, while oxygen is taken from the atmosphere. They produce electricity, heat, and most of them issue pure hot water, so their contribution to pollution is nearly nil.

A common battery generates electricity due to a chemical reaction, and when the chemical materials are spent or saturated the production of energy ends, and they require to be replaced or recharged (when they are rechargeable). Fuel cells, by contrast, do not need any replacement or recharging, and will work as long as they have fuel, which is hydrogen in its different forms. Fuel cells actually consist of the reverse mechanism of water electrolysis. An electrolytic process involves two electrodes, called an anode and a cathode, immersed in water, and when electric power is applied to these electrodes, the water separates into its two components: hydrogen and oxygen.

In a fuel cell, the system consumes hydrogen and oxygen and produces electricity, plus hot water. How does it work?
A fuel cell (Figure 5.4) also has two electrodes also called anode and cathode. Hydrogen is injected into grooves built into the anode, and oxygen is fed into grooves in the cathode. Between these two electrodes, there is an electrolyte (see Glossary), which can be solid, liquid or aqueous, that is made of various chemical elements. The electrolyte can be contained in a matrix or in a membrane. Generally, a catalyst, which is an element, or substance that accelerates a chemical reaction, coats both sides of this matrix.

Hydrogen is the simplest atom, having a proton as a nucleus, with a positive charge, and a single electron orbiting the nucleus has a negative charge. When hydrogen is fed to the fuel cell (left electrode in Figure 5.4), the catalyst, which is usually made of platinum (Pt), produces the separation or ionization of the hydrogen atom. Remember that an ion is an atom, which has either gained or lost electrons.
Since the catalyst separates the electron from the proton, there are now two ions, the positive being called a cation ($H^+$), and it migrates to the cathode through the electrolytic solution. The negative ion, the anion, is the free electron ($e^-$) in a continuous flow, and constitutes the electric current, and it goes to the cathode through an electric circuit and can be used to produce work, such lighting an electric lamp.

The cation from the electrolyte, the anion after doing work, and oxygen from the atmosphere, all meet in the cathode. There, the catalyst helps maintain a reaction that will produce hot water. It is possible to use this pure hot water for different purposes; in some cells, it is recycled to release hydrogen and recommence the cycle.

*5.3.5.1 The fuel cell in automobiles*

Nicolas Otto invented the combustion engine in 1875, and applied it to a motor bike; later, Gottlieb Daimler and Wilhelm Maybach developed an advanced gas engine. Then, in 1885, Karl Benz (the name is perhaps recognizable) built the first practical automobile in history.
The automobile was seen as a curiosity when it arrived at the end of the 19th century: nobody seriously thought that that smelly, noisy and uncomfortable contraption would replace the horse and the horse-drawn carriage of the time. Of course, the automobile brought a revolution in transportation, and from then on, nothing was ever the same.

Table 5.2    Comparison between different types of fuel cells

| Type of fuel cell | Type of electrolyte | Catalyst | Output range | Operating temperatures (in degrees C) | Main applications | Efficiency (%) |
|---|---|---|---|---|---|---|
| PAFC | Phosphoric acid | Platinum | Up to 200 kW | 150 - 200 | Stationary | 40-50 |
| PEM (Proton Exchange Membrane) | Poly-perfluorosulphonic acid | Platinum | 50 – 250 kW | 80 | Cars, Houses | 35 – 40 |
| MCFC (Molten carbonate) | Liquid solution of lithium, sodium or potassium carbonate | | 10 kW – 2 MW | 650 | Stationary | 50 – 55 |
| SOF (Solid oxide) | Solid zirconium oxide with traces of yttrium | | 100 – 200 kW | 1000 | Power houses Housing | 45-55 |
| AFC (Alkaline) | Aqueous solution of alkaline potassium hydroxide | | 5kW | | | 70 |
| DMFC (Direct methanol fuel cells) | Polymer membrane | | | 50-100 | Very small and mid-size applications, such as cell phones. | 40 |
| Regenerative | | | | | | |

Source: This Table was prepared with information from Breakthrough Technologies Institute/Fuel Cells 2000 (see Internet references for Chapter 5)

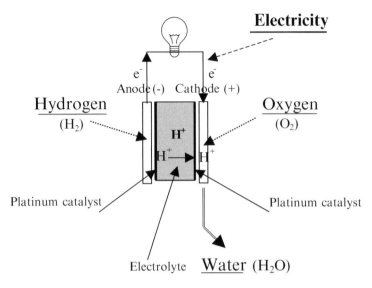

*Figure 5.4* **Diagram of a fuel cell (PEM)**

At present (2004) humankind is at a similar crossroads regarding the imminent replacement of the familiar four-stroke gas engine running cars with a completely new concept: the fuel cell engine. Yet a difference exists, in that people at every level of society acknowledge that the air in cities has become difficult and even dangerous to breathe, that the noise is becoming intolerable, and that fuel costs are becoming prohibitive. Given this, and unlike a century ago, we are now looking at the future, at least in personal urban transportation. It is highly likely that in 20 years' time people will look back and wonder how we could cope with the situation that exists today.

From a technical point of view, this transformation has parallels. The steam locomotive served us well, but then the diesel engine phased it out, and the electric train, in turn, is replacing it as well. Ships were one of the first users of reciprocating steam engines, until fairly recently: the Titanic had two reciprocating steam engines and a steam turbine, using the steam discharge from the former. Steam turbines replaced the steam engines and, in turn, gas turbines substituted steam turbines.

In industry, large food-related industries started with huge reciprocating engines to drive their cooling compressors, to be replaced by steam turbines. In transportation, gas engines propelled aircrafts until the advent of the gas turbine.

All of these large users of primer movers have two common denominators: they use reciprocating engines, and rotating engines replaced them. The advantages of this change are considerable, especially from the point of view of reliability and efficiency. Reciprocating engines have hundreds of moving parts, and change direction constantly: they reach the top of the cylinder, then stop, accelerate until reaching the other end of the cylinder, stop again, then accelerate again, and so on, thousands of times each minute. The rotating engine, whether a turbine or an electric motor, has only one continuous movement, which is rotation.

Without a doubt then, the replacement for the reciprocating gasoline engine of a car should be a rotating one (that is, an electric motor), and the fuel cell will help reach this goal. Unlike the reciprocating engine's operation, this one is very simple: the fuel cell generates electricity that powers an electric motor driving the wheels, and that is it. Interestingly, the first automobiles back in 1888 were electric ones, and battery operated, but they did not survive because of the battery technology available at that time. Electric cars builders then did not heed environmental concerns, but did keep an eye on the price of fuel, which meant that the gas engine came to prevail.

All fuel cells work according the principle sketched out in Figure 5.4. The differences lie in the way that the hydrogen is injected, and in the composition of the electrolyte. It is thus worthwhile to perform a brief analysis of the different alternatives and options that fuel cells offer. Naturally, one important factor to consider is each system's efficiency, in other words, how much, in percentages, of the fuel injected into an engine turns into useful work. This calculation is not difficult, although since this subject is being analysed from the viewpoint of sustainability, it may be worth remembering that computations of efficiency need to consider the entire process.

For instance, gas-engine efficiency is measured according to the ratio between the injected fuel's energy and its output, but one is not considering here the energy needed to extract the oil, to refine and transport it, and to sell the fuel — let alone any environmental costs. In short, the whole process or cycle must be considered. Of course, the same applies to batteries or to fuel cells. Nevertheless, the data consigned in Table 5.2 refers only to fuel cells without considering the energy spent in their production, and without co-generation.

**Latest concerns about fuel cells for automobiles**
Comparing the environmental impact of the different types of fuels and engines requires considering their full cycles: that is, from the source, whatever it might be, to the wheels of a car, bus or truck. This is why it is

sometimes called the 'well-to-wheels' cycle, which can be divided into two: the process to put fuel in the tank (the source-to-tank cycle), and the process to make the wheels rotate and the vehicle move (tank-to-wheels cycle).

The efficiency of fuel cells in the second cycle by far surpasses that of the gas engine, because it loses heat in the tail pipe, in cooling the engine, and energy through friction because of all the moving parts.
There is also to consider loss of efficiency in the combustion process and more losses to operate valves, drive pumps, etc. This is an unquestionable advantage of fuel cells. However, when the source-to-tank cycle is considered some researchers, such as Wald (2004), believe that the convenience is not so considerable, and that drivers will end up putting more $CO_2$ into the atmosphere. To understand this it is necessary to take into account that all fuels exist in nature in the form of fossil fuels (carbon, oil and gas) and biomass, and the same applies to what can be called 'fuels' such as wind, hydro, PV, methanol, ethanol, etc. Fuel cells work, as seen, ionizing hydrogen, so hydrogen is considered a fuel despite a difference when compared to other fuels, *viz.*, that hydrogen does not exist in a free state in nature, and has to be manufactured; this is where the problems arise.

To manufacture hydrogen some fuel has to be used, such as electricity for electrolysis, or methanol or natural gas, or electricity generated by the wind or sun. However, some researchers believe that these fuels could be better used to generate electricity for the grid, instead of to generate hydrogen for cars. Why is this?

Because, if for instance natural gas is used for hydrogen production, more coal has to be burned in boilers to compensate for the absented use of natural gas to generate electricity; and, since coal produces a high content of $CO_2$ in the flue gas, that content is indirectly attributable to fuel cells.

**Different kinds of fuel cells**
As mentioned, there are different types of fuel cells in various stages of development. In general, each of them is appropriate for some specific use/s. To provide just an idea, here is a brief explanation:

*5.3.5.2   PAFC – Phosphoric acid fuel cell*

It has a good output of up to 200 kW, and is used in small stationary power stations; however, they are not suitable for cars as they take some time to warm up. The system needs **pure hydrogen and oxygen**, which could be a drawback.

Actual installations: Holcomb, et al. (see Internet references for Chapter 5) provide details for installing a PAFC in the US, where there are 30 operating units. This fuel cell is supposedly the only available commercial fuel cell.

*5.3.5.3   PEM – Proton exchange membrane*

It has an output between 50 and 250 kW, and is ideal for mobile uses such as cars, trucks, as well as in houses. The system needs **pure hydrogen**, which could be a drawback.
Actual installations: In October 1999, a transit bus rode in Vancouver, Canada, equipped with this technology, and on May 2003, the city of Madrid received its first PEM powered bus, equipped with a 205 kW Ballard fuel cell engine.
Another 10 European cities will receive similar units, and the buses will refuel from 10 hydrogen-refilling stations. At present, most carmakers are considering this technology.

*5.3.5.4   MCFC – Molten carbonate*

Output ranges between 10 kW to 200 KW, which is appropriate for powerhouses. The hot water produced, due to the high operating temperature, can be used as steam to drive a turbine generator. On the other hand, because the high temperatures involved there is no necessity for expensive catalysis materials. It has the **highest efficiency of all fuel cells,** and the system is carbon monoxide tolerant, which means that it is not necessary to use very pure hydrogen.
Actual installations: An estimate put in about 700 the number of large plants installed all around the world, being Japan a leader in this field, as well as the USA. In Europe, the most important market is Germany.

*5.3.5.5   SOFC – Solid oxide*

Output ranges between 100 and 200 kW, which is appropriate for powerhouses as well as housing. It operates at a high temperature; therefore, the generated steam can be used for co-generation with a steam turbine driving an electrical generator. One of its main advantages is that it can **operate with different fuels** such as natural gas, alcohol, diesel, etc. It is not appropriate for car engines because there is a delay in reaching its working temperature.
Actual installations:  De Guire (see Internet references for Chapter 5), reports that Siemens Westinghouse planned to have its fully operational tubular fuel

cell plant by October 2003. She also reports that in Australia planar units have been in operation since 2001. It is interesting to quote that SOFC fuel cells for residential use have a cost between US$ 500 and 1,500 per kW, with a payback period of about 4 - 5 years.

*5.3.5.6 AFC – Alkaline*

This is indeed old technology, but also the system that enabled space missions since the 1960s and it is still used in the Space Shuttle. It requires pure hydrogen and oxygen and is expensive. Its output is about 5kW with a relatively low operating temperature of between 150 to 200 ° C.
Actual installations: On London's streets can be seen an experimental hybrid taxi equipped with this technology. This vehicle offers the Holy Grail of zero emissions.

*5.3.5.7 DMFC – Direct methanol fuel cells*

As it name suggests this fuel cell utilizes methanol (section 5.3.4.1) as the source of energy. The technology is relatively new and it is still in its early stages, however the potential is tremendous and highly sustainable since it uses a fuel derived from vegetable wastes. In this way, the technology can solve two problems at the same time that is to get rid of the 'waste', and to generate electricity. As commented, there are fuel cells that need pure hydrogen. However, there are others that can use a hydrogen carrier and then extract the hydrogen from it, using an element called a fuel 'reformer' (see Glossary). One advantage of the DMFC is that it can eliminate the fuel reformer, increasing the efficiency of the unit.

Extremely attractive is the fact that this system allows the manufacture of tiny fuel cells than can be used for small applications such as laptops, camcorders, digital cameras, and mobile cellular phones. As guessed, the race to produce these micro fuel cells has already started, mainly in Germany and in Japan, and a company is planning to have 100,000 units in the market by 2004. Of course, there will not be something like recharging a battery. Instead, a small methanol cartridge will snap into the unit.

*The Register* (see Internet references for Chapter 5), reported on December 2003 that Hitachi will market a fuel cell the size of an AA battery to be used for PDAs. (see Glossary). The company claims that this device will have enough energy to power a hand held device for six to eight hours. Interesting, the water produced in the reaction is used to dilute the fuel down to a concentration of 3 to 6 per cent. Toshiba, as expected, is also in the race but for the notebook computer market.

### 5.3.5.8  Regenerative

This fuel cell is still in the research stage. An interesting characteristic is that the fuel cell works in a close circuit, that is, the water produced in the electrochemical reaction is further electrolysed to obtain hydrogen. Therefore, these fuel cells have a dual function, as an electric generator and as an electrolysis cell.

**A word of caution.**
Different options and alternatives available for the utilization of non-conventional energy sources have been analysed. Most of then can probably be applied in an area, however, it is necessary to know the potential of the area for a particular undertaking. It is not realistic to think that an area is apt for everything since it is hard to believe that there is sufficient wind to install a wind farm, and at the same time plenty creeks and rivers with the adequate water flow for hydro, and also enough sun and bright days for PV, etc.

Consequently, it is deemed necessary to make an appraisal to determine the physical characteristics of an area regarding frequency and speed of winds, water flow, biomass generation, etc, and then use the most appropriate methodology. In addition, it is imperative to consider two another factors: the scale of the undertaking and the distance that the energy has to be transported. The scale of the undertaking is important for reasons of economies of scale (see Glossary), and the second because the inherent costs of construction a transmission line and the energy losses along the line.

### 5.3.6  *The sea as a source of energy*

There are different methods for using the energy available in the sea, and produced by two diverse effects, however this book only considers those that **have been built and are at present in commercial operation**. The main forms are **tidal energy** that works with the flow and ebb of the tides and **wave energy** that takes advantage of the kinetic energy of waves.

Needless to say, tidal energy is only possible where large tidal ranges exist. Most places have ranges about 7 to 8 meters, but in Canada for instance; the range is between 10 and 12.5 meters. Main places in the world for tidal energy are in Argentina, Australia, Canada, India, Mexico, Russia, South Korea, U.K., and USA. Only one tidal electric generation plant exists in North

America, and it is located in Annapolis Royal, Canada, taking advantage of the tide in the Bay of Fundy, the largest tidal range in the world. Commissioned in 1984 it has an output of 20 MW.

**Tidal energy – The Rance tidal power plant**
In this case, a barrage, that is a dam, is built with tunnels that allow water from high tide to enter a reservoir behind the dam, and in so doing, operating a hydraulic turbine. In low tide, the process reverses and water form the reservoir flows through the same turbine towards the sea. An electric generator attached to the turbine, generates electric energy. The drawback of this type of scheme is that only it works during the tides movement in both directions.

The most remarkable example of this type of generation is the Rance tidal power plant in northern France, which was commissioned in 1966. It has a 330 meters barrage, and a capacity of 10 MW, generated by 24 axial turbines with variable pitch, allowing them to work in both directions (water to the estuary in high tide, and water to the sea in low tide). Naturally, at a certain moment during high tide the level of the reservoir and the tide almost tally, therefore there is little transfer of water, and for that reason the turbines are made to work as pumps sucking water from the sea and discharging into the reservoir, and thus increasing the energy produced when the reservoir discharges into the sea.

There are also projects to utilize underwater currents using turbines as those found in a wind farm (section 5.2.4). In this case, towers are built on the sea floor and each one with two underwater turbines.

**Wave energy – The Inverness wave power plant**
In this type of engine, the kinetic energy of waves produces the rise and fall of a column of water within a conduit, which is connected with the open sea at its bottom. This column of water acts like a hydraulic piston, since during the rise, the water column compresses air above it, and this air is then used to drive a Wells turbine generator (see Glossary). During the fall of water, the water column sucks air, which is again used to drive the turbine, since it can work in both directions.

Therefore, in one instant, the waves raising the water column, creates a high pressure in the air above it and at the next moment, when the wave recedes, the water column lowers, and then sucks air. The air is then alternatively compressed and decompressed, by this oscillating water column (OWC).

An example of this type of pneumatic turbine is the Limpet plant, located in Inverness, Scotland, which was connected in November 2000 to the UK's national grid. The plant is located in the Scottish island of Islay, and according to the Press Release Network, it has a pair of contra-rotating Wells turbines producing 250 kW each. See also "Sciences News" in Internet references for Chapter 5.

The International Energy Agency (IEA) in Paris has published in Internet the ALEP Guidebook (ALEP stands for Advanced Local Energy Planning) (see Internet references for Chapter 5). This document supplies very useful and comprehensive information about energy issues. For instance in http://www.iea-alep.pz.cnr.it/4_Summary.htm, Table 1 depicts an extensive listing of modes to use for planning and analysis purposes and related with energy.

In compliance with the idea already exposed at the beginning of this section that only installed and in operation renewable energy sources would be commented, only two modes of energy from the sea have been mentioned, tidal and wave energy. However, there exist various different schemes to take advantage of the thermal and mechanical energy of the seas. Some of the schemes proposed, are:

- Using the energy of the waves. A device bobbing in the surface, connected to a buoyant platform, can produce electric energy to operate pumps and turbines;

- Using the difference in temperatures in tropical areas and between lower and upper layers of water. Ocean Thermal Energy Conversion (OTEC). There have been feasibility studies made in many countries about this methodology, which first test took place in Cuba in 1929. A good place to get information is http://www.worldenergy.org/wec-geis/publications/reports/ser/ocean/ocean.asp

- Lately, there is research going on to utilize the energy contained in organic matter on the sea floor, when gas methane is released and mixes with salt water to create **methane hydrate**. There is an enormous potential from this source of energy, and some people believe that the available energy is more than twice the energy of all fossil fuels combined. The technology is expected to take about 15 to 20 years to be developed for commercial applications.

**Internet references for Chapter 5**

*Noise from wind turbines*
Source: Department of Trade and Industry, U.K. Government - Noise Working Group, Harwell. U.K. (2003)
Title: *Renewable energy*
Comment: Very comprehensive technical report.
Address:
http://www.dti.gov.uk/energy/renewables/publications/noiseassessment.shtml

Source: Australia Wind Energy Association (AUSWEA)
Title: *Wind farms and noise*
Address:
http://www.thewind.info/downloads/noise.pdf

Source: Danish Wind Industry Association (2003)
Title: *The energy in the wind: Air density and rotor area*
Comment: Comprehensive report in four languages including FAQ (Frequent Asked Questions).
Address:
http://www.windpower.org/en/tour/wres/enerwind.htm
Title: *Renewable energy – Benefiting Minnesota's economic long-term* (2002)
Address:
http://www.mnproject.org/pdf/Renewable%20Energy.pdf

Source: California Energy Commission (2002)
Authors: Dora Yen Nakafuji, Juan Guzman, and Guillermo Herrejon
Comment: This comprehensive report (70 pages) contains valuable information about equipment for and working conditions on large wind generators in five different areas, as well as for small wind generation. It provides information about the physical characteristics of the different types of turbines used. Figure 5.2 of this report is remarkable in showing an almost parabolic growth of energy produced by wind turbines between 1985 and 2001. Another chart shows the evolution of the capacity factor, a measure of efficiency of over 20 percent. There is a very illustrative comment by the authors to the effect that California has more than 10,700 wind turbines, and that the topography of the primary wind resources in California consists of narrow mountain passes leading into hot valleys. There are maps showing the location of these turbines throughout California.
A visit to this site is highly recommended.

Title: *Wind performance report summary (2000-2001)*
Address:
http://www.energy.ca.gov/reports/2003-01-17_500-02-034F.PDF

**Author**: Gyungae Ha – Korea Emergency Management Corporation.
Title: *Ulsan landfill methane as project*
Address:
http://www.pi.energy.gov/pdf/library/EWSL/EWSLkorea.pdf

**Author**: Environment Canada – EnviroZone, issue 38 (2003)
Title: *Turning animal waste into electricity*
Comment: On conversion of manure into heat, electricity, fertilizers, and reusable water. This paper explains the workings of the Integrated Manure Utilization System (IMUS) in the province of Alberta, Canada.
Address:
http://www.ec.gc.ca/envirozine/english/issues/38/feature2_e.cfm

**Source**: The World Bank Group (2004)
Title: *Clean power for small towns in FYR Macedonia-Macedonia Mini Hydro-Power Project*
Address:
http://lnweb18.worldbank.org/eca/eca.nsf/0/571ADB32F2F1B23D85256C1D004AAAE5?OpenDocument

**Source**: European Commission – Energy (2004)
Title: *New and renewable energies*
Address:
http://europa.eu.int/comm/energy/res/index_en.htm

**Source**: Bureau of Energy Efficiency
Title: *12. Application of non-conventional & renewable energy sources*
Address:
http://www.energymanagertraining.com/Book_all/book4_PDF/4.12App%20of%20Non%20conventional.pdf

**Source**: U.S. Department of Energy (2004)
Title: *Make your own clean electricity - Case study: Economics of a home wind energy system*
Address:
http://www.eere.energy.gov/consumerinfo/makeelectricity/eval_wintrb_economics_cs.html

**Source**: World Energy Council (2003)
Title: *The challenge of rural energy poverty in developing countries. Promising technology developments*
Address:
http://www.worldenergy.org/wec-geis/publications/reports/rural/promising_technology_developments/4_4.asp

**Source**: Schatz Energy Research Center (SERC)
Title: *Fuel cells. How the PEM fuel cell works*
Comment: Very simple chemical information for the PEM fuel cell.
Address:
http://www.humboldt.edu/~serc/fc.html

Title: *Project 4: Aragon 2010 including Hamlet (K)* (2002)
Address:
http://www.solar-gmbh.de/eu3/aragon/index.html

**Author**: Craig. Peacock (2004)
Title: *South Australian electricity and renewable energy*
Address:
http://users.chariot.net.au/~cpeacock/#Economics

**Source:** Wind Flow Technology Ltd.
Title: *Wind energy: The perfect solution for New Zealand*
Address:
http://www.windflow.co.nz/backgroundinfo/

**Author**: Tjarinto S. Tjaroko – Asean Center for Energy (2003)
Title: *Applicability of (new) technological approaches / case studies on mini hydro*
Address:
http://www.asemgreenippnetwork.net/documents/tobedownloaded/knowledge maps/KM_applicability_new_technological_approach.pdf

**Source:** Museum Victoria's – Education Gateway (1999)
Title: *Case studies – Pig power*
Comment: Explanation how systems work in a pig farm with 15,000 pigs. This article mentions that this farm's pigs produce the same amount of waste as a city of 40,000. The waste management system created here recycles all the waste and converts it into useable products via an anaerobic digester. It is interesting that water is recovered in different parts of these projects, and used to irrigate the fields close to the pig farms, as well as being used to flush out the pigpens.

Address:
http://www.museum.vic.gov.au/FutureHarvest/case1.html

Title: *Renewable energy: China* (2001)
Comment: The use of PV systems in China, with a market increase of about 20 percent per year.
Address:
http://tcdc.undp.org/experiences/vol8/China.pdf

**Source**: Sustainable Sources (2001)
Title: *Just how much more efficient are GeoExchange heat pumps?*
Comment: This paper makes a comparison between heat pumps and other equipment, such as air conditioning units, gas furnaces, heating oil furnaces and propane furnaces.
Address:
http://www.greenbuilder.com/sourcebook/groundsource/groundsourceeffic.html#AC

**Source**: The Geothermal Heat Pump Consortium (1997)
Title: *Waterfront office building – Louisville, Kentucky*
Address:
http://geoexchange.org/pdf/cs-010.pdf

**Source**: azcentral.com (1998)
Title: *Geothermal heating*
Address:
http://www.azcentral.com/home/diy/geothermal.html

**Source**: Overview of European Geothermal Industry and Technology Prepared by KAPA Systems, Athens, Greece & EGEC (1997)
Title: *Economy of geothermoelectric generation*
Address:
http://www.geotermie.de/egec-geothernet/economics_of_geothermal_electric.htm#_Toc423920728

**Source**: Sinclair Knight Merz Pty, Ltd.
Title: *Geothermal operations and maintenance*
Address:
http://www.skm.co.nz/index.cfm?id=000531DC-2A68-1B1B-9A9D80E5C4250525

**Source**: (PEDA) cited in Bureau of Energy Efficiency
Title: *Application of non-conventional & renewable energy sources*

Address:
http://www.energymanagertraining.com/Book_all/book4_PDF/4.12App%20o f%20Non%20conventional.pdf

**Source**: Breakthrough Technologies Institute (2000)
Title: *The online fuel cell information centre*
Address:
http://www.fuelcells.org/fcapps.htm

**Authors**: F.H. Holcomb, M.J. Binder, N.M. Josefik – US Army (2002)
Title: *Fuel cell technology demonstrations at DoD installations*
Address:
http://www.asc2002.com/summaries/f/FP-10.pdf

**Source**: Sustainable Development International (2002)
Title: *Wind farm on King Mountain*
Comment: The second largest wind farm in the world will have a capacity to power 140,000 houses.
Address:
http://www.sustdev.org/energy/Industry%20News/05.01/24.01.shtml

**Source**: Welcome to the website of the IEA Photovoltaic Power Systems Programme (2003)
Title: *1 MW decentralized and building integrated PV system in a new housing area of Amersfoort - Case studies: Netherlands*
Address:
http://www.oja-services.nl/iea-pvps/cases/nld_01.htm

**Source**: TEKES – Tampere – Finland (2004)
Title: *Solar modules integrated into roof of hypermarket*
Comment: Technical data about this project as well as pictures.
Address:
http://www.tekes.fi/opet/lielahti.htm#Contact

Title: *Solar Thermal Electricity* (1998)
Address:
http://www.teachers.ash.org.au/monkweb/7chap4/7powerplants/solarenergy.htm

**Author**: Christopher Gronbeck (1994)
Title: *Solar thermal case studies*
Address:
http://sol.crest.org/renewables/re-kiosk/solar/solar-thermal/case-studies/trough-power.shtml

**Source**: Goethe – Institute (2003)
Title: *World's largest solar power plant*
Comment: Technical details about the PV plant in Hemau, Germany. This source believes that it will take some time before PV energy becomes competitive in a country with as little sun as Germany. However, it also suggests that this type of electricity will become competitive in the Mediterranean by the end of the next decade.
Address:
http://www.goethe.de/kug/ges/uMW/thm/en50516.htm

**Author**: Eileen J. De Guire - Cambridge Scientific Abstracts (2003)
Title: *Solid oxide fuel cells*
Comment: Good information on fuel cells, especially for SOFC.
Address:
http://www.csa.com/hottopics/Fuecel/overview.html
Ceramic Industry, cited in the above paper
Address:
http://www.ceramicindustry.com/ci/cda/articleinformation/features

**Author**: Tony Smith - The Register (2003)
Title: *Hitachi readies fuel cell for PDAs* (see Glossary)
Comment: This paper deals with what appears to be one of the first methanol cells for small gadgets on the market. A related link within this paper also announces that Toshiba has engineering portable fuel cells for mobile phones that recharge the phone battery without replacing it. There is more related information at this site.
Address:
http://www.theregister.co.uk/content/68/34485.html

**Author**: Ian A. Thain
Title: *A brief history of the Wairakei Geothermal Power Project*
Address:
http://www.geotermie.de/egecgeothernet/ci_prof/australia_ozean/new_zealand/a_brief_history_of_the_wairakei_.htm
Title: *Wave power connection heralds new era*
Address:
http://wire0.ises.org/wire/Publications/PressKit.nsf/H/O?Open&B1EECB3B1E755B0AC125699E003D882E

**Author:** Peter Osborne (2002)
Title: *Electricity from the sea*
Comment: Good information about different devices to use wave energy from the sea.

Address:
http://www.fujitaresearch.com/reports/tidalpower.html

**Source**: Science News
Title: *Oceans of electricity*
Address:
http://www.phschool.com/science/science_news/articles/oceans_of_electricity.html

**Source**: ALEP
Title: *Advanced Local Energy Planning*
Comment: Very valuable information from the International Energy Agency. This paper is packed with information about issues such as the Kyoto Protocol, Energy environmental planning, Technical analysis, and Modeling the energy system, and is mainly geared to energy conservation and the use of non-conventional energy sources. Provides examples of the Linear Programming Optimization Model MARKAL with concrete case studies tom exchange experience and it also provides "Guidebook on Advanced Local Energy Planning" (ALEP).
Visiting of this Website is highly recommended.
Address:
http://www.iea-alep.pz.cnr.it/5_background.htm

Author: David Hall (2000)
Title: *Renewable options 3: Biomass -Bringing biomass up-to-date. Poultry power*
Address:
http://www.peopleandplanet.net/doc.php?id=450

# CHAPTER 6 – MEASURING SUSTAINABILITY

*"What gets measured gets managed. What gets communicated gets understood."* — Steve Percy

To measure something it is necessary first to have it defined; in order to measure sustainability, how can it be defined? Sustainability is a process and, considering the three dimensions it encompasses, it is easy to see that this measurement is not an easy task. In any process, such as an industrial one, there are controls in each of its steps, including the manufacture of its raw materials and components. Inspections are conducted, chemical analyses performed on samples, physical measures taken, and, in each case, the values obtained (indicators) are compared with certain limit values (standards). A piece of equipment or product is usually rejected if it does not comply with these standards, within certain limits. This is the basis of quality control in industry. The set of values for a particular measurement along time constitutes a trend, and one is satisfied when this trend, with its highs and lows, remains between specified limits.

The same concept can be used in measuring sustainability. However, no single number tells us whether or not a process is sustainable; instead, a set of values can be obtained and then compared to standards and thresholds. During a sustainable process, one is interested in seeing that the trend is positive — that is, that it is heading in the right direction — for instance, regarding a decrease of NOx content. The greater the decrease the better, although realistically, nobody expects a zero value for this indicator within a city. Even so, limits can be established and monitored to check if the NOx content remains steady and within the allowed limits over a certain period.

## 6.1 Types of indicators

What is an indicator? Indicators are qualitative or quantitative measures signalling for some condition, for a decision to be taken, to give an early warning, and/or to show the results of a certain action or process.

The following examples show different ways in which indicators work:

- Qualitative indicators
  At a sea-side beach, small triangular flags in four different colours might signal quotients of safety for swimming, reflecting the sea's condition at that time. The flag will not inform people about the wind's force, the water's temperature, or the existence of underwater currents, but will only give qualitative information that reflects the result of an appraisal of a series of circumstances. Thus, a green flag stands for calm seas, a yellow flag advises caution, a red flag denotes rough seas, and a black flag indicate that swimming is not recommended — and swimmers will act according to the information given by such **qualitative** indicators.

- Quantitative indicators
  On a highway, posted speed signs inform drivers about maximum allowed speeds — acting as **quantitative** indicators or measures — which are technically determined as a result of several variables, combined or not, such as the approach of a winding road, sharp downhill slopes ahead, roads narrowing, etc.

- Warning indicators
  When working with a laptop one frequently receives a visual message (or visual indicator) **warning** that only, say, 10 percent of the battery remains, and urging one to change the battery or switch to an electrical outlet. This message appears as a result of an action from somebody: their use of the laptop. If it is ignored, after a while the computer will shut down and one's work may perhaps be lost. This type of indicator can be considered as a lower limit or threshold value, which when exceeded will result in a short while in some consequences.

- State indicators
  Another type of indicator will show the state of achievement relative to a goal, that is, **how close or how far** the result of an action is in reaching it. For instance, it could be a case where the target is collecting a certain amount of money for some humanitarian project within a determined period. The actual amount of money gathered so far will indicate if that target is near, or even if it has been surpassed; it may perhaps suggest that the target is not likely to be met.

- Indicators and time
  Assume, as another example, that some indicator shows the number of people who carpool to their jobs, in response to some transportation plan endorsed by City Hall. Such values, computed

year after year, will allow drawing a line connecting them which will show a **direction,** or a trend. Obviously, an almost horizontal line will indicate that there have been no changes because the plan; if it tilts upwards, it will reveal a trend signalling that indeed the plan is progressing nicely, and vice versa if it points down. This example shows then a pointer which is **temporally related**, indicating a trend through changes in individual values.

### 6.2 Approach for choosing indicators

Indicators can be designed to control conditions that have been set up for the achievement of a certain objective, or, conversely, they can be used to adopt an action based on the information they provide.
To illustrate these two concepts better, imagine a person having to drive 800 km in their car, thinking: *My goal is to reach my destination in 9 hours* — which means an average of 89 km/hr. driving — *yet I also have to stop to refuel, and to stretch my legs, so I'll have to drive a little faster: say, at 95 kms/hr.* The goal will prompt the driver to watch the indicators — that is, the speedometer and odometer — and adjust her driving to the calculated values so as to achieve the desired schedule. This is the **top-down approach,** where the indicators are selected according to the goal.

However, the driver could also take a different approach:
By considering the car's age (one indicator), the condition of its tires (a second indicator), fuel consumption (a third indicator), and the road and traffic conditions (a fourth indicator), the driver may prefer instead to establish a maximum speed limit. That is, the driver may decide that she'd be better off to set the arrival time so as to comply with these safety factors imposed, as it were, by the car, and to abide by a more conscientious manner of driving. This is the **bottom-up approach,** whereby the **indicators condition the goal**. Of course, in either case, the main objective is to reach the destination safely and on time, although with the bottom-up approach its accomplishment is subordinated by the driver to what the indicators say, so most probably he/she will have to drive more slowly, and/or depart earlier than in the first approach.

Let us now see the applicability of the approaches defined above to a sustainable development process:

- Top-down approach: Policy-makers decide on their goals and what indicators they will use to gauge progress.

- Bottom-up approach: Grassroots information is utilized to determine the status of something and to select the indicators to measure it, to gauge variations, and perhaps to limit the scope of the goal.

Which of these approaches is most appropriate to sustainable development processes? One fact peculiar to such endeavours that will often influence outcomes is the need to adjust expectations to existing conditions. This suggests that the bottom-up approach is best. While the goal can be as high and ambitious as desired, reality often shows that due to limited resources all goals should be established to make sure that they are achievable with the safe use of available resources.

## 6.3 Sustainable vs. common indicators

In all of the examples proposed only common or individual indicators have been used: that is, indicators expressing the measurement of a certain issue related to a certain area. One is more interested in **sustainable indicators**, which indicate or relate to other areas or conditions. For instance, when measuring concentrations of SOx (see Glossary) and other contaminants in a city's air, the resulting figure shows the condition of that air, which itself provides valuable information of concern, for instance regarding people's health.

Yet other dimensions need to be addressed, however. That is, it is also important to determine the consequences or effects in other areas within the environment, and out of it. For instance, that pollution has direct effects, such as pulmonary diseases for the population of the city, the potential generation of acid rain, etc.; these influence education, as its amount or value can lead health authorities to close down schools. It may also have an indirect effect on commerce, by leading to traffic measures that curtail cars from circulating in certain areas, etc. In short, there are often consequences for health, education, environment, commerce, etc.: the areas referred to above.

For this reason, and in order to assess or evaluate progress with respect to sustainability, there is no use in using isolated indicators; what is needed is a cluster of them, as well as their evolution in time. The three main domains or areas — namely, society, economy and environment — include hundreds of indicators, although there is a need to develop a manageable set that best represent reality, however imperfectly. Such a set is sometimes called 'headline sustainable indicators', and their number is normally set at between 15 and 20. Sustainable indicators are usually chosen so as to involve the three sustainable domains, although these three classical dimensions are sometimes expanded to incorporate other domains, such as infrastructure, people's participation, government efficiency, etc.

In addition, some indicators must be related to some breakdown or strata. For instance, the ratio between income and house prices depends on the social characteristics of the geographical area under scrutiny. The existence of a good relationship within a certain social stratum does not mean that the whole city enjoys the same satisfactory ratio.
Remitti (2001) offers one of the few examples in the sustainable literature for designing a sustainable indicator by a monetary calculation of the environmental costs of a resource used for a certain function.

## 6.4 Indicator uses

Indicators have many interesting uses, as a few examples will illustrate:

### *Helping decision-making*
People use hundreds of indicators on a daily basis without even thinking about it. For instance the fuel gauge — another indicator — in the mentioned example of a driver. This can indicate that a car is low on gas, and that suffices to alert the driver to take action since it reflects an existent situation with precision, and provides adequate information. In other cases, however, some circumstances can change a decision already taken. Assuming that the same driver was travelling between two points on a highway and at some moment looks at the fuel gauge. This may reveal that the fuel is low, yet the driver may estimate that there is enough fuel left to reach the destination, and decides not to stop to fill up the tank.

If, minutes later, the driver notices that the water temperature — another indicator — is too high, that, in fact, it is in the red zone, that driver is likely to look for the nearest town or gas station to find out about the problem. In that case, a second piece of information unrelated to the first, makes the driver change an intended course of action.

### *Helping to define polices*
Many lodges have a questionnaire in each room that asks guests to fill it in before leaving the hotel, asking for something like this:

*Was the room clean?*
*Was the bed comfortable?*
*Do you consider that the closet, hangers, and chest of drawers were adequate for your needs?*
*Did the TV set work properly?*

*Did you have to ask for anything related to this room?*
*How was the room service?*
*How was the attention you got at the front desk?*
*Was the front desk helpful with any questions about restaurants, museums, maps, addresses, etc.?*
*Did you require a doctor, and asked the front desk to get one?*
*How did they respond?*
*Would you visit us again?*

The hotel collects all this information and extracts some **statistics or indicators** from it and, supposedly, takes measures to improve conditions. Why is this?
Because they have a goal: to generate more business, to get more profits, and this goal can be accomplished best by using the visitor as a very effective and cost-free form of advertisement, since satisfied guests are likely to come back again, and may probably recommend the hotel to friends and associates.

Actually, the hotel's business is to provide guests with a decent room, a clean bathroom, and, technically, nothing else. Therefore, they could have merely asked the guest to express his or her views about it. Yet they also ask about things not directly related to the rental of the room, such as those about the front desk.
Why?
Because if it is true that there is a contract for a room between the guest and the hotel, they are also interested — naturally, on commercial grounds — in that guest's well-being, which is why they want to be as serviceable as possible. That is, to score that goal they must look at the whole scenario, and not at a particular issue. In so doing, they can establish a certain policy regarding advertisement, offering, for instance, something that other hotels are not providing, such as a discount price for excursions to nearby ruins or temples, the free use of bikes for guests to cycle through the downtown, the services of a travel agency and car rentals on the hotel premises, etc. **In other words, these services and their corresponding indicators detect tastes and preferences and can be used to establish policies.**

These basic examples show that more facts are usually needed on a certain issue than just the condensed information provided by just one indicator. This accounts for the interest in looking for indicators that link with each other, since that provides a better picture of all the areas involved.

### *Helping in inputs balancing*
In many cases, there is also need for a measure of complementariness between the different issues. For instance, in a mining process one indicator could express the number of tons extracted per hour, which is essential for economic reasons. Another indicator may show the grinders' capacity in tons per hour,

and, finally, yet another that states the number of people needed to operate the mine and to run the grinders.

Operational personnel get essential information from each indicator — one regarding their own field — although they cannot operate the plant if they do not have the three figures linked quantitatively (ore extracted, the grinder's capacity, necessary people per area).

Why is this?

Because there is a need for all three to be in balance. Otherwise, the operation could lack the ore to feed the grinders because there were not enough miners, or ore might end up stored and accumulating at the grinders' site given the insufficient grinding capacity, or because there are not enough people to run the grinders and other operations.

So it is evident that the operation will be economically and technically feasible only once all the intervening elements are known, so as to balance each resource: labour, ore and grinders.

### *Leading to the discovery of hidden effects*

Since communities have many social areas, such as health, education, entertainment, work, etc., there could therefore be indicators regarding each of these sectors. For example, one indicator could be the ratio of the ridership in the transit system of a city to the number of buses. Assuming that this ratio may be fifteen passengers per bus, a low figure, that would be what this indicator will show, but without reporting on the reasons or the root of this problem. There could be many causes, for instance:

- Too many buses.
- The frequency of service is too high.
- The service is bad.
- The fare is too expensive.
- Patrons have alternative options for transportation.
- People do not have enough money to spend on transportation.
- Trip takes too long because of the poor condition of streets and traffic jams.
- Buses are in very poor shape and spew a lot of gases, which filter into the passenger area.
- Etc.

As seen, an analysis of the reasons for the low value yielded by this indicator could show that they are about areas outside of transportation. There could be links with additional areas of infrastructure, such as inadequate connectivity between areas of the city, or it might be about social problems, and as transportation does not reach poor neighbourhoods because of the danger of vandalism in those areas, or it can go into the environmental sector, because of the emissions. As happens in some countries, it could also be that people are afraid of being robbed while on the bus. In sum, the indicator itself fails to identify the reasons for the problem, and does little to help to solve it.

The ridership-per-bus ratio only refers to one dimension of the whole problem, but one can see that what is needed is something that is also related to all the other aspects. This indicator can therefore lead to a search for other reasons, or the hidden associated aspects that account for the poor ridership. A sustainability indicator does precisely this, so for this reason it is multidimensional indicator.

***Helping to determine weakness in a system***
Assume an indicator that shows the GDP per capita in a city that is the hub for the footwear industry. Several plants in the area use local raw materials, and this region is the main producer for the whole country. Indicators of sustainability may show good links with other areas such as health, education, environment, etc. However, another indicator apparently unrelated to the GDP may signal a dangerous situation and a weakness in the system. This indicator measures the different kinds of industries in the region that constitute its industrial base, and it shows that the industrial activity is highly dependent on the shoe industry. Thus, there are tanneries, plants that manufacture rubber soles, many shoe shops, etc., and, naturally, a strong specialized labour force.

Assume now that new trade regulations allow the import of shoes manufactured elsewhere, and the country is becoming saturated with footwear from other countries at much lower prices.
What will happen?
The footwear industry in the region under analysis will probably be deeply impacted, because many people will prefer to buy an imported shoe at a lower price than a local one at a higher cost. Consequently, many plants will close while others work at a reduced capacity.
All of this is likely to produce massive lay-offs and unemployment, many people will lose their houses to the banks as they can no longer pay their mortgages, sales will drop, generating more lay-offs, and all of a sudden the region's economy will collapse because it was highly dependent on this type of industry.

Unfortunately, this grim scenario is not rare. Thousands of small companies, and not only in developed countries, have disappeared because of the import of much cheaper products from countries like China, the Philippines, Indonesia, etc., which have very cheap labour forces.
In sum, this shows the need for an indicator that relates the economic development of the region that is tied in with other indicators, such as the

mentioned one about diversity ratios. Had the region considered its warning signals, it could perhaps have taken some steps to diversify its economy. This subject was also mentioned in section 1.9, when an indicator was suggested to measure such dependency and a corresponding weakness.

***Unexpected correlation***
In some special cases, knowing the value of an indicator representing one action will facilitate obtaining information about another unexpected action that is completely unrelated to it. Assume for instance an action such as controlling the volume of water in the main water trunks of a utility company that serves an area, and a second action, involving people watching TV. Are the two correlated? Not that anyone might imagine, yet look at the following actual case:

> Some years ago, a young engineer on duty in the control console of a large water utility plant observed that during certain days, in a very precise interval, there was a cyclical demand of water. That is, at certain periods, instruments showed a sudden increase in water consumption, which declined to its original level in a short time of about 5 minutes. This performance repeated again, say, 15 minutes later, and only took place on certain days at a precise time gap such as from 6 to 7 p.m., on Mondays and Thursdays. Since a steady flow of water at that time of the day was an established indicator, its variation warned that something was happening.

> Intrigued, the engineer thought about the phenomenon and realized that it should be related with some particular practice of mass consumption. Then, he remembered a wildly popular TV program aired at the same time. It was very easy for him to find a high mathematical correlation that showed that, indeed, the time of the water increase kept in correspondence with the time of airing the commercials of said program. In other words, people took advantage of these intermissions to go to the bathroom.

> Naturally, thousands of people flushing the toilet almost at the same time produced a surge in water flow that was detected by the water utility instruments. Since this water consumption is related with a human action, knowing the average volume of water per flushing, and the increase in demand of water, it was a simple calculation to determine the number of people using toilets in that interval, and by inference, watching the program, at least in the area of influence of the utility company. Therefore, the determination of number of people watching a TV program (second action) was possible using the flow of water (first action).

To finish this section about indicators uses and properties, it is useful to quote the publication *Expanding the Measure of Wealth* (The World Bank, 1997) which states, "...*the key determinant of a good indicator is* **the link from measurement of some environmental conditions to practical policy** *options"*. The key words in this statement are **'measurement'** and **'practical policy'**.

The former implies that when something is monitored, it must be done so continuously, allowing for the manifestation of significant values, but also to detect a trend.

The latter term expresses that if no action is taken, if no policies are adopted, the exercise is meaningless.

In sum, the attributes of a sustainable indicator involving the three dimensions can be stated as having to:

- Be able to establish links.
- Be understandable.
- Give as much information as possible.
- Show a trend.
- Give enough information to establish policies.

These are tall orders for an indicator to comply with.

## 6.5 Indicator linkages

It is often useful to find the **backwards and forwards linkages** for a certain indicator. They not only show the interrelationship of an indicator with other areas, but will also help to detect some weak link in the chain. Figure 6.1 shows such a linkage for a water-consumption per capita indicator.

The links are not complete, as there could be many more, but the purpose of the drawing is to show how indicators — such as on population growth, water availability in wells and in the aqueduct's capacity, in addition to at the water treatment plant, and average temperature — influence this indicator. In turn, this indicator relates forward to wastewater generation, water related diseases, water pressure, wastewater treatment plants and water losses.

Each of these backwards and forwards indicators is in turn linked with another. This analysis is important because it can illuminate the consequences of taking an action. For instance, increasing water consumption per capita will probably decrease water pressure, and such a decrease will also produce less water losses while increasing wastewater production.

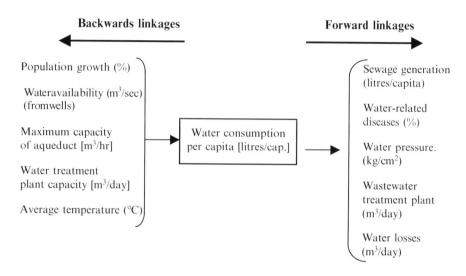

*Figure 6.1*    **Backwards and forwards linkages for water consumption/capita**

## 6.6 Integration of sustainable indicators

Values obtained from indicators seldom belong to just one category, say, the social, as they are often related with others in the environment and economy spheres. From the point of view of sustainability, it is then necessary to have as simple as possible a value relating all the areas involved, that is, an **integration of indicators.** One of the clearest examples is the spawning of the salmon proposed by Seattle Sustainable. They say that the fact that the return of salmon to spawn in their rivers is a good indicator of:

- Good water quality.
- Economic benefits.
- Safety for swimmers.
- Etc.

Naturally, nobody can pretend that a single number can adequately express the often-complex relationships between social, economic, and environmental issues. For that reason, the values given by indicators have to be treated with care, and must always be related to other data. This is why some researchers try to develop **indexes,** that are a final number that aggregates or combines several indicators, giving a general idea or signal of progress — although, naturally, a condensation like this always makes for a loss of information.

## 6.7 Weight of indicators

As a single number, an indicator synthesizes information, showing the result of several factors and interactions. An analyst could know which factors are related to an indicator, but this does not mean that that would enable developing an index that involved all of these factors. This is because it is seldom known what importance or weight is given to each one.

Let us again consider the example in Figure 6.1: water per capita links backwards to some indicators, which in turn are influenced by other factors, such as ground water levels in the aquifer feeding the system, silt and sand in the water, evaporation rates in the open aqueduct connected to the water treatment plant, losses in the network, etc.

To develop an index of the amount of water per capita, it is necessary to combine these indicators, but which weight should be assigned to each one? Two main problems appear here:

1. More often than not, all these linked indicators have different units of measure, so it could be very difficult to find a way to add, for instance, quantities of sand in the raw water, measured in $mg/m^3$, with the capacity of the water treatment plant, in $m^3/day$, and water availability, in $m^3/sec$.

2. Do all the composite indicators have the same weight? That is, does the sand content in the raw water have the same importance as the availability of water? Probably not. So it follows that some sort of weighting is needed to signal the relative importance of each one. Once this is agreed to, how can this weighting be done?

Sometimes, when enough information is available and collected within a certain period, one may use a mathematical technique known as regression analysis (see Glossary and Appendix, section A.6). It is a very well-known and widely used routine that can be readily done with computer software, such as the add-ins incorporated with Excel® and in all statistical packages. It utilizes as input the values logged for each intervening factor or variable during a certain period, say, 10 years.

Therefore, using this set of values for each variable, regression analysis will compute a coefficient for each one, which can be considered as its weighting.

Because the same problem of different units also arises here, these values must be normalized, which is a simple operation, and finally the weight for each factor is obtained. What is the meaning of these weights?

Each one **expresses how much the result for an index changes when there is a variation of a unit value in the variable under study.**

A very well-known application of this method is found in 'Hedonic pricing' (section 7.7), which is normally used to determine the price of a house subject to many different variables, some of which have no market value.

## 6.8 The choice of indicators

It is necessary to select indicators by using either the top-down or the bottom-up approach. Fortunately, many academic journals have published lists of them for every possible circumstance. In other cases, indicators are chosen by decision makers considering specific goals, or by grassroots groups after discussions that consider local and international indicators and standards. Some clarification is needed here, regarding the latter. Some indicators adopt as their yardstick the values of international standards, such as the quantity of drinking water a person needs per day, or the minimum weight for a newborn, because people have the same minimum needs regardless of where they live.

Other indicators, however, correspond and respond to local characteristics, such as, for instance, a city's level of poverty, or its crime rate or its percentage of paved streets. In cases of disagreement between international and local indicators, this author's experience suggests that the latter should prevail.

An example of indicators developed by grassroots groups is shown in Tables 6.1 to 6.4 which reproduces Table 6 'Master list of indicators proposed by workshop participants', from *Measuring Urban Sustainability: Canadian Indicators Workshop*, prepared by David Dilks (1995). The column 'Small groups' identifies the indicators selected in the workshops, while the column 'Survey' identifies indicators chosen by respondents to a survey prior to the workshop.

When indicators under 'Categories/Indicators' are in bold, it means that two or more workshop discussion groups referred them or five or more survey respondents.

*Table 6.1*     Social/Cultural/Institutional indicators

| CATEGORY/INDICATOR | Small groups | Survey |
|---|---|---|
| **SOCIAL/CULTURAL/INSTITUTIONAL** | | |
| 1) Equity/Income/Distribution/Poverty | 1 | |
| • unemployment | 1 | |
| • income distribution (% living below poverty line) | | 1 |
| • of poor [people] living in census tracts with greater than 30% (concentration) | 1 | 1 |
| • socio-economic linkages | 1 | |
| 2) Human health | 1 | |
| • infant mortality/weight | 1 | |
| • incidence of disease | 1 | |
| • number of reported cases of cancer | 1 | |
| • healthy household audit (number that 'pass') | 1 | |
| 3) Education | 1 | |
| • literacy rate | 1 | |
| • % with high school diploma | 1 | 1 |
| • sustainability in school curricula | 1 | 1 |
| 4) Public Safety/Crime | 1 | |
| • walking alone at night | 1 | 1 |
| 5) Community Participation | 1 | |
| • % of population voting in local elections | 1 | 1 |
| • lawn pesticide use | 1 | 1 |
| 6) Heritage/Culture | 1 | |
| • cultural opportunities | 1 | 1 |
| 7) Housing/Shelter Needs | 1 | |
| • accessibility | 1 | 1 |
| • variety (mix) | 1 | 1 |
| • affordability | 1 | 1 |
| • quality | 1 | |
| 8) Government/Public Services | 1 | |
| • hard services provided | 1 | |
| • soft services provided | 1 | |
| • availability/accessibility of public services | 1 | |
| • ability of community to provide public services | 1 | 1 |

Source: Environment Canada and Canadian Mortgage and Housing Association
Reprinted with kind permission of Environment Canada and CMHC

*Table 6.2*     Environment indicators

| CATEGORY/INDICATOR | Small groups | Survey |
|---|---|---|
| **ENVIRONMENT** | | |
| 1) Air Quality | 1 | 1 |
| • exceedance of standards | 1 | 1 |
| 2) Water Quality and Use | 1 | |
| • surface water quality | 1 | 1 |
| • ground water quality | 1 | 1 |

| | | |
|---|---|---|
| • treatment (before and after use) | 1 | 1 |
| • recreational use | 1 | |
| • % of population drinking bottled water and/or using water filters | | |
| • **water consumption** | 1 | 1 |
| 3) Soil Quality/Contamination | 1 | 1 |
| 4) Ecosystems/Green space/Biota | 1 | |
| • **access/distance to green space** | 1 | 1 |
| • classification | 1 | |
| • % of land base that is green space | 1 | |
| • green space per capita | 1 | |
| • **total amount of natural space** | 1 | |
| • ecology | 1 | |
| • ecosystem integrity | 1 | |
| • % of bird species that would be present if whole area remained natural | | |
| • presence of indicators species | 1 | |
| | 1 | |
| 5) Land Use/Urbanization | 1 | 1 |
| • **density (change in net residential density)** | 1 | 1 |
| • **mixed use** | 1 | |
| • urban form | | |
| 6) Energy & Resources Consumption/Conservation | 1 | |
| • **energy consumption** | 1 | 1 |
| • non-renewable energy use per capita | 1 | |
| • land consumption | 1 | |
| • product consumption | 1 | |
| • per capita consumption | 1 | |
| • efficiency | 1 | |
| 7) Solid Waste | 1 | |
| • **Generation** | 1 | 1 |
| • Disposal | 1 | |
| • **Diversion** | 1 | 1 |

Source: Environment Canada and Canadian Mortgage and Housing Corporation (CMHC)
Reprinted with kind permission of Environment Canada and CMHC

*Table 6.3*  **Economy indicators**

| CATEGORY/INDICATOR | Small groups | Survey |
|---|---|---|
| **ECONOMY** | | |
| **Potential Economic Indicators** | 1 | |
| • **employment (including diversity)** | 1 | 1 |
| • disposable income | 1 | |
| • real purchasing power | 1 | 1 |
| • public debt | 1 | |
| • dependency ratios | 1 | |
| • office and retail availability | 1 | |

Source: Environment Canada and Canadian Mortgage and CMHC
Reprinted with kind permission of Environment Canada and CMHC

*Table 6.4*          **Infrastructure/Influencing factors**

| CATEGORY/INDICATOR | Small groups | Survey |
|---|---|---|
| **INFRASTRUCTURE/INFLUENCING FACTORS** | | |
| 1) Population | 1 | |
| • growth | 1 | |
| 2) Transportation | 1 | |
| • **modal splits** | 1 | 1 |
| • expenditures | 1 | |
| • **commuting distance/time/mode** | 1 | 1 |
| • vehicle kms. Driven per year | 1 | 1 |
| • energy/pollution | 1 | |

Source: Environment Canada and Canadian Mortgage and Housing Corporation
Reprinted with kind permission of Environment Canada and CMHC

Clearly, these Tables are an excellent reference for choosing indicators since they not only cover the three dimensions of sustainability, but also include some preferences (which are noted in bold face) regarding the more important issues. It is also interesting to observe some linkages within the same category and between inter-categories, for instance:

- **Unemployment, income distribution, and percentage of poor living in census tracts greater that 30 percent,** in Table 6.1, appears to be more important than all the other indicators, except for **accessibility,** variety, and affordability of housing. It is obvious that there is a linkage between unemployment and income distribution, and the conditions to gain access to house ownership, which of course is also linked with economic conditions in the region.
- People are concerned about **exceedance of standards** in air quality (Table 6.2), which relates also to health problems, and perhaps to economic issues.
- The state of **surface and ground water** concerns people, as does **water consumption.** This last relates to social issues such as public health.
- People consider that **access/distance to green spaces** and its **total amount** are good indicators of quality of life.
- **Population density** is another important issue that is also linked with social factors (such as discrimination), and affordability of housing. Of course, it is also associated with transportation.
- **Energy consumption** is another big issue, with links to environmental concerns about air quality as well as to land use.
- It appears that **solid waste generation** has a large impact, together with its **diversion**. It could suggest that it could be associated with packing, amongst other issues, and recycling.

- **Employment** (Table 6.3) is of course a major concern, and is naturally connected with social issues.
- **Population growth** (Table 6.4) appears to draw attention, and this is obviously related to social and economic issues.
- Finally, in transportation, there is concern about **modal splits**, that is, how people travel (owning a car, by bus, train, etc.). No doubt this issue could be considered jointly with the environment, air pollution, economy, and social aspects.
- The last concern appears to be **commuting** and **distances** that people have to travel.

## 6.9 Multipliers

This very important economic concept is used to measure the economic impact that a new activity or an increase in a certain activity can have on the local economy. Simply put, it is a ratio between an initial change in an activity and the total change on the whole local economy produced by this initial change. The name 'multiplier' is used because initial changes can multiply along a local economy with something like a domino effect.

Let us take the example of a plant producing fruit and chocolate candies in a certain region. The manufacturer hires local people and also purchases local sugar — produced by a sugar beet plant — and other local inputs from the regional economy and outside of it. This local purchasing of fruit, sugar, water, energy, etc., have spin-off effects in the region, prompting the development of suppliers, who in turn hire more people. The sugar-producing plant will then have a forward effect related to its customer, and a backward effect in relation to its suppliers.

Suppose now that the plant receives an order from an outside buyer for $1,000,000 worth of merchandise, and that this industry has a multiplier of 1.69. What does this mean? It means that if the plant sells candies worth $1,000,000 to buyers outside the region, the total sales, output, or business activities throughout the region will increase by $690,000. Therefore, an initial change of $1,000,000 will trigger a total change of $1,690,000.
How can this be explained?
The plant will have to hire more workers to handle this order, so these people will be the direct beneficiaries. That is, they will receive compensation for their work, translating into a **direct effect**. In addition, and because of the increase in production, the plant would have to buy more sugar, to pay for more transportation, to use more electric energy, to buy more paper for wrapping its products, etc., so that initial change will also involve suppliers, who are then the indirect beneficiaries. This is called an **indirect effect**.

When the people employed by the candy manufacturer, by the sugar beet plant, by the transportation company, etc., are paid, they, in turn, will spend money in purchasing goods and services such as electric energy, food, entertainment, etc. Thus, the initial change has produced another effect, which is called **induced effect**.

Now, these effects take place within the regional economy and outside of it, as there are many inputs from places that are external to the region. Say, for instance, that of every dollar entering the region for this sale, only $0.39 is re-spent there, while the balance $0.61 is spent elsewhere. The manufacturer's re-spending will involve, for instance, payment to other firms within the region for inputs such as paper wrapping for candies, to another company to provide security services, etc. That manufacturer spends the $0.61 in purchases of inputs from outside the region, such as colorants, flavours, chocolate, etc. This also includes taxes paid, as well as financial obligations. This spending is called a **leakage.**

This is the first **round** of re-spending. Now, a second round of spending will also take place, through the purchases that the local suppliers have to make, in turn, to provide the candy manufacturer with goods and services. Again, a part of their purchases will be from within the local economy, but another part will be outside of it. Let us say that $0.18 remains in the area, and $0.21 'leaks' outside of the region.

In a third round, the supplier to the candy manufacturer's supplier will also make purchases within the region and outside of it, so that, say, out of the $0.18 spent in the area, only $0.08 remains, and $0.10 is leakage.

In a fourth round, out of that $0.08 only $0.03 remains in the local economy, while $0.05 leaks out; and finally, in a fifth round, out of $0.03, only $0.01 remains and there is a $0.02 leakage.

Table 6.5 shows these values and the cumulative re-spending in the region.

*Table 6.5*   Rounds, respending and cumulative respending

| Round | Respending ($) | Cumulative respending ($) |
|---|---|---|
| Initial entering the region | 1 | 1 |
| First | 0.39 | 1.39 |
| Second | 0.18 | 1.57 |
| Third | 0.08 | 1.65 |
| Fourth | 0.03 | 1.68 |
| Fifth | 0.01 | 1.69 |

Figure 6.2 shows, for each round, the relationship between re-spending and leaks.

Each activity possesses a multiplier within a region, and the question that now arises is, how is this value obtained? Acquiring indicator values is based on Input/Output inter-industrial analysis, which was developed by the 1973 Economics Nobel Prize winner, Wassily Leontieff (Leontieff, 1951).

The calculation is not difficult when the needed data are available. A model called IMPLAN, developed by the University of Minnesota, calculates multipliers, generating three series of multipliers, namely, for direct, indirect and induced effects.

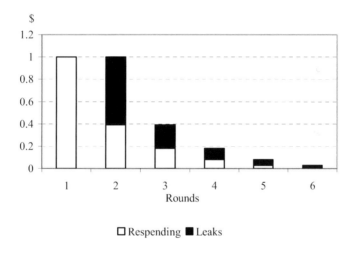

*Figure 6.2*  **Rounds, respending and leaks**

Figure 6.3 displays the flow of the economy in a region, taking the hospitality industry (hotel, motels and cabins) as generator of change.

Hotels pay their employees, which constitutes the direct effect.

The indirect effect takes place when hotels buy linen, energy, food, furniture from suppliers (commerce), or when they contract construction people for expansions or new premises.

Hotels' employees, as well as the supplier's employees, also spend money in the area in the form of goods and services such as electricity and water, which is the induced effect. The construction industry, for instance, purchases energy, pays for sewage and water, and hires people. In turn, the utility company providing energy hires more employees to match the demand from other sectors. Their people, in turn, buy electricity, food, etc., producing a ripple effect. These turnovers are called rounds, and they contribute to each successive round of the local economy, although the leakage is reduced with each round until it practically disappears, as explained above.

When using the **Input-Output model** (see Glossary) for calculation, it is necessary that it capture only inter-industrial relationships of goods and services. Another more elaborate model called Social Accounting Matrix (SAM) also takes into account non-market financial flows of goods and money, providing a better idea of the whole relationship. The description of these two models is beyond the scope of this book, and is only mentioned to let the reader know about their existence. These are economic models, and there is abundant literature in books and in the Internet about them.

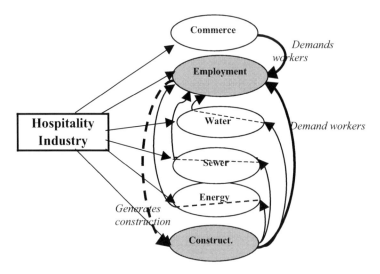

*Figure 6.3* **Economic flows in a region with tourism as main driver**

### 6.10 Framework for indicators

What is a framework for indicators? It is a theoretical structure, used for:

- Systematising data and information;
- Selecting indicators;
- Making explicit the existing interactions between different issues.

Several different frameworks have been developed, each one addressing a specific purpose, but probably the most used framework is the OECD approach (see Glossary) known as P-S-R, or Pressure, Stress, Response. It was developed to link Pressures on the environment by pointing out the human actions or factors ('stressors') that exert pressure on the environment. The Stress refers to the effects produced by these stressors, and the Response is what is being done to correct these effects.

Frameworks closely relate with a desired objective or purpose. Thus, the P-S-R is mainly concerned with the state of the environment due to human actions. The United Nations has also developed a Human Development Index (HDI): as it name implies, this deals with variations in human development. Other frameworks exist for quality of life, for accounting purposes, for policy development, etc. For instance:

Sachs, *et al.* (Sachs, 1998), uses:

- Material input.
- Primary energy consumption.
- Water use.
- Land use.
- Selected emissions.

The World Resources Institute takes into account pollution (Hammond, *et al.*, 1995), and therefore considers:

- Pollution.
- Depletion of resources.
- Risks to the ecosystem.
- Environmental impact on human welfare.

The World Bank considers: (Serageldin, *et al.*, 2000)

- Economic capital.
- Natural capital.
- Human capital.
- Social capital.

The OECD framework developed a cause-effect relationship framework. An example that considers different indicators is illustrated in Table 6.6.

A very useful reference is the analysis by Alberti (1996) on a list of United Nations Centre for Human Settlements (UNCHS) urban indicators, which includes a working list of indicators of sustainable development from the United Nations Commission of Sustainable Development (UNCSD), following the P-S-R framework.

*Table 6.6*        **Comparison of indicators in the PSR framework**

| Pressure | Stress | Response |
|---|---|---|
| Sewage | Percentage of population connected to the sewage network | Percentage of population that will be connected according to new plans |
| Domestic waste | Kg. of domestic waste produced per capita | Percentage of domestic waste recycled Percentage of composting |
| Land use | Population density per square km. | New bylaws to increase building construction in the affected areas |

## 6.11 Thresholds

In a natural system, a threshold can be defined as the point, level, or critical value where there is a remarkable change in the fundamental characteristics that until that point were intrinsic to that natural system. The problem in many cases is that one does not know how the ecosystem reacts to certain perturbations or impacts, or, in other words, one does not know the rate of the response of the ecosystem to some impacts, since it could be linear or non-linear.

As an example, let us consider soil contaminants. Soil generally contains minerals such as potassium, nitrogen and phosphorous, mainly from fertilizers. When it rains, surface water carries these elements to rivers, with serious consequences to human health and the ecosystem (section 3.5.3). It is therefore fundamental to know what are the **thresholds** of those elements; to be more exact, what are the numerical values above which there is an increase in the danger of this runoff happening.

While it may seem that indicators and thresholds are the same thing, this is not really the case, since a threshold determines a point where remarkable changes occur, while an indicator weighs up the performance of something. For instance, an indicator can tell us how frequently a certain pollutant is above a given threshold, and one of the most appealing aspects of indicators is that they can condense a great deal of useful data.

A threshold can express values according to any unit of measure; Table 6.7 provides a few examples.

*Table 6.7*      **Examples of threshold units**

| Area | Units |
|---|---|
| **Environment** | |
| Urban solid waste | kg / person-day |
| Maximum content of CO in streets in 8 hour period | 9 µg / m$^3$ |
| Paper recycling | % |
| Particles in suspension | mg / m$^3$ |
| **Infrastructure** | |
| Street flooded with heavy rain | % |
| Dwellings connected to drinking water | % |
| Traffic flow | vehicles / hour |
| **Transportation** | |
| Expenditure in road infrastructure | $ / capita |
| **Social** | |
| Households below the poverty line | % |
| Total number of housing units | houses / 1000 population |
| Median usable living space per person | m$^2$ |
| **Government** | |
| Wages of local government staff to Local expenditures | % |

## 6.12 Carrying capacity

'Carrying capacity' can be broadly defined as the environment's ability and potential to indefinitely sustain the use of a renewable resource.

The foremost example is the Earth's capacity to sustain life, which some estimate at 10 billion people, at least with the knowledge we have today. Back in the 18$^{th}$ century, the economist Thomas Robert Malthus alerted about the danger of population growth compared with food production, a theory that fortunately did not materialize in view of the increase in food supplies brought about by new developments in agriculture.

This carrying capacity refers not only to the production of food, but also to the problem of water, since the world is getting thirsty; there is also the quandary of land availability. For example, the walled city of Delhi, in India, has a density of 166,300 persons per square kilometre! (see *The Global Development Research Centre*, under Internet references for Chapter 6). This creates vast demands on all orders of life.

Food, water, land, clean air and resources; humankind has to deal with the carrying capacities of all these different areas, and life on the planet now depends critically on a good deal of knowledge about them, and of their magnitude. As long as society respects their carrying capacities, the earth's resources can last for a long time indeed; but when these levels or thresholds are breached, renewable resources begin to decline. Some resources can react and make a comeback if left unexploited for a certain period, while for others, unfortunately, this is not so. It is possible sometimes to guess about the response of the system when a threshold or tolerance limit is breached, but at other times it could be very difficult, to say the least. In addition, it is often impossible to identify properly the secondary or even tertiary effects that this breach can trigger. For instance, overfishing in an area will most likely alter the ecosystem, since it will probably disturb the food chain.

For another thing, does society really know the consequences of the continuous logging of forests? This has been going on for centuries, but what is nature's precise threshold? What will its response be? Some known penalties have taken shape as erosion, desertification, rain generation, the disappearance of species, but what is the final outcome? In addition, how does the logging of forests interrelate with other ecological systems?

Intimately related to this subject is the concept of resilience, discussed in Chapter 1.9, which can broadly be defined as the capacity or tolerance to sustain damage. One parallel might be, a boxing match, where both fighters have the capacity or resilience to absorb or sustain damage, but only until a point or level where nothing is left and a fighter reaches a breakdown. It is useful to consider this in analogy with the environment: it too has a resilience to withstand damage up to a certain point, when it is no longer able to recover or provide a resource.

### *6.12.1 Carrying capacity in the environment*

Real-life examples of breaching the threshold or carrying capacity in the environment are:

1. The decline in production of the Ogallala aquifer in the USA, up to a point that it will be depleted, in some areas, by 2025. Can it recover? Probably, but in a geological time, not in ours. Why has this happened? Because the rate of extraction of water through wells has exceeded the replenishment rate of the aquifer. In other words, the farmers did not respect the extraction rate established by nature, and the carrying capacity of the aquifer was surpassed.

2. The cities of Mexico, Shanghai and Bangkok are sinking due to the same action, which is the rate of water extracted from their undergrounds exceeds the replenishment capacity of their aquifers.

3. Many researchers blame over-fishing for the disappearance of the 'anchoveta' (anchovy), a fish that was very abundant off the Peruvian

coast. Industrial fishing for fishmeal processing plants began in the 1950s, making Peru the world's largest fishing nation by volume (see Muck (1989), Greenpeace International, in Internet References for Chapter 6). This very publication also contains this significant paragraph:

*A group representing FAO and the Peruvian Government's Ocean Institute (Instituto del Mar del Perú) estimated* **the maximum sustainable yield at 9.5 million tonnes per year** *which included nearly two million tonnes estimated to be consumed by sea birds. In 1970, the group issued a warning: if the catch remained above* **9.5 million tonnes,** *the fishery would be in danger of imminent collapse (Instituto del Mar 1970)... The Peruvian Government turned a deaf ear... In 1970, a harvest of* **12.4 million tonnes was allowed,** *followed by 10.5 million tonnes in 1971 (Paulik 1981). Yet by 1973 it had fallen to two million tonnes and by 1983 to a mere 100,000 tonnes.*

The breach of the 9.5 million tons threshold led the fish to vanish, at least in that area.

4. African elephants were facing extinction from a lack of food due to continuous human encroachments of their natural habitat. The magnificent Kruger National Park, South Africa, with a huge area of almost 20,000 $km^2$ (slightly smaller that the area of the State of Israel) stretching 360 km from north to south, was designated a protected area for the African elephant. Fortunately, the number of animals increased up to about 12,000. However, considering that each animal consumes up to 450 kg of vegetable matter and about 200 litres of water each day, the Park is clearly very close to its carrying capacity — if that has not yet been surpassed — from the point of view of providing food. This is why the government now faced some very difficult options: kill some animals (culling) for the rest of them to live. Is this the only solution? Perhaps some other alternatives could be explored, such as the sterilization of some female elephants, in order to keep a balanced growth.

5. Section 2.12.3.2 discussed the town of Clearlake as an example in the use of geothermal energy. They have been extracting steam for electric generation since the 1980s, although population growth and increased electrical consumption have contributed to breaching their reservoir's carrying capacity. Fortunately, in this case, the previous capacity can be restored by re-injecting wastewater, since the steamfield's natural replenishment must be measured in geological time.

6. The River Rhine in Germany was abundant with salmon until the late 1950s. Many undertakings as well as river contamination led to the disappearance of the fish, which is just another example of a threshold breached. Pollution came from treated and untreated sewage, agriculture, industrial discharges, and contaminated ground water, running to a total of more than 100,000 tons per year, and from accidental discharges of toxic products. Chemicals and organic wastes polluted water such that it depleted its carrying capacity to the point of no longer sustaining life, partly because too much oxygen in the water was taken to oxidize waste.

Thirty years later, a massive international movement began with the aim of cleaning the basin, and the rehabilitation project that was launched included engineering works to allow the fish to swim upstream for spawning. Today, the fish are back, and this particular endeavour led to a repair of the damage done. It is interesting that the economic output of the Rhine Basin — which is probably the most industrialized area of the world — has remained unchanged, proving that sustainability can be achieved without damaging economic growth.

### *6.12.2 Carrying capacity in the social fabric*

From the social point of view, one could measure a society's resilience to accepting certain government policies. Of course, democratic mechanisms allow electing new governments at election time, but in some cases people's patience may also reach a point of exhaustion. Within democratic channels, they may decide to express dissatisfaction through massive demonstrations — such as happened in 2003 in many countries around the world, protesting the war in Iraq, or the long-term strike in Venezuela which asked for changes in the government.

In such cases where people's thresholds are surpassed — after their tolerance or resilience are drained — they may rebel against local authorities. For example, this may occur when people complain for years about some measures not taken, for instance, to build sewers, and City Hall has not tried to solve the situation. In this case, once the level of tolerance is breached, people may decide not to pay taxes until their problem is settled.

Let us now analyse another aspect of this concept by considering the following scenario: An entrepreneur developed a beach resort in the early '70s by taking advantage of the white, extensive and fine beaches, and began by building some few cottages, installing a couple of small shops, and planting hundreds of pine and eucalyptus seedlings. Many liked the quiet, the wild atmosphere, and built weekend houses there, overlooking the dunes and the sea, within walking distance of everywhere. Tourism increased and the road to the village was paved and regular bus service started. The pioneer visitors continued going, to meet old friends while the village expanded with the arrival of more tourists, and with the development of new buildings and utilities. There was room for everyone, the beaches were clean, and there was plenty of space for walking, fishing, playing beach sports, even walking to the nearby beach.

As time went by, a casino was built, streets were paved, a large hotel was built, and more shops and a supermarket gave the village the air of a small town. Little by little, something started to change. Old timers and new visitors realized that it was becoming hard to get a hotel room, that the beaches were crowded, that there were queues to get served in restaurants, and that the town was now noisy day and night. A formerly leisured three hour stroll to the village now took considerably longer due to the number of vehicles on the road. Others noticed that the dunes no longer existed because of construction, and that the air was polluted with car fumes. From the economic point of view, it was a bonanza, but little by little, the village was getting fewer and fewer visitors. Is it hard to guess what happened?

**The quality of life the village offered in the past had changed, and for the worse.**

In other words, the social carrying capacity of the village that allowed for good service, a friendly little village atmosphere, clean, pine-scented air, time for everything including beach sports, etc., had been breached because of overcrowding; people began visiting other places in search of a more relaxed atmosphere.
The notion of carrying capacity here is clearly the same as that seen previously, and the concept also applies to many other things, such as public administration, the police force, and so on.
Therefore, when planning a sustainable city, for instance, it is necessary to determine **what is the social carrying capacity of the city that will be able to support the new lifestyle, whatever that it might be?**

### 6.12.3  *Carrying capacity in the economy*

Economic resilience and thresholds are also often found in practice. If the economy of a country is performing badly, with ongoing devaluations of currency — on top of social problems arising from strikes, picketing of factories and roads, etc. — then this could lead to a loss of resilience in commerce.

When merchants can no longer keep their shops and outlets open, and entrepreneurs must close down their factories, thresholds are being breached, as a result of which bankruptcies, lay-offs, and unemployment often follow.
All these examples in three different areas of sustainability have been selected to show the importance of these two concepts: resilience and carrying capacity, and the need to consider them in any study, program, or sustainable process.

After analysing the concept of carrying capacity and its relationship with a threshold or level of use or exploitation of a resource, it must be asked: how is that level established? In some cases, it is possible to determine it approximately, such as for instance in groundwater extraction, provided that the aquifer's recharging rate is known and predictable. But in many other circumstances this level is unknown, and it is often not possible to discern how close or far from that threshold a commercial or industrial operation is. Worse yet, sometimes it is also uncertain if a renewable resource whose threshold has been breached will ever experience a come-back.

In some cases, such as that of the Ogallala aquifer, there probably will be a return, but on a geological time scale, not a human one; therefore, even if people know this, it will not help. In other cases, such as the depletion of anchovies off the Peruvian coast, it is unknown when, or whether or not the fish will be seen again; it could take several years, decades, or never occur. For both these cases, the determination of threshold values is of paramount importance, requiring as many studies as possible and, of course, not something to be left in the hands of decision-makers or grassroots groups. This determination has to derive from serious studies and research from recognized institutions.

Because of this uncertainty, the one thing that can be done calls for using indicators, which will play a fundamental role. In the case of the fish, for instance, a trend in the values of indicators showing the rate of capture in successive years can show declining values serious enough to decree a ban in the activity, so as to give nature time to recover. In fact, the use of thresholds is customary in the mining industry, where there might be two thresholds: the minimum price of metal, making mining and processing profitable, and decreases in the gold content in a ton of ore. Either of these factors can exert enough pressure to stop operations. Perhaps, as does happen, the operation can be restarted years later, once the price of gold is high enough, or if new technical developments make it profitable to exploit mine sites formerly considered to have unprofitably low levels of gold content.

## 6.13 Selection of a set of final indicators

An issue that pops up very often is how to select indicators, as there could be hundreds of them. In another words, on what basis are indicators selected? This question does not have an easy answer. First, one needs to select indicators representing issues in the three dimensions of sustainability, otherwise, the system will be measuring only one, as is the case of the GDP that only calculates economic growth. Some features related to a final set of indicators need to be examined:

1. *Quantity of indicators*
   Indicators are used to gauge the progress or the status of something, and to derive conclusions. This is why it is impossible to deal with hundreds of indicators; what one needs is a final set of, say, 15 or 20, and this selected set has to be representative of the whole scenario.

2. *Selection according compliance*
   One way to select indicators is to prepare a table with indicators in columns and selection criteria or goals in rows, and then to check off each indicator according to the compliance with each criterion. Then an addition of the number of check-marks in each column will provide a number that indicates how well each indicator performs regarding the criteria — and the greater the number of check-marks, the better.
   This system is often used, but if we chose a final listing of 15 indicators, for instance, and selected them in accordance with their decreasing total value per column, then there would be no guarantee that the three dimensions would be equally represented, as most indicators could end up in only one of them.

3. *Importance or weighting*
   Some indicators are more important or carry more weight than others. For instance, 'derelict areas in a city' is more important than 'the number of meetings at City Hall that the citizens attend'.
   While the quality of a city's life can be severely affected if the percentage in the whole area is significant, the quality of life of the community will not be affected if some people did not attend, say, two meetings. This is why it is sometimes convenient to assign a weighting, for instance, using a 1-to-10 scale, to gauge the indicators based on the decision makers' perception of their importance.

4. *Objective*
   The selection of indicators depends on the goal, as explained in section 6.10, about frameworks.

5. *Understandable information*
   Indicators are designed to convey information, but to be effective and useful this information has to be easily understood by everyone. For instance, to measure the contamination of a body of water due to organic matter, scientists use the $BOD_5$ concept (see Glossary), and in the scientific community everybody understands its meaning and its values. This is a widely used concept, however, is too technical and its meaning is not accessible to everyone. Perhaps it would be easier to use some other measure, such as a sampling of the number of fish per $m^3$ of water, and to compare its value with previous measurements, since if the fish can live, the water is good enough.

6. *Importance of information*
   Indicators should communicate significant information. In education, for example, if one wants to gauge the quality of the school environment, an indicator expressing the average age of school buildings is relatively important. But an indicator of the average number of students per teacher, or of the average number of students per classroom would definitely be far more important to the central issue.

7. *Reliability*
   This relates to the accuracy of the information provided. If the values for air pollution in a certain part of the city are taken once a week, and on Sundays, they are not very reliable as they do not reflect the reality. The samples obtained must be of a quality and taken at intervals such as to represent average measurements.

## 6.14 Monitoring progress

Different procedures to measure sustainability have been analysed, although once a project has started, it is necessary to measure its progress, lest it become a merely academic exercise. Monitoring is usually done through the use of indicators, and for monitoring purposes these are used in connection with:

- **Target values**
  Here, a particular indicator institutes comparisons with some established target value for a certain time; in other words, the indicator serves as a **measure of performance.** If the target for dwellings connected to a sewer was established as 68 percent by the year 2004, and the actual figure at that time is of 55 percent, obviously the city has fallen short of its target.

**Administrative indicators** are used for administrative work, measuring how well a government office, for instance, is performing its functions, plans, and programs. This could be a measure of the average number of days it takes to solve routine issues, such as authorising a construction permit, or the reaction time to decide on problems presented by citizens. This type of indicator actually belongs to a new category called 'Governance indicators' that has been promoted by the World Bank. As a matter of fact, Daniel Kaufmann, *et al.* (see Internet references for Chapter 6) developed a methodology that considers an aggregate of 31 different Governance indicators, including Government Effectiveness, Rule of Law and Graft.

- **Threshold values**
  Thresholds can have minimum and maximum values, or both simultaneously to indicate a range. A minimum threshold might identify, for instance, the minimum income people need to earn in order to pay all their expenses. A high threshold value could be the maximum amount of dioxins allowable from an incinerator. A gap or range between two threshold values pertaining to a certain criterion indicates its minimum and maximum levels: for example, a poll can show that between 34 percent and 51 percent of the dwellings surveyed in a poor neighbourhood has a family size of between four and five persons.

- **State of the environment report**
  A state of the environment report (SoE) is issued to inform citizens about the condition of the environment, the way it is improving or worsening, or if it is just being maintained without changes or to provide reasons for any changes.

## 6.15 Indicators for the city

Table 6.8 shows a listing of indicators that can be used in an urban environment. It compares baseline conditions with goals.
It is important to bear in mind that one is interested in measuring progress related to sustainability, so selected indicators should reflect the above-mentioned comparisons values related to economics, social and environmental issues, considering linkages as well. Thus, an indicator showing a positive increase in one area can also indicate a positive decrease in another. For instance, a river clean-up can account for a rise in fish population — leading to an increase in economic benefits for anglers — while at the same time

confirming a decrease in human diseases linked to the human consumption of water.

Sometimes, what appears to be the betterment of an area may, with a more thorough study, only be shown to **transfer the problem to another area**. For instance, the air can be cleaned up with measures to clean flue gases from an industrial smokestack, but it is also necessary to find out what happened to the dust that was collected in cyclones, filters, and scrubbers. If it just ended up dumped on the ground, then the industry has only changed the recipient of the contamination from the air to the ground.

Table 6.8    Some indicators to be considered for the city

|  | Baseline values | Goal values |
|---|---|---|
| ***Carrying capacity in each area*** | | |
| Social | | |
| Ecological | | |
| Economic | | |
| Administrative | | |
| | | |
| ***Social problems in the city*** | | |
| Percentage unemployment | | |
| Drug consumption | | |
| Ratio of crime / population | | |
| Number of children working the streets | | |
| Ratio of homeless people / population | | |
| Water consumption per capita | | |
| Percentage of children with low birth weight | | |
| Percentage of single mothers | | |
| Percentage of households managed by women | | |
| Ratio of car ownership / population | | |
| Gini index of distribution | | |
| How is the relationship among the different ethnic groups of the city? | | |
| | | |
| ***Safety*** | | |
| Crime rate | | |
| Road accidents | | |
| Prostitution | | |
| Trend of thefts | | |

|   |   |   |
|---|---|---|
| **Economic infrastructure** |   |   |
| Ratio of population / main economic activities |   |   |
| Ratio of economic output per activity / total economic output |   |   |
| Ratio of local suppliers/input to large local plants |   |   |
| Ratio of total income in different areas of the city |   |   |
| Number of paved streets per area |   |   |
| Number of houses connected to sewage |   |   |
| Percentage of population with access to clean water |   |   |
| Percentage of total output that the city exports to another areas or countries |   |   |
| Percentage of raw materials the city imports from other areas or countries |   |   |
| Percentage of value added per industry |   |   |
| What plans are there to revitalize run-down areas of the city? |   |   |
| What use is to be given to land not in use near railway tracks? |   |   |
|   |   |   |
| **Infrastructure** |   |   |
| Potable water quality |   |   |
| Losses in the network |   |   |
| Quality of treated sewage discharged |   |   |
| Percentage of houses connected to sewage |   |   |
| Percentage of houses with cesspools |   |   |
| Percentage of streets in bad repair |   |   |
| Percentage of time lost because of traffic jams |   |   |
| Percentage of approval of transportation system |   |   |
| Fare structure |   |   |
| Number of times per year streets are flooded due to heavy rains |   |   |

| | | |
|---|---|---|
| Average travel time in the transit system | | |
| Average car occupancy | | |
| How is the state of repair of public buildings controlled? | | |
| How is the state of repair of road and rail bridges controlled? | | |
| What plans are there to repair and rebuild city assets, such as buildings, sidewalks, tunnels, bridges, etc.? | | |
| What is the frequency of garbage and recyclables collection? | | |
| How does the city manage the heavy traffic of trucks? | | |
| | | |
| **Structure of City Hall** | | |
| Efficiency | | |
| Good governance | | |
| Percentage of corruption cases in the last 10 years | | |
| Percentage of public works or projects initiated and finished in one administration | | |
| Number of community centres | | |
| Ratio of municipal employees / population | | |
| Number of meetings between citizens and the City Hall on city issues | | |
| Number of citizen suggestions that have been put in practice | | |
| Trend in taxes collected | | |
| How is the communication between City Hall and citizens? | | |
| How is the system to hire personnel? | | |
| What is the average distance between the barycentre of an area and the next fire station? | | |
| | | |
| **Economic infrastructure** | | |
| Ratio of population to main economic activities | | |
| Ratio of economic output | | |

| | | |
|---|---|---|
| per activity related to total of the city | | |
| Ratio of local suppliers and their input to large local plants | | |
| Ratio of total income between different areas of the city | | |
| Number of paved streets per area | | |
| Number of households connected to sewage | | |
| Percentage of population with access to clean water | | |
| Percentage of total output that the city exports to other areas and/or countries | | |
| Percentage of raw material, sets, assemblies, etc., that the city imports from another areas or countries | | |
| Percentage of value added per industry | | |
| **Health care** | | |
| Ratio of population / hospital beds | | |
| Ratio of population / doctors | | |
| Ratio of population / administrative personnel | | |
| Ratio of population number of ambulances | | |
| Average waiting time for emergency services | | |
| Average waiting time for an ambulance | | |
| Percentage of infectious diseases | | |
| Ratio of number of walk-in clinics / population | | |
| **Primary and secondary education** | | |
| Location of schools related with population density | | |
| Average age of school buildings | | |
| Average distance a child must walk to school | | |
| Ratio of children / floor space, in $m^2$ | | |
| Number of drop-outs before | | |

| | | |
|---|---|---|
| finishing school | | |
| | | |
| **Tertiary education** | | |
| Ratio of university students / population | | |
| Percentage of students completing a career | | |
| Percentage of graduates finding a job in the city | | |
| Average of people from secondary level who enter to the tertiary level | | |
| | | |
| **Housing** | | |
| Percentage of people owning a house | | |
| Ratio of household income / rental | | |
| Ratio of household income / mortgage | | |
| Number of houses without basic infrastructure (water, electricity, sewage. pavement) | | |
| Percentage of people with clear titles | | |
| Number of people per house | | |
| **Environment** | | |
| Average air quality at soil level (NOx, $CO_2$, CO, $SO_2$ particulate) | | |
| Average soil contamination | | |
| Degree of contamination of rivers and lagoons (in $BOD_5$) | | |
| Percentage of solid waste land-filled | | |
| Percentage of solid waste with in-source recycling | | |
| Percentage of solid waste in landfill recycling | | |
| Ratio of litter on streets / person | | |
| Number of $m^2$ of green space per person | | |
| Ratio of kilometres of bicycle paths / population | | |
| How does the city control that cars are running in good condition, regarding noxious emissions? | | |
| How does the city manage | | |

| | | |
|---|---|---|
| hazardous wastes from hospitals and industries? | | |
| How does the city manage recyclables? | | |
| Does the city have a green belt? | | |
| Ratio of new construction / decrease of agricultural area | | |
| How is the average density in the city? | | |
| How is noise abated in urban highways? | | |
| What measures are being taken to increase wildlife in the city parks? | | |
| What are the regulations and penalties for killing an animal? | | |
| ***Not satisfied needs*** | | |
| Social area | | |
| Environment | | |
| Economics | | |
| ***Energy*** | | |
| Percentage of electric consumption generated by non-conventional sources | | |
| Percentage of electric energy generated by hydro | | |
| Measures taken to reduce emissions in the local power plant | | |
| Is there any chance of producing acid rain with the combination of emissions from the power plants and other utilities? | | |
| What indications are there of energy-saving measures? | | |
| ***Transportation*** | | |
| Mode of transportation | | |
| | Bus | |
| | Dedicated highway for buses | |
| | Streetcars | |
| | Subway | |
| | LRT | |
| | Bike | |
| | Boats, ferries | |
| Is there a regional system? | | |
| Ratio of kms of urban | | |

| | | |
|---|---|---|
| highways / population | | |
| What are the measures taken to reduce traffic accidents? | | |
| **Geography** | | |
| How well is the city interconnected? | | |
| How clean is the water in canals, rivers, creeks and lagoons? | | |
| How clean is the harbour? | | |
| **Cultural** | | |
| What programs exist for people to attend theatres, festivals, concerts, etc.? | | |
| How many museums does the city have? | | |
| How many art galleries does the city have? | | |
| How many libraries does the city have, and what percentage of the city budget is spent on this activity? | | |

As a closing comment, a remark on the efforts by United Nations-Habitat to create what they call a Global Urban Observatory (GUO) may be of interest. This is a global monitoring system with a network of regional, national, and local urban observatories with the mission of compiling information on urban indicators and promoting the creation of Local Urban Observatories (LUO) at the local level (see Internet references for Chapter 6).

Section 6.15.1 illustrates the selection of urban indicators for the city of Guadalajara (Mexico). This analysis is interesting not only because of the size of the city, which is about 6 million, but because it involved the city itself along with eight satellite cities. The consideration of the eight satellite cities facilitated developing a strategic plan than considered the city as well as its region. Thus, projects such as sewage or main roads to connect several cities began being coordinated as if they were intended for a single city.

Indicators played a substantial role here. It is interesting to note that when these selected indicators were used to choose projects to be executed, some projects were rejected because they did not comply with the threshold values (section 6.11) established for them; in other words, because they were found to be unsustainable.

### 6.15.1 Case study: Selection of indicators for the city of Guadalajara, Mexico

In the year 2002, a two–year-long study developed for the city of Guadalajara, Mexico, was released with the title "Study for the urban development of Greater Guadalajara, using sustainability indicators".

Guadalajara, Mexico's second city, has a population of about 6 million living in Guadalajara and its eight satellite cities. These cities in their environs greatly differ in population, economic growth, and industrial activity.

The study involved developing a methodology to detect problems in the area, propose solutions, analyse limitations, determine sustainability indicators, and select urban projects subject to the availability of resources and certain values expressed through various indicators of sustainability.

The project encouraged citizen participation at each step of the process, which included seven meetings and a survey, as well as participation by municipal officials.

The initial set of 76 indicators that was developed after discussions
with all the parties involved comprised the following areas:

- Environment.
- Society.
- Economics.
- Social-Economic-Environmental.
- Social-Economic.
- Citizens' participation.
- Municipal administration.
- Municipal organization.
- Infrastructure.

After considering communication issues and the citizens' understanding of the indicators and their interaction, the decision was to have only 20 final indicators. Therefore, 76 initial indicators were considered and being subject to 28 different criteria, involving sustainable goals, selection criteria, participating areas and adopting the OECD framework approach (section 6.10).

A multicriteria analysis procedure (section 7.3) was chosen to select these 20 indicators, after an optimization process enabled maximizing the information content.

The benefit from this project was that the whole region was considered as a whole, that is the city and its region. Besides, it

> *allowed to introduce sustainability restrictions into the strategic plan process for the city.*
> *For more information on this project, see "Project 222" (see Internet references for Chapter 6).*

**Trends**

It is always useful, provided that enough information is available, to draw a diagram that shows **trends** on different subjects. This generally involves graphic displays of the evolution of an indicator as a function of time. A trend can be positive, when it points in the right direction; otherwise, it can be negative. For instance, assuming that City Hall is monitoring a number of industries that emit liquid discharges into a river. Table 6.9 provides this information, and Figure 6.4 graphically shows an evolution in the number of polluting industries, with actual values against the target goal. Not all the targets have been reached, but the final observed values are close to the target for each year.

*Table 6.9*   Target and observed values of various polluting industries

| Year | 1998 | 1999 | 2000 | 2001 | 2002 | 2003 | 2004 |
|---|---|---|---|---|---|---|---|
|  | Baseline |  |  |  |  |  |  |
| Target values | 305 | 280 | 265 | 240 | 215 | 160 | 135 |
| Observed values | 305 | 290 | 280 | 240 | 226 | 197 | 161 |

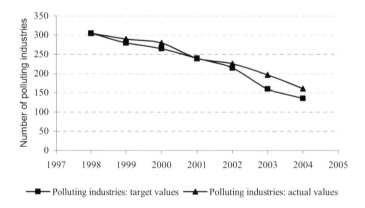

*Figure 6.4*   **Comparison between targets and observed number of polluting industries**

Graphs showing trends are very useful because they indicate the evolution of, say, a pollutant. As an example, assume that an indicator shows a trend about number of people that smoke cigarettes and another indicator reveals the number of people affected by lung diseases; then a correlation analysis can be derived from both set of figures indicating the relationship between them, and this information can be useful for prediction purposes

The trend line's inclination shows the rate at which the observed phenomenon changes. An almost horizontal trend line, whatever it might represent, obviously shows no changes through time. This may even reflect a healthy situation — as, for instance, when the line indicates the grades of high school youths after measuring their general knowledge (provided, of course, that their levels are satisfactory).

Sometimes, a trend line is compared with a horizontal line that represents some desired condition at a certain time. For instance, a horizontal line may indicate the maximum allowed amounts of sulphur oxides (SOx) in the air. Its comparison with a trend line highlights out the progress achieved in reaching that objective, or a lack of progress.

## Reporting

Table 6.10 depicts a form recommended for presenting annual reports about sustainable issues. Naturally, information is condensed here to provide people with a quick grasp of what have been achieved, and where the community stands in the process with that sustainable issue. It goes without saying that this is only an example, and an annual report should cover many more issues.

The column of 'comments' provides a basic and necessarily brief explanation on the performance of each corresponding activity. However, it is understood that this explanation should include a detailed clarification of the reason or reasons for the stated performance. Please notice that 'Pos.' = Positive trend and 'Neg.' = Negative trend.

Trends are a fundamental tool for appraising progress. They allow determining, for instance, if a community has progressed in arresting the destruction of the environment since measures were first taken. In this connection, the Washington-based *National Center for Economics and Security Alternatives* published the *'Index for Environmental Trends'* (see *Rachel's Environmental and Health News*, in Internet references for Chapter 6).

Their study researched 21 indicators on environmental quality, providing brief discussions about them, and summarising their data into a single value, called an 'Environmental index'. For instance, for Air Quality, the study employed six different measures: SOx (sulphur compounds), VOC (volatile organic compounds), NOx (nitrogen compounds), CO (carbon monoxide), particulate, and $CO_2$ (carbon dioxide). This research was not limited to the US, as with some trends such as air quality, the study condenses information gathered in nine countries.

Table 6.10  Annual progress report on sustainable issues

| Projects | Units | Baseline year 2001 | Target year | Value for target year | 2002 | 2003 | 2004 | 2005 | Trends | Comments |
|---|---|---|---|---|---|---|---|---|---|---|
| **Social and community** | | | | | | | | | | |
| Community centres – libraries | No. | 5 | 2006 2008 | 7 9 | 5 5 | 6 6 | | | Pos. | On schedule |
| Construction of bikeways | Km | 5 | 2003 | 12 | 5 | 5 | | | Neg. | Construction has not started yet because of legal problems securing right of way |
| Households connected to sewer | % | 76 | 2004 | 85 | 76 | 76 | | 85 | Pos. | Work is currently a little behind schedule, but will be finished by early 2005 |
| **Economics** | | | | | | | | | | |
| Increase in Regional Gross Product | % | 27.5 billion dollars | 2002 2003 2004 2005 | 3.5 3.0 2.9 2.9 | 3.7 | 3.1 | 2.8 (esti mate) | | Pos. Pos. Equal | The meeting of targets is attributable to the installation of two large industries in the area |

| Projects | Units | Baseline year 2001 | Target year | Value for target year | 2002 | 2003 | 2004 | 2005 | Trends | Comments |
|---|---|---|---|---|---|---|---|---|---|---|
| **Energy** | | | | | | | | | | |
| Electricity spent in municipal buildings | Thousand of kW/hr | 783 | Reduction of 10% by 2003 | 705 | 779 | 749 | | | Pos. | Although the trend is positive the target has not been met because……… ………… |
| Fuel used by garbage trucks and street cleaning equipment | Thousands of litres/year | 1,818 | Reduction of 12% by 2003 | 1,599 | 1,890 | 1,674 | | | w/o notice-able change | Increased expenditure in the purchase of new trucks not approved because of a steep rise in costs |
| **Environ-ment** | | | | | | | | | | |
| Air contami-nation | Average number of days w/clean air, as per standard | 257 | 2002 2003 2004 | 250 300 365 | 240 | 310 | | | Pos. | Targets met |

| Projects | Units | Baseline year 2001 | Target year | Value for target year | 2002 | 2003 | 2004 | 2005 | Trends | Comments |
|---|---|---|---|---|---|---|---|---|---|---|
| Transportation | | | | | | | | | | |
| Replacing old buses by eco-efficient units | No. | 67 | 2002 | 29 | 13 | 7 | | | Neg. | Technical problems arose with the fuel cells, which performance was not as expected and by potential operational risks. New targets need to be established |
| Light Rapid Transit | Km | 0 | 2005 | 16 | | 7 | 12 | 16 | Pos. | 12 km will be completed and put in service by Fall 2004 |

**Internet references for Chapter 6**


**Author**: Maureen Hart (2002)
Title: *Sustainable Measures*
Comment: This is an excellent site to look for sustainability community indicators. It offers listings of indicators for Economy, Education, Environment, Government, Health, Housing, Population, Public safety, Recreation, Resource use, Society and Transportation, and includes a breakdown of issues within each category.
Reading of this paper recommended.
Address:
http://www.sustainablemeasures.com/

**Source**: UN Department of Economics and Social Affairs – Division of Sustainable Development
Title: *Indicators of sustainable development* (2004)
Comment: The Commission of Sustainable Development (CSD) provides a wealth of information about indicators on this site, in the social, economic, environmental, and institutional fields. Look also for Guidelines for Testing and Reporting using indicators of sustainable development (ISDs).
Visiting this site is recommended.
Address:
http://www.un.org/esa/sustdev/natlinfo/indicators/isd.htm

**Author**: Adam Mannis, University of Ulster (2002)
Title: *Indicators of sustainable development*
Comment: The author starts with an opportune statement when he says, "The word for *indicator* in Arabic is *pointer*. Indicators point to a desirable outcome, to 'which way is up' in the policy arena". The paper continues with the Human Development Index, which is composed of the three following indicators: longevity, educational attainment, and standard of living. Each one is explained in detail, and the author also discusses its construction, and an example taken from two countries. The author incurs then into sustainable indicators as per Chapter Agenda 21 (section 1.13), using the Pressure-State-Response framework (section 6.10).
Visiting this site is highly recommended.
Address:
http://www.ess.co.at/GAIA/Reports/indics.html

**Authors**: Tony Jackson & Peter Roberts (2000)
Title: *A review of indicators of sustainable development:
A report for Scottish Enterprise Tayside*
Comment: This comprehensive study deals with a subject seldom found in the sustainability literature: the strong and weak interpretation of sustainability (section 1.3). It also discusses subjects such as the development of models of indicators for sustainable development, mainly following the OECD approach (section 6.10), the OECD criteria for selecting environmental indicators, the application of sustainable indicator models in policy evaluation, and many other related subjects. Despite being a paper addressing an audience of economists it is very clear and concise.
Visiting this site is highly recommended.
Address:
http://www.trp.dundee.ac.uk/library/pubs/set.html

**Source**: Marbek Resource Consultants (1996)
Title: *Performance indicators for environmentally sustainable transportation – A discussion paper*
Address:
http://www.tc.gc.ca/programs/environment/SD/pdfdocs/eperform.pdf

**Source:** Sustainable Seattle (1998)
Title: *1998 indicators of sustainable community report*
Comment: The City of Seattle is one of the world's leaders in sustainable development. In this paper, they explain the use of 40 selected indicators and also their trends in issues about the environment, population and resources, economy, youth and education, and health and community. Also includes a useful section with FAQ.
Address:
http://www.sustainableseattle.org/Publications/40indicators.shtml

**Authors**: Bambang Juanda and Upik Rosalina Wasrin
Title: *Selection and modeling of sustainable development indicators: Indonesian case*
Address:
http://www.sarcs.org/documents/SDI%20Hanoi%20Presentations/Bambang%20Juanda_Indonesia.pdf

**Authors**: S.L.A. Ferreira and A.C. Harmse (1998)
Title: *The social carrying capacity of The Kruger National Park: Policy and practice*
Address:

http://www.parks-sa.co.za/conservation/scientific_services/ss_ferreira.html

**Authors**: Gretchen C. Daily and Paul R. Ehrlich (1992)
Title: *Population, sustainability, and earth's carrying capacity - A framework for estimating population sizes and lifestyles that could be sustained without undermining future generations.*
Comment: Includes an analysis of the current human population situation on earth and the carrying capacity today, and a useful classification of resources, with a definition of their maximum sustainable use. Discusses carrying capacity from the social point of view, with spatial and temporal implications.
Address:
http://wildlife.wisc.edu/courses/360b/earthpop.htm

**Authors**: Julian Dumanski and Christian Pieri
Title: *Keynote: Monitoring progress towards sustainable land management*
Address:
http://spc3.ecn.purdue.edu/nserlweb/isco99/pdf/ISCOdisc/SustainingTheGlobalFarm/K014-Dumansky.pdf

**Source**: Swale Borough Council (2004)
Title: *Monitoring progress – Sustainable indicators*
Comment: Introduces the concept of 'Best value indicators' acting as a main yardstick against which performance can be judged. Note especially the example of how some impacts are interrelated with others in space and time.
Address:
http://www.swale.gov.uk/index.cfm?articleid=1092&articleaction

**Author**: P. Muck, cited in Source: Greenpeace International (1997)
Title: *Industrial 'Hoover' fishing – A policy vacuum*
Comment: This paper deals with the very serious issues involved in overfishing, and the consequent depletion of fish. Several cases are analysed, and solutions and recommendations are proposed.
Address:
http://archive.greenpeace.org/comms/cbio/hoovrpt.html

**Source**: The Global Development Research Center
Title: *Urban environmental management*
Comment: Under the title 'Urban Communities and Participation', and within this under the heading 'Notes on Quality of Life', different definitions are given of this very important concept. It is worth mentioning the extensive listing included about Quality of Life attributes.
Address:
http://www.gdrc.org/uem/

**Author**: Tom Bauler
Title: *Concept, application and validation/efficiency of an environmental information system – Indicators for sustainable development in an inter-regional context*
Comment: According to this author, *"SD (Sustainable Development) is a dynamic process of 'social learning' that corresponds to a continuous and iterative process of choice between economical efficiency, social equity and environmental sustainability: thus the necessity to rely on flexible and dynamic evaluation tools such as indicators for sustainable development"*. This statement defines the process extremely well.
Address:
http://www.ulb.ac.be/ceese/tom/tom_phd_issues.htm

**Author:** Rosalyn McKeown
Title: *ESD toolkit*
Comment: Thresholds of education and sustainability. A thorough analysis of this important subject.
Address:
http://www.esdtoolkit.org/discussion/

**Source:** London Sustainable Exchange (2003)
Title: *More affordable homes possible if thresholds abolished, says report*
Comment: Expresses that more housing could be built in London if threshold values are lowered. This is a good example of how established thresholds can have social and economic consequences. The paper claims that lowering thresholds would allow for the construction of 10,000 new houses in London per year.
Address:
http://www.lsx.org.uk/news/_page360.aspx

**Source:** The Sustainability Report - Affiliated with the Institute for Research and Innovation in Sustainability (2000)
Title: *A brief story of sustainable development*
Comment: The paper refers to the Brundtland report (section 1.2) when it states, '*There are thresholds that cannot be crossed without endangering the basic integrity of the system. Today we are close to many of those thresholds.*'
Address:
http://www.sustreport.org/background/history.html

**Source**: Tecnológico de Monterrey – Campus Guadalajara (2002)
Title: *Project 222: Urban development study of the extended urban zone of Guadalajara, according to indicators of sustainability, México.*

Comment: This study developed a methodology to select projects in different areas of the city of Guadalajara, complying with sustainability objectives.
Address:
http://www.partnerships.stockholm.se/new_tavlande_index.html

**Authors**: Peter Hardi, Stephan Barg, Tony Hodge and Laszlo Pinter
International Institute for Sustainable Development (1997)
Title: *Measuring sustainable development: Review of current practice*
Comment: Very thorough description of indicators and models.
Reading is recommended.
Address:
http://strategis.ic.gc.ca/pics/ra/op17-a.pdf

**Source**: Seattle Office on Sustainability and Environment (2003)
Title: *Moving towards sustainability 2002: An annual progress report on the City of Seattle's environmental action agenda*
Comment: Very good information about how to write a progress report on sustainable issues. Provides brief and good information about city's assets and operations. Very illustrative paragraphs on 'Causes for concern....', and 'Reasons for optimism'. See especially Table 1 'Annual electricity use in representative city buildings', Figure 2 'Annual consumption of water in representative city buildings', as well as the other many Tables and data provided.
Reading highly recommended.
Address:
http://www.cityofseattle.net/environment/EAAReport2002.pdf

**Source:** EUROSTAT (2001)
Title: *Development of headline indicators in Australia*
Comment: Excellent document with abundant information on this subject, including a summary of 'headline' sustainable indicators.
Visiting this Website is recommended.
Address:
http://www.unece.org/stats/documents/2001/10/env/wp.15.e.pdf

**Source:** Communitybuilders.nsw (2004)
Title: *Sustainability Indicators - a community builder's toolkit*
Address:
http://www.communitybuilders.nsw.gov.au/getting_started/statistics/s_tkt.html

**Source**: U.K. government (2002)
**Title:** *Sustainable development - the UK Government's approach*

Comment: Produces a listing and explanation of 15 headline indicators, together with assessment of progress when compared to initial baseline values.
Address:
http://www.sustainable-development.gov.uk/ar2002/

**Authors:** Daniel Kaufmann, Aart Kraay, Pablo Zoido-Lobatón (1999)
Source: The World Bank Development Research Group
Title: *Policy research working paper - Aggregating governance indicators*
Comment: Very comprehensive study including a large data sample from 160 countries.
Visiting this Website recommended.
Address:
http://www.worldbank.org/wbi/governance/pdf/agg_ind.pdf

**Source:** United Nations Human Settlements Programme (2003)
Title: *Global Urban Observatory*
Address:
http://www.unchs.org/programmes/guo/

Source: Environmental Research Foundation (1998)
**Author:** Peter Montague
Title: *Rachel's environmental and health news*
Address:
http://www.rachel.org/bulletin/pdf/Rachels_Environment_Health_News_508.pdf

# CHAPTER 7 – SUSTAINABLE IMPACT ASSESSMENT (SuIA)

The SuIA acronym is used to differentiate Sustainable Impact Assessment from Strategic Impact Assessment (SIA).

The six previous chapters have commented on fundamental concepts and initiatives such as the Brundtland report (section 1.2), the Bellagio Principles (section 1.14), and the 'ecological footprint', while discussing the curse of our age — waste and pollution — and proposed some concrete solutions (elimination of waste, recycling, etc.).
An analysis of the feasibility of establishing sustainable processes for a diversity of human activities has also been the subject of this book's material, along with different steps that industry can take to reverse various damaging processes. Energy has taken up a whole chapter, as befits such an important subject for our civilization, and techniques and tools have also been raised as the effective ways to measure sustainability.

It appears, then, that all the elements needed to determine the state of the environment and society are in place, as well as some relevant economic factors. This chapter will now addresses the issue of **how to use these tools** and how to **determine the values** that they require, so as to obtain the sets of results that can alone enable assessing the effects of any remedial actions that are taken.

## 7.1 Urban and regional sustainability

Sustainability has been examined taking into account activities in the three fundamental dimensions. However, if one reflects that a region is some sort of envelope or frame where all of these activities are performed, it is then natural to address the subject of sustainability at the regional level. Figure 7.1 shows the area of a city with its geographical limits, with a superimposition of the three dimensions of sustainability that apply to it.
Urban sustainability is extremely important because most of the world's population now lives in urban areas; therefore, the social, economic and environment problems generated by that population are considerable, and far more significant than those in rural areas are. Instead of city limits, regions

316  Introduction to Sustainability: Road to a Better Future

should be considered as formed by the city and its suburban areas, which usually constitutes a vague zone with blurred limits.

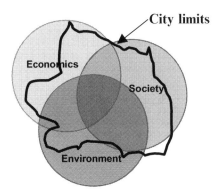

*Figure 7.1*  **The city and the three dimensions of sustainability**

To study urban sustainability, a series of steps needs to be followed. These steps are:

### 7.1.1  Assets inventory

This involves determining the resources the city has, in human and natural capital, its prospects, the health and diversification of its industrial base, its universities, and, of course, its government, including decentralization and communication with the city's inhabitants. A city could be endowed with intangible assets such as heritage, proximity to beautiful beaches or imposing mountains, near natural parks, etc., in addition to being beset with negative aspects, such as slums, poverty, poor transportation, deficient infrastructures, etc.

### 7.1.2  The baseline concept

This is a fundamental concept that calls for an inventory of the present-day situation before changes are made. It covers not only the environment but also the economic and the social fields. In the **environmental arena**, it will detail the scenario at present, including such intangible aspects like a nice view, local weather, air, soil and water quality, etc. This information will be used as a snapshot of the city's conditions, or as a reminder of how it was before actions were taken. Old photographs from the 50s and 60s showing, for instance, buildings covered with grime and soot, and chaotic traffic, can bear witness to a situation that has often been superseded; unfortunately, the reverse is also true, and one can observe forests that no longer exist as they have been logged to make room for dwellings.

From the **economic angle,** it is necessary to record the main economic indicators, such as local GDP, number of industries, diversification, decentralization, etc. From the **social point of view** information is needed about the way the population live in the area, how much household incomes are, the value of the ratio linking household incomes to rent, the percentage of underweight babies, the quality of the health service, etc.

The baseline is important because it will be used as a yardstick to measure changes in all of these areas.

Difficulties in a community seldom pertain to just one of the areas depicted. Say, for instance, that the problem is poverty. Although it lies in the social area, most probably it has links with economics, because one of the reasons for poverty could be the region's high unemployment. There could be shantytowns located on hills in the outskirts of the city, and because the local trees were logged for heating fuel, poverty turns out to be related to environment, since the land has been eroded due to runoffs from rainfall.

The city should thus be considered with all its different inputs and outputs, reflecting the interaction of the environment, society, and economics, and aimed to the goal of becoming a sustainable entity.

Marginal notes
The following observations can be useful in this regard:

### a. On strengthens and weakness

It is important to list and analyse the city's **strengths** in terms of, for instance:

- The region's industrial base.
- The number of universities and graduates.
- The transportation system.
- Its connectivity with other urban centres.
- People's participation in government decisions.
- Social conditions, especially for those in the lower strata.
- The availability of land, water and energy.
- The local weather.
- Etc.

Also to list such **weaknesses** as:

- The lack of industrial diversification.
- Deficiencies in its specialized labour.
- Weak social institutions.
- Poor transportation.

- Unfavourable climate.
- Undeveloped road infrastructure.
- Poverty.
- Discrimination.
- Etc.

### b. On the need for a goal
What is a goal?
The dictionary (Shorter Oxford English Dictionary, 2002) defines a goal as the object of one's ambition or effort, and also as a desired end or result.
In sustainability, this 'goal' could be to ensure for people a good quality of life, and to utilize existing resources in such a way that future generations will be able to enjoy them. Then, of course, it is essential to define what 'quality of life' is. This can be expressed as a 'goal' for social equity, for everyone to have shelter and food, access to education, good health care, a fine and pleasant environment, good job conditions, etc.

'Targets' should probably also be established; this involves deciding, for instance, that the final goal should be reached in ten years' time, along with setting intermediate targets for short- and medium-term goals. Setting up a goal is easier said than done. Members of a group can think in very different ways due to a diversity of wishes and dissimilar values. To take an everyday example, assume that a family of four has decided to go to the beach for their summer vacation. They can reach their destination by using the family car, the bus, the train or air transportation.

Each of these options has its advantages and disadvantages, compared to the others and gauged against any single criterion — which could be, for instance, travel time. The driver, since it is long drive, might prefer choosing the bus, the train or the airplane, in order to relax. Others in the family may prefer the car because of its comfort and its availability at the trip's destination. Others will choose the train, since there is more freedom of movement during the long trip. The family 'team' can thereby gauge each alternative or option using a certain numerical scale, seeking an average value. However, the goal of the trip remains to be defined precisely, and this an even more complicated issue. Most probably, the older generation may want to sit in easy chairs by the beach with a good book, sunbathing and chatting with friends. Meantime, one of the youngsters may have as an objective to go disco dancing every night, and chat with friends from last summer's visit to the area.

The other youngster may be hoping to scuba dive the summer away, and barbecuing with friends on the beach. So there are three different objectives, although they do not need to be harmonized since each one can do what they like best — as the ultimate goal for everyone is to have a pleasant vacation time, at least as each one defines it.

Each one will probably be able to place a value on the time spent in their preferred activity, such that one will be able to say, "*I will spend 70 percent of my time scuba-diving*". And if, at the end of their vacation, the family members are asked about how well they complied with their objective, the parents could say, *"we hoped to spend at least 50 percent of our time on the beach, but because of the weather (rain and wind) we had to remain indoors, so maybe we spent only 35 percent of our time in our preferred activity"*. In other words, each one can make an evaluation of their results.

This analogy can be used to analyse what happens when a community decides to have a sustainable neighbourhood. Some members can agree, and others disagree, on where they want to become — that is, their destination or goal — but once this goal is established, the objective or means to reach it could differ according to the different societal groups and levels. However, the analogy stops here, because a community needs **common objectives,** and not a mixture of them. This leads to the fact that it is necessary to identify the agreed goal, the means by which to reach it, and the criteria to gauge the results within a context, which is also called a frame.

It is probably possible to establish deadlines for targets that are a measured partial result taking place at a certain time. These could be thresholds and/or concrete actions completed or in the process of being completed, and the evaluation will consist of comparing **actual** results accomplished with the **target** threshold or achievements.

### *c. Note on information gathering*

This is one of the most cumbersome aspects of such projects, since a great deal of data is usually required. However, in most cases, the information sought already exists; the problem is often that it is perhaps archived in many different locations, offices and dependencies, and also that people sometimes consider that they own the pertinent records, reports or figures such as statistics, and do not want to share them. Some recommended steps for such cases are:

- Determine the amount and level of detail of information needed, and list the locations where it is.

- This information sometimes does not exist, or is too old. For instance, this could be the case with data on social aspects, such as the need for adequate shelter. In such cases, it is probably most expedient to make a survey to determine the present conditions. A survey is a very useful and reliable tool, provided that the number of people interviewed is significant (that is, a true sample of the population), and that the questions are clear or are without ambiguities and are not too numerous. A questionnaire therefore needs to be prepared, such as that shown in Table 7.2.

- In other cases, some questions get mixed answers because there is no straight answer but a series of facts related to the question. One example could be the community's feeling about the necessity to undertake certain infrastructure work: although all answers respond to some need, there could be a discrepancy about the seriousness of a determined situation, and, of course, disagreements about one's comparative importance (see, for instance, example in section 4.2). In this circumstance, the purpose will be served best through meetings that involve all levels of the community, and with a moderator conducting the sessions. These will facilitate identifying problems, which can then be analysed, discussed, and voted on. Everyone will thereby have an opportunity to express their ideas and comment on pre-established issues.

This procedure is especially important because people very often put forward suggestions not considered by the decision makers and analysts. Many problems or inconveniences generated by a project, alternative or policy, are known far better by the people who will suffer as a result of them; therefore, they are bet able to provide very useful input. It is important to underscore, again, that a cross-section of the community must be represented at such meetings in order to achieve social equity, and of course, with the presence of city government experts who can clarify on the spot whether or not a proposed scheme is feasible. These meetings should not only be concerned with necessities, as they must also reflect citizens' opinions about the often intangible aspects that are valued by the community, such as preserving quietness, views, the community's social fabric, etc.

It is useful to invite to these meetings people who have actual experience in the subject under discussion. For instance, in a program for upgrading slums in the outskirts of Mexico City, the meeting organizers invited people living in former slums that had previously been upgraded. These people brought very valuable experience to the table, helping with the understanding of problems that came up unexpectedly, with niceties about financial difficulties, etc.

- City Hall probably already has considerable information immediately available through their Geographic Information System (GIS).
  This is a good tool, and since it is visual, and it could be invaluable for showing the community different aspects of infrastructure — such as the problems to steep or unsuitable terrain and for dwellings built on them, or the erosion that indiscriminate logging will produce in a certain area. Information is often available about buried utilities, water flows in trunks, type of crops and size of cultivated area, location of forests, creeks, etc., and their interactions with a proposed project, which can often be very helpful. If one of the projects is, for instance, to build an interurban road, GIS can help to geographically identify sensitive areas, agricultural zones, protected areas, etc., leading to the road's correct layout in order to thereby minimize damage.

## 7.2 Agreeing on the goal

What is the goal of sustainable development? What does the community want to accomplish?
This subject touches on many diverse areas of human activity, so it is practically impossible to deal with every specific case, and one must be

content to just specify some general norms and procedures. A few examples of activities will be discussed to illustrate the procedure

One very important concept mentioned in Chapter 1 that now needs further clarification is the significance of economic growth and its relationship to sustainable development. Is it possible to have both at the same time, or this is only wishful thinking? To illustrate this very important question, assume that a government has singled out three alternative small towns in different areas as candidates for the terminal of a gas pipeline coming from gas fields located in the south of the country. It is requesting opinions from each area people about this undertaking, which will weigh into the process to select a place. The three pre-selected towns are on the seashore, two of them 70 km apart from one another, and the third one 193 km from the closest of these.

The idea is to develop in one area a petrochemical complex that utilizes this natural gas as raw material. All three towns have small fishing harbours, and the selected town will see it expanded to accommodate ships for exporting liquefied gas.

This undertaking will clearly lead to economic growth in the selected town, which will take the form of more employment, better roads, improvements in transportation connections, perhaps even a new airport, etc. However, is that really what each town wants? If these communities are looking for **economic growth,** it is probably without thinking in terms of sustainable development. Why is this so?

Because linked with these advantages each town will also have air, water and soil pollution. Most probably, smells will become part of their daily life, and their beaches will eventually be contaminated with residues. Current crime rates, which are practically non-existent, will probably rise, and their real estate taxes will also rise, due to the ever-increasing value of the land. The greater commercial activity will also bring workers from other parts of the country, putting a strain on their schools and hospital capacities. Naturally, the lifestyle for the old residents would change.

In other words, each community will be better off economically but worse off from the point of view of sustainability, **if the social and environment aspects of sustainability do not accompany the economic growth.** To make the decision, trade-offs must be considered, such as (this analysis is for one town, but it applies to them all):

- The **peaceful life** of the small town will be lost forever _but_ their **standard of living** will rise.
- The **environment will also lose** some of its quality _but_ it will gain some others since the complex will locate where **mosquito-infested marshes now** lie.

- Prosperity will create **new opportunities for young people** and probably stop their migration to other places, *but*, at the same time, the **cost of living will rise** and more land will be used for dwellings.
- There will be **demand for more specialized workers** that probably will be filled with the training of existing labour in the region, *but* then there could be a **scarcity of workers** for agricultural-related activities.
- Because of the construction of the petrochemical plant, there will also be the installation of plants working with natural gas and the products manufactured by the plant, such as plastics for different uses. Therefore, there will be a sharp **increase in industrial activity** in the selected town, *but* subject to the fortunes of a single industry. If the market prices or if demand falls, there will be a **strong slump** in the selected town.

Therefore, even if the goal is economic growth it also should include a view of its sustainable development; whatever the goal, this has to be clearly stated and a plan prepared to reach it. This is essential; each community has to decide **what they want and the means to have it**, and once this is done, the citizens and authorities must make a serious effort to determine their resource inventory, and how to use it better.

## 7.3 Understanding the problem

How can they solve this very complex problem? In other words, how can they achieve an equilibrium between economic growth and sustainable development, which are apparently contradictory forces? It is this author's view that a solution can be found when **trade-offs** can be calculated and appraised. Only with a knowledge of these trade-offs will the policy-makers be able to make sound decisions about what should be done, what projects or plans have priority, etc.

The community has to develop a **model,** a **framework**, that will give them responses to many questions, and which will allow them to make several analyses that consider diverse options and alternatives. A model can be prepared using mathematical tools, with the advantage that the result of these studies is quantitative, so easier to understand and evaluate. The tool proposed in this book is called 'Multicriteria analysis', and is currently used in everyday activities for many different situations. What does 'multicriteria' mean, and what is its relationship with this study?

To reach a certain sustainable goal **plans, projects** and **procedures** (activities) obviously have to be established, and some of them will compete for certain resources and be subject to diverse **restrictions** or **criteria**. The way to evaluate these conflicting activities is to gauge them by analysing their compliance to a set of criteria; hence the name 'multicriteria'.

The use of this approach will require defining a series of concepts, as follows:

- Goals and objectives.
- Plans, projects and procedures (activities) for reaching that goal.
- Criteria and indicators to gauge activities that lead to the achievement of the objective.
- Weights to be assigned to these criteria.
- Thresholds, norms, or levels for the criteria and indicators, against which results can be compared. Application of multicriteria analysis to find answers.
- Monitoring progress.

## 7.4 Resources inventory

### Factors to consider

Whatever the activity, to design a sustainable plan some objective clearly has to be developed based on the assets of the site. If, for example, a community wishes to use renewable sources for generating energy, then this stage is responsible for making an inventory of all the resources the community wants to develop, and the feasibility that the area offers for each one. For instance, although solar-based energy is obtainable anywhere in the world, a community should find out the suitability of their area.

Why ask this question, since energy is available to anyone?

Because it could very well be that the geographical location of the community makes this venture possible, but not practical.

The reason is latitude, which is very important since the intensity of the heat, normally measured at noon, decreases with increasing degrees of latitudes from the Equator, whether North and South, due to the inclination of the sun's rays. So a solar plant built in the community may provide people with energy, but not in the amount expected nor as advertised in a manufacturer's technical literature.

This means that two solar-based plants identical in size and construction will have different outputs if one is located in the Sahara and the other in Norway. Therefore, if this alternative is an option, any analysis must consider the geographical location of the project.

The same holds regarding wind energy, or biomass or hydro, since they all depend on geological, geothermal, geographic, hydrographic, and meteorological factors. This search for information applies to any project, such as a pipeline, a road, a national or a thematic park, a refinery, etc. Here, it is necessary to consider the **comparative advantage** (see Glossary) of a site in relation to another.

Even if the community is in a hilly country with apparently good conditions for hydro generation, there is a need to determine its potential. It can be appraised considering facts such as rainfall or snowfall regimes, the average water flow in creeks and rivers, if the streams freeze in the winter, determining the amount of detritus carried by the streams, etc. Such analyses will reduce the range of possible options.

In the of the construction of a lengthy pipeline between the oilfields and a harbour for instance, which can mean hundreds or thousands of kilometres, the main project of constructing the pipeline is usually broken down into several alternatives, mostly involving the consideration of some alternative routes between the two points.

As there are other considerations aside from minimum cost and distance, many facts need to be taken into account, including the sustainability of each alternative. Here a study should bear in mind, besides the cost and distance, some additional factors, including:

- Kind of areas crossed by the pipeline (deserts, populated areas, marshes, protected areas, agricultural zones, etc.).
- Risks involved, such as sabotage on some part of the route;
- Rivers to be crossed.
- Tunnels to be built.
- Environmental impacts during construction and operation.
- Risk of leaks in zones with geological faults prone to tremors and earthquakes;
- Etc.

In industrial projects, information must be collected regarding for instance:

- Environmental impacts on air, soil and water.
- Risk to the population in case of accidents.
- Economic impact to the region.
- Influence of this industry in the regional gross product that will be attributable to it. This is socially important, as was seen in section 1.9.

In the case of municipal works, such as the expansion of a sewer system, the erection of a water treatment plant, the construction of a new landfill, the construction of a waste incinerator, etc., other factors that must be considered include:

- Land use.
- Sustainability of the water source.
- Capacity of the wastewater treatment plant.
- Geological conditions of the site, concerning the construction of a landfill.
- Prevailing winds, in the case of the incinerator.
- Distances.
- Etc.

Natural parks are not an exception. Besides cost considerations, it is necessary to take into account:

- How the presence of visitors will affect the ecosystem.
- How many hectares of forest will be cleared to make room for roads and trails.
- Risk of fires produced by cigarettes.
- Etc.

In other words, the inventory of resources, whatever the nature of the project(s) must **list all the potential alternatives, costs, advantages, disadvantages, geographic, hydrographic and geological aspects, land availability, etc.**

**Assets to consider**
Assets are naturally varied, but as an example, some assets are listed involving a few activities. For instance:

*Tourism*
In tourism, assets involve a wide range of tangible and intangible resources. Many areas are beach destinations and preferred for marine activities, for example, swimming, yachting, scuba diving, surfing, fishing, etc. Others rely on the beauty of their mountains and lakes, as well as such related events as hunting, trekking, climbing, fishing, skiing, etc. Many have artistic and architectural treasures, museums, concert halls, palaces and castles, etc., their main asset centering on cultural activities, tours, etc.

Others profit from their virgin forests and national parks. An examination of assets is necessary to determine their best use and the work required for their restoration to their original beauty. For instance, in a city with heritage treasures such as Prague or Saint Petersburg, this could involve cleaning ancient buildings covered by soot, and/or the repair of damaged structures.

*Industry and urban centres*
In general, industrial sites have a certain main activity and diverse ancillary lines of business usually connected with it and with the local city. Assets include people's capacity, the industrial heritage or expertise in certain areas of human action, universities and technical institutes, research centres, etc. Some examples are the cities of Toulouse in France, home of the Airbus, or Everett in the USA, headquarters of the Boeing Company, both leaders in the aircraft industry having a considerable pool of very skilled people, in a high-tech industry.

*Cultural centres*
Cities like Oxford, England, Boston, the US, and Paris, have old and renowned universities whose assets are not only the ancient buildings which are probably themselves very valuable, but also the prestige, their research centres, and the commercial activities that they engender.

*Research centres*
Probably the best-known examples of such centres are Silicon Valley in California, the CERN facility straddling the Swiss-French border in Geneva, Switzerland, which is home of the Super Proton Synchrotron (SPS), the Lawrence Livermore National Laboratory, in Livermore, California, etc. The assets here are the know-how, the installations (unique in the world, as in the Swiss case), the research institutes, the pool of 'brains' working in the area, the high standard of living, etc.

In all of these cases, the assets should **be inventoried, their future prospects evaluated, their needs appraised, and their ways to reach sustainable development defined.**

### 7.5 Plan to accomplish the objective

When a community has the basic information, it can go ahead with the process. It needs to develop an action plan, that is prepare the detailed information about programs, projects, etc., and, once this is done, to schedule their execution, normally using what is called a Gantt Chart (See Table 7.1). In this example, this chart only depicts the title of the big issues to be treated, and their timing. Naturally, much more detail should be allowed for each one.

### 7.6 People's opinion

Technical information about the elements of a project is important but people's opinion must also be considered. As mentioned, this is done through polls and surveys that yield numbers that as an average represent a subjective issue. For instance, in the construction of a highway, a questionnaire is prepared and a poll conducted on a cross-section of the population affected by the project. One of the questions could be for instance: *"How do you think the traffic noise from the highway will affect you?"*

*Table 7.1* **Bar chart for a sustainable process**

| Projects, plans and programs | Responsible | 2004 | 2005 | 2006 | 2007 | 2008 |
|---|---|---|---|---|---|---|
| Transportation issues | Eng. Dept., M.U. & R.L. | ■■■ | | | | |
| Road infrastructure | Technical Dept., R.L. & M.O. | | ■■■■■ | | | |
| Social programs | University, R.T. & I.E | | | ■■■■■■ | | |
| Community budget | Financing Dept. & P.R. | | | | | ■■■ |
| Downtown revitalization | Chamber of Commerce & R.U. | | | ■■■■■ | | |
| Incineration plant | Eng. Dept. and S.L. | | | | | ■■■ |

The person consulted is then asked to place a value on the issue of between 1 and 10, the greater the worse (or the better, since this is only a convention). When the poll concludes, an average is found about the effect of the noise produced by the project and that value is inserted on a table or matrix, corresponding to a criterion than can be named, for instance, 'Effect of noise'. The questionnaire should also include other related queries, in this case, regarding aesthetics, about the effect of the highway on shortening travel times, the benefits of the project, problems that might be created (such as by the separation of the population on either side of the road), etc.

However, the questionnaire should not be very long because people will probably get tired or bored with it, so eager to get rid of the interviewer as soon as possible by answering questions without much analysis or thought. This is why the preparation of the questionnaire is a task for specialists who can get a maximum of information within a reasonable number of questions yet without ambiguity. Table 7.2 shows a sample of such a questionnaire.

*Table 7.2*     Sample questionnaire for a highway project

| Questions | Marks | Comments |
|---|---|---|
| 1. What is your general opinion of the highway? | 1 2 3 4 5 6 7 8 9 10 | |
| 2. Do you think it will improve your commuting to downtown? | 1 2 3 4 5 6 7 8 9 10 | |
| 3. What is your perception about the noise? | 1 2 3 4 5 6 7 8 9 10 | |
| 4. How do you think the traffic noise on the highway will affect you? | 1 2 3 4 5 6 7 8 9 10 | |
| 5. The highway will divide the neighbourhood into two zones. Would this affect you? | 1 2 3 4 5 6 7 8 9 10 | |
| 6. In this area there will be two underpasses connecting the two zones. Do you think this is enough? | 1 2 3 4 5 6 7 8 9 10 | |
| 7. Do you believe that the highway will obstruct the view of the lake? | 1 2 3 4 5 6 7 8 9 10 | |
| 8. How will the highway affect you? | 1 2 3 4 5 6 7 8 9 10 | |
| 9. What is your feeling about the highway changing the rural atmosphere in this neighbourhood? | 1 2 3 4 5 6 7 8 9 10 | |
| 10. Do you believe this work will have an economic effect in this area? | 1 2 3 4 5 6 7 8 9 10 | |
| 11. When the highway is in operation the tram service to downtown will be discontinued and replaced by rapid bus service. How do you feel about this? | 1 2 3 4 5 6 7 8 9 10 | |

## 7.7 Criteria and indicators to gauge projects

Criteria are normally used to gauge projects, plans, and alternatives, i.e. to evaluate the extent of the contributions of a project to achieving the objective, bearing in mind a particular criterion.

There are different types of criteria, for instance:

a) To gauge several projects, one criterion could have to do with the degree of annoyance (measures on a 1-to-10 scale) produced by traffic; another could be about total funding available, and yet another could be concerned with travel times.

b) Another type of criteria deals with indicators (section 6.1). For instance, a criterion could gauge reactions to air pollution, in which case a maximum value or threshold there could be agreed to or established. Another could examine labour costs, to establish a minimum amount or threshold to be paid to workers.

c) It is also advisable to use sustainable indicators (section 6.1) in order to ensure that the alternatives also comply with criteria pertaining to sustainability issues. For instance, in a mining and metal refining operation a sustainable indicator such as degree of contamination of water from the floatation process (see Glossary) can have a chain effect on wildlife populations, especially for birds, when water from the process is stored as 'tails' in ponds, and a further effect may be aquifer contamination.

In these cases, the analysis is also useful for suggesting that remediation measures be adopted. For example, in this industry, it is mandatory to make sure that industrial operations will not harm birds; this is accomplished either by using wire or plastic mesh on the pond. Of course, in order to avoid polluting the underground water, the bottom of the pond must be lined with a high-density polyethylene sheet that rests on a compacted clay layer. Accidents do happen, however, as seen in the following note. In this case, had the operator had an indicator showing the structural strength of the dam, and/or sensors as indicators in its interior, a disaster could probably have been avoided.

> Note:
> In August 1995, a leak in a tailings pond dam for a gold mining operation in Guyana leaked millions of litres of slurry containing cyanide into the Essequibo River. The incident led to a massive contamination of the river, including a huge loss of life to fish, domestic animals, and local wildlife. Since the river runs through a jungle whose local native populations live along its shores, these natives had to be warned about the danger. It is likely that the extent of the damage to people will never fully be known.

d) Some criteria usually relate to things that do not have a market value but a qualitative and subjective worth, such as watching birds, going to the beach, or enjoying sunsets, and it is possible that a project will alter that worth.

Assume, for instance, that a highway project will follow a lake's shore, and a populated hilly area lies on a gentle slope to the lake. Once the project's intention is known, there will probably be an uproar from these residents since the highway will alter the aesthetics of the site. However, the construction has a high priority because it will connect two very important commercial centres that are linked at present by an urban road with very heavy traffic that produces delays, bottlenecks and accidents, as well as being the access to a first-class ski resort.

As expected, a survey conducted in the area shows a strong opposition to this project, and the objections are that the project will produce a loss in property values, and that it will only benefit the centres located at both ends of the highway while bringing no benefit to the affected area. Several alternatives routes are then proposed, including tunnelling in some sections of the route, and even using a trench system.

Now, how does one put a value on a criterion such as 'Loss of property value'? There is a need for a figure that shows how each alternative will affect the intangible benefits that the population enjoys at present, such as an unobstructed view of the lake, and/or a direct access to the beach. Several techniques exist to make this non-market value appraisal, but one of the most used is a method called 'Hedonic pricing'. In essence, this is a statistical method that considers how the prices of real estate properties have behaved in the past due to similar undertakings in other sites; in other words, it tries to determine the dollar loss in property values produced by a loss of an aesthetic non-market value.

The core of the method rests on the assumption that the final price of a house is the consequence of a series of factors that intervene in the generation of that final price. If it is possible to determine the weight of each factor involved, then it will be feasible to get the final price for the property. In this way, it is sometimes possible to place hedonic values on different projects that involve the criterion 'Loss of property value'.

To obtain the values of the weights or coefficients attached to the factors, this method uses regression analysis (see Appendix, section A.6). This way, one can assign to each alternative, and for the 'Loss of property value' criterion, a measure of the damage, measured as a percentage.
The information necessary to feed the regression model comes from copious data that the Real Estate industry has usually compiled for several decades in various parts of a country.

A whole set of techniques known as 'Cost-Benefit Analysis' can be used for the determination of coefficients and thresholds (see Internet references for Chapter 6, and Munier, 2004, in the Bibliography).

e) Risk criteria
   These are criteria concerned with the existence of risk arising from the execution of each of the alternatives. Suppose that a project calls for a construction to take place on geologically unstable soil such as, for instance, houses built on a steep mountain slope, partially supported by

high concrete pillars or stilts (as seen in parts of the Santa Monica Mountains in northwest Los Angeles). This instability could stem from the frequency of tremors or earthquakes in the area, leading to the association of risks with these constructions.

There could be also risks connected with erosion in soils, producing landslides with heavy rainfall, such as the hundreds of landslides that occurred in Sierra de Avila, north of Caracas, in 1999, which destroyed thousands of houses and apartments. Sabotage risks are also a possibility, such as with the explosion of a pipeline in Iraq, in a site 140 kilometres northwest of Baghdad, in August of 2003.

Other risks include the theft of heavy machinery and materials from construction sites, and risks related with social crime — as seen in the molestation, attack and rape of women in villages close to construction sites, which happens especially during the construction of pipelines. It is also important to consider health risks as shown by the Bhopal disaster in India (see Glossary).

Risk is very difficult to quantify, although it is feasible sometimes to establish an estimated value of occurrences, a threshold. For instance, in a waste incineration project it is possible to determine, in accordance with experience in other places, that over a 10-year period there could be on average 1.5 per cent of people affected by emissions. This average or estimate can be refined by considering local conditions, such as prevailing winds and their force, technically improved incinerators, higher combustion temperatures, etc.

It is also possible to work with low and high estimates for the resulting average, and using the standard deviation (see Glossary).

## 7.8 Application example: A community looks for a sustainable energy option

Considering all of the above, the best way to clarify these concepts is through the following example, which considers a community that wants to utilize its natural renewable resources for energy generation and to complement the power they receive from the national grid. Their objective is not strictly environmental, as they also want to minimize their energy costs with renewable sources. The community must obviously analyse and evaluate its resources, their potential, how to use them better, and what are their future prospects.

## 7.8.1 Goals

The initiative above pertains to an area abundant in small rivers, creeks and lakes, with forests and a high population density living in a cluster of small towns. There is a strong sense of community in the area, and because of that, a community-based group has been formed involving local entrepreneurs, academics, diverse professionals, engineers, city planners and local municipalities. The general feeling is that something must be done to improve their environment, although there is no clear idea of what should be done. There is however, a consensus of what their priorities should be:

- Minimising the damage to the environment.
- Cutting their dependency on fossil fuels.
- Decreasing the cost of energy.

The community does not yet know how to approach the problem, but it has begun forming several commissions to study different aspects of the issue, such as:

- o Resources abundantly and reliably available.
- o Technical and financial viability of different options.
- o People's perception.
- o Environmental problems.
- o Similar undertakings within the country and in other parts of the world.
- o Visits to libraries and Internet sites to gather information.

## 7.8.2 Resources inventory

From the point of view of technical and financial viability, and considering their resource availability, the following potential sources have been identified:

- *Small hydro.* Many water streams and creeks exist in the area, with adequate water volumes and a sufficient slope to generate water head, yet the need to build roads and trails for construction and maintenance must be considered, as these might raise some environmental problems.
- *Wind energy.* The potential exists, since this is a windy area. Many farms could use this type of energy with turbines installed on their land, and any excess energy could be sold off by injecting it into the grid. They consider, as an example of a similarly small development, the 75-kW wind turbine erected by the community of Bro Dyfi, in

Wales (see 'First community wind turbine launched in the UK', in Internet references for Chapter 7).
- *Biomass:* The area shows some potential for this, because of the wood wastes from forests and sawmills.
- *Biogas:* There are many landfills in the area, so the utilization of methane gas for heat and energy production is feasible; however, their being located over a rather large area makes it expensive to transport the methane gas to a central generation unit. Nevertheless, this is a viable alternative.
- **Solar based**. Meteorological statistics show that the area has a large percentage of sunny days per year, and as it is not located at a high latitude, its solar energy is considerable, especially in the summer when the weather is hottest. If this option prevails, the construction of a solar energy facility using mirrors is contemplated.

**Technical aspects**
The technical characteristics of each option, in accordance with their availability in the area, were translated into technical coefficients, such as cost/kWh, reliability (in %), energy production in solar plants, slopes, wind speed, length of piping and electrical wire needed, quality of methane, etc.

**Location**
There were several potential locations for the power plant using methane from landfills, and different places to install two or more large wind turbines (the number of wind turbines was another variable), as with the biomass utilization plant. In the case of installing a solar-based facility, a suitable area would be available. Regarding the installation of hydro –units, several places were available, along with different options regarding number and sizes of water turbines, water volumes and slopes.

*7.8.3 Criteria*

Using the same example, ten criteria were used by the community. The criteria can be technical, or true values dealing with objective issues, or estimates, that is, educated guesses as well as subjective appraisals. The list of criteria prepared for this application is:

- Generation potential.
- Land use.
- Dependability.
- Environmental impact.
- Human impact.
- Social impact.

- Maintenance cost.
- Return on investment (ROI).
- Sharing land with other users.
- People's opinions.

Alternatives and criteria were related through the use of a **matrix,** or a table that had all the alternatives in columns while the rows detailed all the criteria. A number at the intersection of a column and a row indicated the value of the relationship between an alternative and a criterion.

Suppose for instance that the criterion of 'Dependability' is used to gauge two alternatives — such as 'solar based' and 'small hydro'. If a number such as 11 is placed at the intersection of the 'Dependability' row and the 'Solar based' column, this indicates that this alternative is good for only 11 hours a day. If its intersection with 'Small hydro' has the number '24', this alternative is good for 24 hours each day. So, from the dependability point of view, hydro and biomass are better than solar-based, as they contribute the most for energy generation.

Yet it is important to bear in mind that this advantage of some sources over others is only valid for this criterion. Since the objective is to minimise costs for energy production, subject to a set of criteria, this analysis must be performed for each of them. For instance, under 'People's opinion' the small hydro alternative contributes the highest value, while the Biomass alternative shows the lowest contribution.

This kind of analysis where projects or alternatives are subject to a set of criteria to reach a certain objective has the name 'Multicriteria analysis'. If there are also different objectives, then their name would be 'Multi-objective analysis'. The value at the intersection of an alternative and a criterion that indicates this relationship is called a **coefficient.**

**Criteria selection**

According to the established objective-and-resources inventory, criteria involving social, economic, and environmental aspects have to be developed. How is this done?

For each alternative, it is fundamental to analyse its positive and negative impacts on the environment, its economics, and the society involved, and then to create the criteria that ranks the different alternatives. Very often, this appraisal comes from expert opinion; that is, from suggestions by experts knowledgeable about social, economic, and sustainable effects. For instance, a wind farm can have various positive impacts, measured as cheap energy, income for the community (if surplus electricity is sold), substitution of a local diesel power plant, etc. But there could also be negative effects, such as noise, visual pollution, killing of birds, etc.

The experts can then establish a set of criteria that apply to most of the options. For instance, one criterion could be called 'cost of energy in kWh', with coefficients assigned for each alternative.
Another criterion could be 'noise', with a coefficient measuring the noise produced by each alternative that would be placed at each intersection.
Another criterion might be 'land use' and the corresponding intersection denoting how much land each alternative demands.
Many different criteria may come into play, such as 'return on investment', or 'people's feelings about the initiative' or 'danger to wildlife', etc. But the important thing to bear in mind is **that all of these criteria can be used** for the four alternatives, or for some of them.

Most of these are usually technical criteria, but it is necessary also to consider suggestions and ideas from grassroots groups, i.e. to ask for the opinion of people who directly affected by the projects considered.
Another source used to select criteria is the Delphi method, whereby information about the objective(s) and alternatives is sent to different specialists, some perhaps thousands of kilometres away. These experts, unknown to one another, have a time limit to analyse and study the problem and send back their opinions. The procedure is reiterated until a reasonable agreement is reached. This system has its pros and cons, but it is widely used.

Once the community has a variety of criteria, they can use them all, although a selection is normally made to choose a **set of criteria** that best represent the problem. Besides, more often than not, two or three criteria refer to the same issue, and it is convenient to avoid repetitions.
Each situation is different, although from the point of view of sustainability calls for considering criteria that encompass sustainable issues within the three dimensions. In this example, the criteria include only economic, environmental and social issues but not a sustainable one, since all these projects are environmentally sustainable; that is, they all work with renewable resources. Nevertheless, there is emphasis in economic issues because their objective is an economic one, and on social issues, because the population will be affected by whatever decision is taken.

More complications are yet to come. Observe that some criteria call for a maximum effect, while others appeal for a minimum effect. Table 7.3 shows criteria chosen in this example, and different situations to which they apply.
If all the alternatives are sustainable, in the sense that they use renewable resources, why do they need so many criteria if the objective is to minimize the cost of energy production?
The community could just select the alternative with the lowest generation costs; however, the reason for including more criteria to rank projects or

alternatives lies at the core of the concept of sustainability, as it refers not only to economics but also to social and environmental issues.

*Table 7.3*     **Criteria conditions**

| Attributes/Criteria | The maximum effect the better | The minimum effect the better |
|---|---|---|
| Generation potential | X | |
| Land use | | X |
| Dependability | X | |
| Environmental impact | | X |
| Human impact | | X |
| Social impact | | X |
| Maintenance cost | | X |
| Return on investment (ROI) | X | |
| Sharing land with other uses | X | |
| People's opinion | X | |

*Notes on damages*
Unfortunately, electric generation has in the past only taken into account the economic aspect, and this gave rise to the construction of huge nuclear, hydro and fossil fuel powerhouses. Some of these undertakings, especially the **fossil fuel-operated plants,** have caused very serious damage to the environment, such as through the production of acid rain (section 2.8), global warming, and the reduction of the ozone layer. From the social point of view, these plants are largely responsible for lung diseases and the deterioration of many works of art. Examples exist everywhere, but perhaps most remarkable are the Ruhr area in Germany and the Katowice and Krakow regions of Poland.

*Large hydro-generation installations* and the dams they require are responsible for altering the ecosystem, since the dams act as a barrier in two ways: they impede some species from swimming upstream to spawn, and they prevent the mud's biota from being transported by the water to reach areas downstream. This last is the case with the Nile River's rich sediments, which, before the Aswan Dam's construction, fertilized the soils in Egypt's lowlands, as discussed in section 5.1.
As if this were not enough, the lake behind dams usually takes up a large tract of valuable land, and whole villages sometimes need to be flooded and their inhabitants moved to other areas. Other options, such as generation using the wind's dynamic energy, can environmentally affect a landscape or produce noise that disturbs people and wildlife. Some animals are very sensitive to noise, and can react to it violently.

*Nuclear plants* are efficient and enjoy a good safety record, although the problems with disposing of their radioactive residue still remain unsolved, both as to where to deposit it and how to neutralize its very harmful effects. From the social point of view, the Chernobyl disaster — due to human error — is a reminder of the consequences that failure or mismanagement can have on the population.

This illustrates the necessity of gauging a project or alternative not only from the perspective of its economic performance, but by also considering the environmental and social aspects, which is done using several criteria as in this example.

### 7.8.4 Criteria weights

Most multicriteria methods assign a relative weight to each criterion; these weights come from expert opinion and reflect how much more important one criterion is than another. This weighting is a correct procedure and its use should be encouraged. Besides, these methods identify which criterion — due to its importance — deserves a better appraisal through its coefficients and calls for a more thorough estimate. However, and without denying their usefulness, it is necessary to reckon that they are subjective and cannot be replicated: that is, it is only by chance that another team of experts may give them the same weights as another team.

A methodology developed by Milan Zeleny (see Appendix, section A.1) permits calculating weights based on the information provided in the performance matrix (section 7.8.6).

Projects and alternatives can also have weights. For instance, if in a set of projects one refers to an area prone to periodic flooding, while another refers to a new black coat for a road, both projects it obviously do not have the same importance; as a consequence, it would perhaps be wiser to assign different weights to them. This does not mean that projects with greater higher weights will automatically be selected, as they must also comply with the restrictions imposed by criteria; but whatever method is used to make a selection will have to consider this weighting.

### 7.8.5 Threshold selection

Threshold is defined in section 6.11. Logically, the nature of thresholds is the same as the nature of the corresponding criterion that they limit; that is, coefficients in a criterion and a corresponding threshold have the same unit of measure. For instance, if one criterion is concerned with the toxic emissions produced by each alternative, expressed in $mg/Nm^3$ (mg per normal cubic meter), then the threshold must have the same units. However, in some cases, the value assigned to the threshold by the analyst will allow the determination of the number of units that each alternative will have; in other words, it will allow the determination of the value for the alternative.

### *Notes on threshold selection*
For example, a threshold in an educational strategic plan could be the total amount of square meters a city possesses in all its classrooms and all its schools. This threshold corresponds to a criterion called 'Floor space'. Assume now that there are several projects to build and/or improve elementary, junior and high school buildings. For each of these projects there is a coefficient that reveals how much area is assigned to each student, comprising his or her desk, chair, aisle, (a proportion of) the teacher's desk, cabinets, tables, cupboards, shelves, etc.

This value is a standard, posited for argument's sake, to be 2.5 m$^2$/student. Another standard, which this time considers didactical reasons, saying that there should be no more than (or a maximum of) 40 students per classroom. This information, put in mathematical terms in order to analyse units, is:

$$2.5 \text{ m}^2/\text{student} \times 40 \text{ students}/\text{room} \geq 100 \text{ m}^2/\text{room}$$

That is, the average classroom size should be at least (that is, greater than) 100 m$^2$/room.

Suppose now that a survey conducted in elementary schools shows that, on average, there are 77 m$^2$/classroom. This means that schools are short of room space, and some values either have to be relaxed, or, from the sustainable perspective, it turns out that more funds need to be allocated to these schools so as to increase their available classroom space. In this case, there is an indicator (77m$^2$/room) that informs about the current situation, and a threshold (100 m$^2$/room) stipulating for us how it should be.

This is a sustainable threshold since it embodies the economic (funds) and social (students) aspects. Another threshold, for instance the maximum allowed concentration of NOx (see Glossary) in the air in certain areas of a city, establish limits to the number of cars in an area during certain hours. In this case, the concentration is expressed in mgNOx/Nm$^3$. Consequently, when the average car's emission (i.e. a coefficient) in mg/Nm$^3$-car is known, it will be possible to assign a value to the number of cars that ought to be allowed to circulate.

This issue of NOx concentration and its decrease has been addressed by some large cities by a system that apparently works, simply by dividing the car's population in two. Half of the population is allowed to circulate on Mondays, Wednesdays and Fridays, while the other half can do that on Tuesdays, Thursdays, and Saturdays. Inspectors checking the last digit of their licenses, even or odds, do the control. Does this system work?

Yes, to a certain extent, since the number of cars circulating on any given day certainly decreases; however, there will be more that the quantity allowed.
Why is this?
Because of people's idiosyncrasy; some of them solve the problem by owning two cars, one with a license plate ending in an even number and other in an odd one; so it appears that the only solution for these cases is education and some consciousness-raising. This actual case is mentioned to make the reader aware that a solution that seems straightforward and simple to implement will not always work and that careful and thorough consideration must be given in studying these situations from different angles.

Another type of threshold could indicate, for instance, some value for the community's wishes. For example, for safety reasons a community might want to have a police car patrolling the streets between certain hours in a proportion of one vehicle for every 250 inhabitants, which then becomes the safety threshold for a criterion that might be called 'street safety'.

### 7.8.6  Gathering the information

All the elements that play a role in the study (alternatives or projects, criteria, thresholds and coefficients) have been briefly discussed, and they make up a database called a **performance matrix**, whose layout is depicted in Figure 7.2. This choice of name is rather obvious, since the matrix shows the performance of the different alternatives according to the criteria selected.

Nothing has yet been said about how the coefficients are obtained. In this example, as in any other scenario, this is probably the most difficult and time-consuming task. It will be the topic of the next discussion, where the information-gathering task is broken down into the following stages:

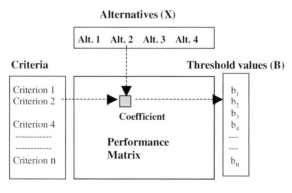

*Figure 7.2*   **Layout of a performance matrix**

## 7.8.7 *Coefficients*

Coefficients represent the relationship between an alternative and a criterion, and they can have different origins since there are different types of coefficients, as follows:

*Technical data:*
These coefficients pertain to technical and objective data. This information is available from tables, manuals, technical publications, from other similar applications, from technical specifications supplied by equipment manufacturers, and from the Internet. Examples of technical data for a wind project are curves depicting the output of the electric generator coupled to the wind turbine and as function of wind speed. Figure 7.3 shows such a curve.

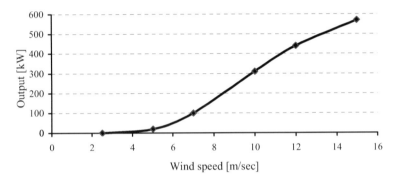

*Figure 7.3*     **Curve of turbine output as a function of wind speed**

*Econometrics data*
Other data comes from econometrics or from very well-known economic tools, for example, on the value of the multiplier effect (section 6.9) for a certain industry, or the return on investment (ROI) for each alternative. This last, which involves very important information, can be easily obtained when data on initial investment, operating costs, savings in the electric bill or profit from the sale of electricity, are used to calculate the ROI over a certain period. The ROI is expressed as a percentage and obtained when the net present value, that is (- initial investment- expenses +benefits - residual costs), is computed for different interest rates. The ROI corresponds to the percentage value of the interest rate found at the intersection of the Net Present Value curve with the percentage axis; or, in other words, when the discount values for gains vs. investment and expenses are equal. (See Appendix section A.2.)

*People's opinion*
People's opinion is considered to be fundamental, and to this end personal and telephone surveys are conducted, as well as several meetings in each area with all the parties involved. From this point of view, a person is consulted about their opinions and feelings regarding each option on different subjective issues, such as:

- Aesthetics for each of the potential sources.
- Noise production.
- Environmental impact.
- Land use.
- Cost and ability to pay to develop each alternative.
- Economic benefits that this undertaking could bring to the area.
- Benefits as a pilot test for the rest of the country.

These opinions are quantified on a scale of from 1 to 10, the higher the better.

### 7.8.8 Alternatives selection

The problem now is to select the option or blend of options or alternatives that will produce the most sustainable effect. To do this, the performance matrix is built with the alternatives in columns and the criteria in rows, the idea being to use these rows or criteria as benchmarks to select the alternatives. This is a typical application of multicriteria analysis, and is solved by using the methodology discussed in section 7.8.9. The method allows for the analysis of different 'what if…?' scenarios, thereby helping the decision-makers to reach well-grounded, rational assessments.

### 7.8.9 Solving the problem

To solve the problem of selecting the most rational, sustainable alternative, Mathematical Programming (MP) is used; it is a very well-known and proven procedure for optimizing an objective subject to many criteria. There is no need to dwell on this technique's theoretical aspects, since there are literally hundreds of books and articles published on the Internet that explain the technique at all levels of difficulty.

It is also possible to utilize a large array of algorithms (or 'codes') that are likewise available on the Internet. This book uses the algorithm called 'Solver', which is found as an add-on in the Excel® spreadsheet program. Readers who have it installed on their computer should look under 'Tools', then under 'Complements', to mark the 'Solver' box. This is a very powerful software for solving linear and non-linear problems, and provides very valuable additional information. 'Solver' feeds from data contained in the performance matrix, as built by the user.

Having already identified the potential sources of renewable energy, and determined criteria, coefficients, and thresholds values, the community has therefore been able to construct the performance matrix depicted in Table 7.4.

*Table 7.4*  Performance matrix

|  |  |  | Solar based | Small hydro | Wind energy | Bio-mass |
|---|---|---|---|---|---|---|
|  | Cost | $/kWh | 0.037 | 0.041 | 0.035 | 0.045 |
| **Attributes** | **Area** | **Units** |  |  |  |  |
| Generation potential | Economics | MW | 1.6 | 1.82 | 3.1 | 2.1 |
| Land use | Environmental | Ha | 40 | 3 | 29 | 45 |
| Dependability | Economics | Hours/day | 11 | 24 | 9 | 24 |
| Environmental impact | Environmental | % | 3 | 5 | 10 | 20 |
| Human impact | Social | % | 0 | 0 | 5 | 12 |
| Social impact | Social | % | 2 | 0 | 20 | 48 |
| Maintenance cost | Economics | $/kW | 0.00043 | 0.00017 | 0.00072 | 0.00068 |
| Return on investment (ROI) | Economics | % | 4.8 | 6.2 | 5.8 | 5.1 |
| Sharing land with other uses | Economics | Ha | 0 | 41.6 | 28 | 0 |
| People's opinion | Social | No. | 4 | 9 | 7 | 3 |

According to their search, four main renewable sources have been singled out:

- A desert region that is large enough to install a **Solar-based** facility (section 5.3.3);
- A hilly area situated at a much higher elevation that is suitable for **small-hydro electric units** (section 5.2.6);
- An off-shore location with shallow waters where **wind turbines** can be installed to generate electricity (section 5.3.1);
- A heavily populated region where gas generated by **biomass** (section 5.3.4) from landfills can be collected in enough volume to produce energy.

The unit cost for each alternative is known in $/kWh.

At present, they assume that developing these renewable sources will account for a considerable portion of energy consumed. In order to evaluate projects or alternatives, 10 criteria or 'attributes' are used by them.

Explanation of Table 7.4.

The unit cost to produce each type of alternative is broken down as follows:

*Production cost* = Capital cost + Land leasing + Operating cost + Working capital + Disposal cost.

The attributes are:
*Generation potential*
This refers to the potential that each alternative has in the region for generating electricity, considering:

- Land availability.
- Water availability.
- Slopes.
- Rainfall and snowfall regimes.
- Winds regime.
- Biomass generation.
- Intensity of solar radiation.
- Number of days with clear skies per year.
- Distance to the electrical grid.
- Etc.

The generation potential for each of the different alternatives is:

- *Solar based.* Depends on: Intensity of solar radiation + number of days with clear skies per year + availability of suitable land to install the mirrors, sun tracking equipment, heat storage and heat transmission unit, turbo-generators, and transformers, + distance to the electric grid, etc.

- *Small hydro.* Depends on: Average water flow in small rivers and creeks + adequate slopes + access to the area + distance to the electric grid, etc.

- *Wind energy.* Depends on: Wind regime + wind average speed + frequency and force of tornados and hurricanes + land availability + distance to the electric grid, etc.

- *Biomass.* Depends on: Characteristics of landfills + distance to the electric grid, etc.

*Land use*
Refers to the amount of land needed for each alternative, considering the economies of scale. Measured in hectares.

*Dependability*
This is about the availability of the resource, and is evaluated in number of hours per day. For instance, solar energy is usable while the sun shines, and heat accumulators can perhaps be utilized for a couple of hours after sunset.

*Environmental impact*
Refers to the damage produced to the environment. For instance, wind turbines can potentially kill birds, generate noise — which can disturb people and wildlife — and modify a site's aesthetics. Biomass is capable of producing dangerous emissions. Hydro units need access roads or operation trails, sometimes in the middle of a national park or a natural forest. Measures are given in terms of a percentage of rejection.

*Human impact*
There could be risks to human health, as in the case of using biomass. It measures as a percentage of rejection.

*Social impact*
This refers to the ways that a project can influence people's lives. For instance, the impact produced by wind turbines. It measures as a percentage of rejection.

*Maintenance costs*

- Solar based:
  Implies costs from damage to mirrors + maintenance of sun tracking equipment + electricity needed to operate the tracking system + turbine and generators maintenance. It measures in $/year.

- Small hydro:
  Entails maintenance cost for hydro-turbines + electrical poles + substation + access roads + cleaning of garbage grates, etc. It measures in $/year.

- Wind energy:
  Involves maintenance of wind turbines (blades and nacelles) + gearbox + generator + electric batteries, etc. It measures in $/year.

o   Biomass:
    Means maintenance of methane purification equipment + turbines/reciprocating engines + electric generators. It measures in $/year.

*Return on investment (ROI.)*
This is the rate of return on all investments and expenses for each alternative, considering the following over a 20-year planning horizon:
- Investment - operating cost - working capital - residual cost (cost of dismantling and disposing of the unit at the end of its useful life) + energy savings + income for energy sold to electric utility + social interest rate, etc. Measured in percentages.

*Sharing land with other uses*
Solar energy prevents using the land where the mirrors are located, since the mirrors, tracking equipment and installations, cover it.
Unit of measure: Hectares; the same for biomass. However, wind energy generation allows for using the land around the towers provided that the blades are at a convenient height. The same goes for hydro, although if a dam is built to create a reservoir, the land occupied by the lake should be considered. Measured in hectares.

*People's opinion*
This expresses, on a 1-to-10 scale, how people perceive each alternative without technical considerations. Of course, there could be many other attributes to consider: for instance, the effect on taxation, since on the one hand, land value is enhanced by the construction of these undertakings, but, on the other, the government normally relaxes taxes with the generation of electric energy. Measures given as a cardinal number.

Another aspect, albeit not depicted in Table 7.4, could be the multiplier effect (section 6.9) created by these new technologies: in new jobs for the manufacturing and maintenance of this equipment, increased levels of education needed to manufacture and operate them, the development of ancillary industries, etc.

The performance matrix is a **mathematical model** trying to stand for a reality; as such, it must be as representative and accurate as possible, and from that point of view, Table 7.4 is incomplete since some fundamental technical restrictions must be established, such as:

1. In the criterion 'Generation potential', one must set up a **minimum total production output** measured in megawatts (MW); otherwise a project could not be economically justifiable

because of economies of scale (see Glossary). This consideration leads to considering that perhaps it is not fitting to develop only one source of energy, but it may instead be preferable to develop two or even three of them combined.

2. Minimum generation for each type of option.
   Because of economies of scale, there is a minimum value or threshold for the energy generated by each option/equipment. This means that a minimum output has to be established because equipment is manufactured with a certain minimum power. Thus, Table 7.5 is built.

*Table 7.5*                      **Threshold values by option**

| Option | Minimum generation capacity [MW] | Comments |
|---|---|---|
| Minimum generation for solar | 0.10 | This value is low because there are not enough large areas available for installing an array of mirrors close enough to human settlements |
| Minimum generation for hydro | 0.40 | Corresponds to the minimum generation value to offset land use, wiring, etc. |
| Minimum generation for wind | 0.37 | Corresponds to the minimum generation value to offset infrastructure needed |
| Minimum generation for biomass | 0.12 | Corresponds to the minimum generation value to offset equipment needed |

3. As mentioned, equipment comes in a large variety of sizes and performance standards, so the study should consider this to allow the model select the most appropriate combination or mix of equipment for the minimum total output established.
   The table must thereby be expanded with columns for the different outputs available for each alternative, with the corresponding unit cost values of production in $/kWh, plus maintenance costs, etc. In other words, columns should be created (with a column to correspond for each size available per alternative or option) to lead up to something like Table 7.6. Of course, a real life situation could have many more columns and many more rows.

*Table 7.6*    **Performance matrix considering several sizes per source**

|  | Solar based | | | Small hydro | | | Wind turbines | | |
|---|---|---|---|---|---|---|---|---|---|
| Size | | | | | | | | | |
| Cost | | | | | | | | | |
| | | | | | | | | | |
| Attributes | | | | | | | | | |
| Generation potential | | | | | | | | | |
| Land use | | | | | | | | | |
| Dependability | | | | | | | | | |
| ---------- | | | | | | | | | |
| ---------- | | | | | | | | | |
| ---------- | | | | | | | | | |
| ---------- | | | | | | | | | |
| People's opinion | | | | | | | | | |

Thresholds established for each criterion are depicted in Table 7.7.

*Table 7.7*  Threshold values by criterion

| Criterion | | Value |
|---|---|---|
| Generation potential | > | 2.15 |
| Land use | < | 45 |
| Dependability | > | 9 |
| Environmental impact | < | 20 |
| Human impact | > | 0 |
| Social impact | < | 48 |
| Maintenance cost | < | 0.0068 |
| ROI | > | 4.80* |
| Land shared | > | 1 |
| People's opinion | < | 9 |

\* Appendix section A.2 explains how to obtain this value

**Explanation of coefficients**

Table 7.6 includes data from Tables 7.4 and 7.5. Columns show the different alternatives, while rows depict the attributes, conditions, or criteria that are used to rank the alternatives.

The values in the 'Dependability' row, for instance, indicate the corresponding availability in number of hours per day. Thus, '11' shows that solar-based electric generation is good for 11 hours each day (when the sun is shining, and including two hours of stored heat at sunset). Small hydroelectric generation is available over the 24 hours, wind electric generation is available, on average, for 9 hours, and biomass used to generate electricity is available over 24 hours.

*Table 7.8*   **Performance matrix with the incorporation of generation limits**

|  | Solar based | Small hydro | Wind energy | Biomass |
|---|---|---|---|---|
| Cost ($/kWh) | 0.037 | 0.041 | 0.035 | 0.045 |
| **Attributes** | | | | |
| Generation potential | 1.6 | 1.82 | 3.1 | 2.1 |
| Min. gen. for solar | 1 | | | |
| Min. gen. for hydro | | 1 | | |
| Min. gen. for wind | | | 1 | |
| Min. gen. for biomass | | | | 1 |
| Land use | 40 | 3 | 29 | 45 |
| Dependability | 11 | 24 | 9 | 24 |
| Environmental impact | 3 | 5 | 10 | 20 |
| Human impact | 0 | 0 | 5 | 12 |
| Social impact | 2 | 0 | 20 | 48 |
| Maintenance cost | 0.00043 | 0.00017 | 0.00072 | 0.00068 |
| ROI | 4.8 | 6.2 | 5.8 | 5.1 |
| Land shared | 0 | 41.6 | 28 | 0 |
| People's opinion | 4 | 9 | 7 | 3 |
| Expressed in percentage | 1 | 1 | 1 | 1 |

The Rate of Investment (ROI) criterion will provide an indication of the expected return for each kind of project. The origin of these values can be found in Appendix, section A.2)

The 'Social impact' criterion shows, in its row, another type of coefficient expressed in percentages. That is, only 2 percent of those polled believe that solar-based energy will have an impact on their way of life. There is no value for hydro, as people do not see how this option may have an impact on their lives, as long as it produces energy.

This percentage jumps to 48 percent for biomass, because people do not like the idea of having a landfill in the neighbourhoods. For wind-generated energy, 20 percent are against the alternative, because people do not feel comfortable with the noise or the aesthetics. However, to an extent this is a matter related to geographical and density aspects. It is geographical because the wind turbine can be located in many different parts such as farms, offshore, in mountains, etc., and it is related with density due to the fact that it could also be positioned in deserted areas.

People in urban areas are not likely to support the idea of a wind generating electricity farm on the outskirts of their city or offshore due to visual considerations. However, in rural areas people's feelings are probably different, since the owner of the farm gets extra income from the installation of a wind turbine on their land, by receiving a fee for the use of the land. Besides, in rural areas it can be very advantageous for a farmstead to generate its own electricity using smaller wind turbines, showing substantial savings in their electrical bill and perhaps even a profit if excess energy is sold to an electric utility.

As mentioned before, the criterion 'People's opinion' expresses subjectively, on a 1-to-10 scale, how people generally feel about a project. In this example, the winner is the small hydro plant, and it is easy to understand why people gave it the highest marks, as it is:

- Practically hidden.
- Noiseless.
- Reliable.
- Not a land user (in the case of run-of-the-river installations), that is, when there is no need for a dam. In this type of scheme, water is extracted from the river at a certain point and channelled or piped to a hydro unit that discharges it into the river at a lower level. The scheme therefore uses the difference in levels between two points of the river. The river's flow is not altered, and a dam is not put in place as a barrier.

### 7.8.10  *Objective of this exercise*

The goal of this exercise was presented in section 7.8.1 as a broad and imprecise desire: **To have the environment improved.**
Now that the community has studied and discussed the whole plan, they have established objectives that are precise, and that require compliance to reach the goal. What are this exercise's objectives?
The objectives are several, as follows:

- To generate electricity from their own renewable sources.
- To minimize the use of fossil fuels.
- To minimize their electricity bills.

Naturally, it would be ideal to have a blend of projects of different sizes and outputs that can achieve the simultaneous objectives of:
- Reducing the electricity bill.
- Minimizing the damage to the environment.
- Minimizing the use of fossil fuels.
- Maximizing the return of their investment.
- Maximizing the reliability of electrical generation.
- Etc.

These, clearly, are conflicting objectives, since maximizing reliability, for instance, could also damage the environment — in the case of a preference for fossil fuels — and may entail higher costs. Other restrictions can arise, as solar-based and wind-based energy generation might not be complementary, as both use the same tract of land (so it is one or the other). These constraints must therefore be taken into account in the study.

### *7.8.11 The database*

All of the elements of the problem, as follows, have now been defined:

- **Objective**: To minimize the cost of energy generation.
- **Projects:** Four different alternatives, two of them being incompatible (all of them are environmentally sustainable).
- **Criteria:** 10 different criteria for maximization and minimization of economic, social, and environmental issues.
  Most people feel that criteria should be assigned weights to express people's perception of their relative importance.
- **Coefficients:** At the intersection of a project and criteria, these express the relationship between them.
- **Thresholds:** One is assigned per criterion. The relationship between the coefficients in the rows and the thresholds is denoted with the sign '$\geq$' for 'equal to or greater than', the sign '$\leq$' for 'equal to or lesser than', and the sign '=' stands for the same as or equals'.

Combining all these elements Table 7.9, the database, is obtained.

With the information from the database the problem can be stated in a mathematical form that relates the coefficients, the thresholds and the signs. For instance for the criterion 'Generation potential' and the threshold '2.15 MW', the expression for that row will be (see Table 7.9):

$$1.6 \, x \, (\text{Solar based}) + 1.82 \, x \, (\text{Small hydro}) + 3.1 \, x \, (\text{Wind}) + 2.1 \, x \, (\text{Biomass}) \geq 2.15 \text{ MW}$$

If we designate $x_1$ as the solar-based alternative, $x_2$ as the small hydro, $x_3$ as the wind alternative, and $x_4$ as the biomass alternative, then the above equation becomes:

$$1.6\, x_1 + 1.82\, x_2 + 3.1\, x_3 + 2.1\, x_4 \geq 2.15 \text{ MW}$$

*Table 7.9*  The database

| Attributes | Solar based | Small hydro | Wind energy | Biomass | Values from computation | | Thresholds from Tables 7.5 and 7.7 |
|---|---|---|---|---|---|---|---|
| Cost ($/kWh) | 0.037 | 0.041 | 0.035 | 0.045 | | | |
| Generation potential | 1.6 | 1.82 | 3.1 | 2.1 | 2.32 | > | 2.15 |
| Min. gen. for solar | 1 | | | | 0.10 | > | 0.10 |
| Min. gen. for hydro | | 1 | | | 0.40 | > | 0.40 |
| Min. gen. for wind | | | 1 | | 0.38 | > | 0.37 |
| Min. gen. for biomass | | | | 1 | 0.12 | > | 0.12 |
| Land use | 40 | 3 | 29 | 45 | 21.62 | < | 45 |
| Dependability | 11 | 24 | 9 | 24 | 17.00 | > | 9 |
| Environmental impact | 3 | 5 | 10 | 20 | 8.50 | < | 20 |
| Human impact | 0 | 0 | 5 | 12 | 3.34 | > | 0 |
| Social impact | 2 | 0 | 20 | 48 | 13.56 | < | 48 |
| Maintenance cost | 0.00043 | 0.00017 | 0.00072 | 0.00068 | 0.000466 | < | 0.0068 |
| ROI | 4.8 | 6.2 | 5.8 | 5.1 | 5.78 | > | 4.80 |
| Land shared | 0 | 41.6 | 28 | 0 | 27.28 | > | 1 |
| People's opinion | 4 | 9 | 7 | 3 | 7.02 | < | 9 |
| Expressed in percentage | 1 | 1 | 1 | 1 | 1.00 | = | 1 |

| RESULT | 0.10 | 0.40 | 0.38 | 0.12 |
|---|---|---|---|---|

That is, the production capacity of $x_1$, plus the production capacity of $x_2$, plus the production capacity of $x_3$, plus the production capacity of $x_4$, **should be equal to or more than 2.15 MW**, i.e. this value is the **minimum production level to achieve.**

The same can be done for the other criteria.

The objective (z) can be expressed as:

$$z = 0.037 \, x_1 + 0.041 \, x_2 + 0.035 \, x_3 + 0.045 \, x_4 \text{ (Minimum)}$$

That is:

The cost per kWh using solar-based means, times the percentage of use of solar-based, plus
the cost per kWh using small hydro means, times the percentage of the use of small hydro, plus
the cost per kWh using wind means ,times the percentage of using wind, plus the cost per kWh using biomass means, times the percentage of using biomass, should be a minimum

An algorithm (see Glossary) in Mathematical Programming (MP) can now be applied, and the result will indicate the optimum mix of alternatives or options for this example. This result, shown in the last row in Table 7.9, indicates that the best alternative or energy source, considering all the restrictions imposed, is small hydro, followed very closely by wind, and then by biomass and solar-based. These values also represent percentages, therefore it means that the optimum solution includes 40 percent of energy produced by small hydro, 38 percent by wind, 12 percent by biomass and 10 percent by solar based.

The shadowed column portrays the thresholds imposed to the model. The column 'Values from computation' depicts values computed by the model in compliance with the thresholds considering the mathematical signs.
The four criteria 'Min. gen. for solar', 'Min. gen. for hydro', 'Min. gen. for wind' and 'Min. gen. for biomass', indicate, as mentioned before, that those sources must produce, because of economies of scale, a minimum of energy, as shown by these signs '>'. If these minimum values were not considered, it would be possible for the model to choose an option that is too small; because of this, it is possible that no equipment is available on the market.

This exercise is static in that it establishes policy regarding energy generation over a period of, say, 20 years. However, it could very well be that future technical advances change the values of some coefficients, such as the efficiency of solar-based energy increasing in output while its costs reduce, as has already happened with photovoltaic systems (section 5.3.2) and wind turbines (section 5.3.1).
For this reason, and considering that long-term planning can involve 40 or 50 years, and a 10-year period for each plan, it is convenient to have five planning schemes that can be developed following this method. Then,

undertaking again a new study each decade, as in the example, by replacing values according to current best estimates.

*Main characteristics*

This model produces two sets of results. The first set supplies the information that one is looking for, that is, **the blend of alternatives or options** that **optimizes** the objective.
The second set supplies the values of the weights calculated for the criteria. These weights are known as **shadow prices** or **marginal costs,** which have precise meanings. These marginal costs indicate how much the objective for a unitary change will vary within the threshold value. Understandably, this has a tremendous value for doing a sensitivity analysis that will determine how the objective will vary for different values of the thresholds; this is very useful when the values of some thresholds are not known with precision. It is possible to perform this analysis on one criterion or on all of them simultaneously.

The second set also supplies what are known as 'opportunity costs', that is, the abandoned opportunity when resources are used in some projects instead of others. For example, if the model selects, say, project or alternative 2, but for whatever reasons the community wants to develop alternative 3, the opportunity cost will tell them how much it will cost – either monetarily or in some other terms – to choose that non-optimal alternative.

What happens when, besides many criteria, there are also different objectives? Very often an objective can be considered to be a criterion, and vice versa. For instance, in the solved example, the objective was minimizing energy production costs; if another objective is considered, such as minimizing the environmental impact, then the objective will be, as per Table 7.9, as follows:

$$z = 3 x_1 + 5 x_2 + 10 x_3 + 20 x_4 \text{ (Minimum)}$$

In addition, the cost values that were an objective will be now a criterion of the '≤' type; but how does one manage when there are two simultaneous objectives? Obviously, the substitution above cannot be used.
Yet it is possible to utilize another tool in the MP family to perform the job. It is called 'Goal Programming' (GP), and it works as explained for the MP exercise, only in GP the objective also incorporates positive or negative 'deviations' for a stated objective.
The GP algorithm is not included in the Solver method of Excel®, because it is sold separately.

*Special characteristics*
Some special features of the MP method are:

- Complementarity
  Sometimes in a set of projects or alternatives, two or three are complementary, in that if one of them is selected the other should also be chosen.
  For example, hydro-projects are usually built in 'cascade', that is, several dams are erected along a river, and then, using the difference in slope, the discharge of the hydro-turbines from the first dam is made to feed the lake formed by a dam constructed for a second undertaking that is usually many kilometres downstream.
  In this case, it is obvious that both of the projects, the one upstream and the one downstream, are complementary, since altering the operating conditions of one dam will affect the other.

- Standalone
  Sometimes two or more projects will not allow for another project nearby. For example, assuming that three options have been studied for a road between two points, A and B, the selection of one option will understandably preclude selecting either of the others.

- Selection with integer results
  In many cases, a fractional result makes sense, as occurs when one ranks projects or alternatives, and the ranking is expressed as a relative percentage. For instance, the results could be as depicted in Table 7.10.

*Table 7.10*          **Results in percentages**

| Projects | | | | |
|---|---|---|---|---|
| 1 | 2 | 3 | 4 | 5 |
| 0.21 | 0.19 | 0.19 | 0.23 | 0.18 |

These fractions can be used to establish a hierarchy of projects, in this case the most significant project being number 4. Assume now another example where there are three projects as follows:

- Project 1: School construction;
- Project 2: Sewer construction;
- Project 3: Paving route 926/A.

The result could be as depicted in Table 7.11.

*Table 7.11*     Construction project

| Projects | | |
|---|---|---|
| School construction | Sewer construction | Paving route 1209A |
| 1 | 2 | 3 |
| 0.40 | 0.28 | 0.32 |

In this case, it would appear that project 1, 'School construction', is the most important. It probably is, but this result is of little help since it makes no sense to build 40 percent of a school, 28 percent of a sewer and 32 percent of a road, unless these projects are included in, say, a two-year construction plan. In that case, these percentages could indicate the progress in construction over the first year, and the balance to be completed during the second.

In general, one does not work with such fractions or decimals, so integer results are normally needed. In this case, it is necessary to work with integer programming, and most especially with binary results. In this last case, the result could be something like what Table 7.12 indicates.

*Table 7.12*     Binary results

| Projects | | |
|---|---|---|
| School construction | Sewer construction | Paving route 1209A |
| 1 | 2 | 3 |
| 1 | 0 | 1 |

This result shows that for whatever reasons, generally a lack of funds, only projects 1 and 3 can be executed, and project 2 will not be constructed.

## 7.8.12 Conclusion

As mentioned, information-gathering is a long but very important process, and one not to be taken lightly as the outcome of the study depends on such input. It is not possible to standardize the length of this information-gathering process since it depends on the following factors, among others:

- Number of projects or alternatives.
- Number of criteria.
- Complexity of each issue.
- Existing statistics.
- Enough funds available for the research.
- Time needed to prepare questionnaires, to perform a Delphi analysis, to determine hedonic prices, etc.

What is encouraging is that most of this information usually exists somewhere, so the problem becomes finding out where it is, and having it computed. This usually involves many people, according to the type or blend of types of projects, and it essentially involves constituting a team of specialists that includes engineers, economists, social workers, statisticians, architects, city planners, health care professionals, education workers, stakeholders, etc.

Last but not the least, it is necessary to consider that this involves a **feedback process**, which is why there should be top-down and bottom-up flows of information linking decision-makers and the population. Being at a high level, decision-makers do not know all the particulars of a problem, and those at the grassroot level are ignorant of all the intervening factors, mainly those that are economic and political.

**Internet references for Chapter 7**

**Author**: Peter Bartelmus (1999)
Title: *Sustainable development – Paradigm or paranoia?*
Comment: Medullar discussion on this vital theme.
Address:
http://www.wupperinst.org/Publikationen/WP/WP93.pdf

Title: *Sustainable development*
"…we are confronting a situation of survival of our planet as a whole"
Address:
http://csars.calstatela.edu/boone/geog322/geog322/viana/sustainable.html

**Author**: Sharon Beder (1994)
Title: *The hidden messages within sustainable development*
Comment: This author introduces two paradigms, namely a shift in priorities where the environment is part of the economic system, and the primacy of the market to allocate them. Consequently, management of the environment is an economic problem.

Address:
http://www.uow.edu.au/arts/sts/sbeder/esd/alternatives.html

**Source**: Report – The Maine Center for Economic Policy (2001)
Title: *Sustainable development*
Comment: An excellent paper. Under the title "Integrating the practice of sustainable development", there are nine essays on nine different and varied subjects such as Energy, Forestry, Fisheries, Health Care, Community Development, Food Security and Living Wages. According to the paper, each essay is an exploration of what sustainable development could look like in Maine, and what actions they need to take now to promote a sustainable state. This reading is recommended.
Address:
http://www.mecep.org/report-sustain/report.htm

**Source**: Tellus Consultants Limited
Title: *Visions and indicators*
Comment: Information to develop indicators. There is a very good list of indicators for a sustainable community, and a consistent treatment of the subject of targets.
Address:
http://www.tellusconsultants.com/visions.html

**Authors**: Stanley Guy and David Rogers (1999)
Title: *Community surveys: Measuring citizen's attitudes towards sustainability*
Comment: This is a sensitive issue since the result of the survey can strongly influence further actions. However, it is not an easy task, since questionnaires have to be carefully drafted to avoid ambiguous answers. According to what these authors say, *"This paper describes how community surveys can be used to assess individual and public support for sustainability"*. This paper includes sample questionnaires for economic, social, and environmental sustainability.
Address:
http://www.joe.org/joe/1999june/a2.html

**Source**: University of British Columbia, Vancouver, Canada (2001)
Title*: Fraser Basin/Georgia Basin sustainability tools and resources (STAR)*
Comment: Useful explanation about information gathering in an actual case, including inventory components. *"This information gathering will determine the first set of additional website/database components featuring tools and resources to support sustainability activities"*.
Phase 3, which lays out a scheduled set of deliverables, is recommended reading.

Address:
http://www.pkp.ubc.ca/publications/star_proj.htm

**Source**: Murdoch University
Title: *The tragedy of the commons* (1968)
Comment: One of the first, 'classics' about the use of resources that belong to everybody.
This reading is recommended.
Address:
http://wwwscience.murdoch.edu.au/teach/biotech/tragedy/tragedy.htm

**Source**: Edited by Julian Dumanski, RDV, World Bank
Title: *Guidelines for conducting case studies under the international framework for evaluation of sustainable land management*
Comment: Useful information about thresholds.
Address:
http://www.ciesin.org/lw-kmn/slm/guideslm.html

**Authors:** A.J. Smyth and J. Dumanski
Title: *FESLM: An international framework for evaluating sustainable land management*
Address:
http://www.fao.org/docrep/T1079E/t1079e00.htm#Contents

**Authors:** Alice Born, Claude Simard, Robert Smith (2001)
Title: *Technical guidelines for indicator selection*
Address:
http://www.nrtee-trnee.ca/eng/programs/current_programs/SDIndicators/ClusterGroups/Cluster Group_BackgroundDocuments_TechGuidelines_e.htm

**Source**: peopleandplanet.net – People and renewable energy (2003)
Title: *First community wind turbine launched in UK*
Address:
http://www.peopleandplanet.net/doc.php?id=1924

# CHAPTER 8: CASE EXAMPLE - A COMMUNITY IN SEARCH OF ITS FUTURE

## 8.1 Background information for a process

This chapter's case example deals with urban sustainability, and will serve to apply the concepts discussed in the preceding seven chapters. It is necessary to bear in mind, however, that the sustainability process is not a straightforward one. Since it involves the environment, economics and mainly people, there are many facets, ideas, intermediate targets, etc., that must be taken into account, considering that each one interacts with the others in complex ways. It is also worth noting that the issues are not treated at one single level but on several, including those pertaining to grassroots, stakeholders, decision-makers, policy-makers, local and national governments, as well as that the process is looped in that decisions, suggestions, additions, amendments, etc., move back and forth not just horizontally (at the same level) but also vertically (between the various listed levels).

As discussed in section 7.2, it is difficult to agree on the population's goal or goals, or to pinpoint a prevailing aspiration, as well as the intermediate targets to be met, and when and how to reach them. Another difficult issue has to do with measuring the goal's actual achievement. When can the community say that it has reached its purpose? Is it possible for them to use some international yardstick to determine if a society has reached a sustainable way of living?

Hardly.

Every community is different from others, in that each has its own problems and needs, resources and drawbacks. One way to solve this issue is probably to establish some general parameters about what the community needs, independently of the present situation. Their members can say for instance, *"One of our problems is that part of the population is living in poverty, with makeshift houses and with few opportunities to find stable jobs; therefore, one of our objectives should be to improve living conditions in this sector"*. From here on, this will most probably be a high priority area to consider.

Without necessarily being in disagreement, others may hold, for example, that the region is in a risky position since not enough diversification exists in the local economy — a problem that might also take a high priority. However, when studying these two issues, it is perhaps possible to find a common ground between them. For example, they may submit a potential solution that involves not only improving living conditions but also capacity-building for this people, and for the rest of the city's inhabitants. This would provide the opportunity to learn something new while at the same time establishing a base pool from where new industries can draw. This good initiative might perhaps be complemented with political and economic decisions from City Hall to attract new industrial and commercial ventures by lowering taxes and granting some additional benefits.

It is therefore likely that **synergies** may be detected once problems and objectives are identified. After all, synergy means the existence of interests common to different participants, as in this case. From this point of view, it would also perhaps be wise to look for **synergies among all the city's problems** — and the community is likely to find them — because a city is a socio-economic-political and ecological system that may be described as many different parties, yet that also interact in order to reach a common goal.
The goal or goals could already have been defined, although it is also necessary to consider that there is a need for **some sort of measure** of how **much of a goal** they want to achieve. In other words, how do they quantify the amount of their efforts to improve the condition of these low-income people?

Are the decision-makers satisfied that these low-income people will end up having a basic infrastructure? Is that enough?
Somebody may respond, *"Well, at least they will live better than before"*. True, but this is only an improvement of their current situation, and does not achieve social equity.
Consequently, one policy that the community can establish regarding this problem is to progressively allow these people to enjoy the same benefits, in quantity and in quality, that are offered to the rest of the city's population. This means access to a nearby healthcare centre, to transportation connecting their area with the rest of the city, access to a community centre where citizens can talk and be heard by the local authorities, and gain access to telephone lines, computer terminals, etc.

Fine, but now that they have reached some apparently reasonable agreement, how can they measure the quantity and quality of services for all the population? For instance, while they can say that at present there is a ratio of 400 hospital beds per inhabitant, just what does this mean? Is it too many, enough, or too few? The search for an answer invites one to look for

recognized standards in this regard. Indicators have been compiled by institutions, such as the United Nations, which give a figure of what is considered desirable.
All right, now they can use that figure, whatever it might be, and say *"we are short of beds or we have an excess of beds relative to international standards"*, and thereby know just where they stand. Is this assessment enough?

Well, yes and no.

Yes, in that there is now a yardstick, the international standard, against which the ratio of available beds can be appraised. However, they should also consider local situations, at the local and national levels. Why is this?
Because, unfortunately, not all countries have the same standard of living and resources, so standards and indicators should be adjusted to local conditions, needs and problems. To clarify this point, consider that in some countries fresh water, in quantity and of good quality, is taken for granted, so its consumption easily meets international standards. Others who are not as fortunate have to be content with a fraction of this, simply because they lack enough water sources. Therefore, as long that they receive water of good quality and in adequate quantities to satisfy their hygiene and cooking needs, well, their standard is as good as the other countries', where water is plentiful (which, one might add, are scarce indeed these days!).

So they now have the synergies, and have considered international standards adjusted to their local needs. They still also need additional things, such as:

- To determine how to measure progress.
- To make sure that one indicator is linked to others, that is, darting across economic, social and environmental issues and their interactions.
- To decide on the timespan to complete the process.
- To settle on how to keep people informed, and what kind of information to furnish. That is, it is necessary to agree about the state of reporting on the sustainable environment.

Now, this problem being solved, a new one arises: It is necessary to establish some general grounds about **what a priority undertaking is** and which of alternatives should be postponed. More precisely, once all their goals are defined and the undertakings for how to reach them are identified, making a ranking or selection becomes crucial, subject to several constraints, such as:

- Not enough funding is available for all projects or undertakings (an almost universal condition).
- The span of time usually covers many years, so funding allocations have to be specified on a yearly basis;
- Targets need to be established, maybe in phases. For instance, section 4.2.1 discussed the construction of a bus system in Bogotá, and how they phased in the undertaking regarding the extension and expansion of the service in a timely fashion, along with the allocations of funds for it;
- Sustainability is a dynamic process in that objectives, priorities, and even goals may change over a period. So the sustainability plan must consider that perhaps in five years time it could be quite different. Naturally, nobody can know that, but changes can be possibly foreseen, and the mechanism to prepare for and deal with any contingency plans has to be established to face these unforeseen circumstances;

Establishing a time-plan or schedule is also essential. The reason for this is not only economic, as has to do mainly to a natural or technical sequence of events, because some activities must precede others. Consider, for instance, this situation: Because of traffic congestion problems at an intersection where five avenues converge, City Hall has opted for the following solution:

- Construction of a roundabout at street level.
- Excavation of a tunnel with a north-south orientation on the second level, beneath the roundabout.
- Excavation of a second tunnel with a north-east/south-west orientation on the third level, beneath the other tunnel.

Of course, the massive excavation is the main task to be undertaken. Before beginning it, however, information must be gathered about the location and condition of any buried utilities, such as sewer, water, telephone lines, etc. Once this is done, usually with the help of GIS, studies must be conducted about rerouting utilities, replacing cables and water and sewer trunks in poor condition, etc. There are also permits to get, drawings to prepare, guidelines to be issued, etc.

That is, a series of activities must be accomplished before, during, and after the excavation, and all of this work has to be coordinated. Therefore, and especially in complicated undertakings, it is essential to have a bar-charts of activities, or, in the more complex and elaborate works, some development of a critical path schedule (see Glossary).

In general, there could be the feeling that, considering the magnitude of the problems involved, the achievement of sustainability for a city can only be feasible for small cities. This is not so, and the case of Curitiba (section 3.3.1) provides a good example. More recently, in 2002, London's Mayor set up a commission to determine ways in which London could also become a sustainable city.

## 8.2 Introduction to the sustainable initiative for a community

The community of Meadow Forest with a permanent population of about 160,000 is nestled in a mountain valley along the mouth of a river draining into a deep lake. The area has lush vegetation, and offers facilities for hiking, trekking, fishing, swimming, and off-road vehicles recreation. It gets a lot of tourism year-round; this is its main activity, while the others are a couple of high-tech light industries and some reduced production of cottage industries. Clearly, tourism brings many benefits to an area, but many problems also arise. For this reason, the population wants to take steps to have a sustainable city in the near future. One the first things the citizens did was to analyse the advantages and disadvantages of tourism in general, taking as a guide Table 8.1; then they prepared their own appraisal.

*Table 8.1*   Advantages and disadvantages of tourism for a region

| Domain | Advantages | Disadvantages |
|---|---|---|
| SOCIAL | | |
| | Creates employment opportunities, which increases the standard of living. People with more disposable income purchase appliances, cars, houses, make improvements, etc. | Because land price increases, it is more difficult for low-income people to rent and/or buy a house |
| | Promotes construction of houses for low-income people if part of the city's revenues are channelled to social projects | Prices in resorts tend to be higher that in other places. For this reason, the cost of living increases for everybody |
| | Helps health care for people who can buy prescriptions as well as better food | Real estate taxes tend to increase |
| | Enhances education, since people have the economic means to pursue other studies | Unless schools and hospitals improve their capacity, there could be problems in capacity to provide good services |

| | | |
|---|---|---|
| | For foreign visitors, tourism allows a healthy interchange of social and cultural issues, such as learning the history of the area, listening to its music, and enjoying its dances, and the opportunity to learn a new language | There is a potential rise in crime for the area |
| | Will encourage the development of artistic activities, induced by a public with free time and eager for entertainment and cultural events | |
| | The presence and action of man always has an impact on the environment, and tourism only increases it. There could, however, be some advantages, such as the planting of trees in a semi-desert area, the reinsertion of some animal species, initiatives to stop erosion, etc. In some areas, health aspects may be improved, such as, for instance, the eradication of mosquitoes | The ecology of the area is always impacted one way or the other. It can severely affect sensitive areas. The Altamira Caves, in Spain, famous the world over for paintings from between 16,000 and 9,000 BC, are open only for small parties due to a fear of damage to the paintings from humidity and temperatures brought by people. In other caves, paintings that survived thousands of years have deteriorated rapidly from presences of visitors |
| | Some areas that could greatly benefit are the wildlife population by the creation of sheltered zones, by protecting them from poachers and through provision of guaranteed food. The Kruger National Park in Africa offers such an example (section 6.12.1) | Tourism is responsible for the consumption of large amounts of water, and the need to treat the corresponding wastewater |
| | | There is usually an increase in air contamination due to cars, and a rise in the consumption of fossil fuels due to increased energy consumption |
| | In this area, tourism has its greatest influence. It can create human settlements with all amenities and life comforts even in deserted areas, such as in Cancun and Las Vegas | Sharply increases land values, making it more difficult for small business to start |

| | Business flourishes not only due to increasing number of patrons in hotels and restaurants, but also because many people save money all year long to spend it in a couple of weeks at some resort | Can ruin local producers when large amounts of produce and fruit are bought by hotels and restaurants from outside the area |
|---|---|---|
| | Tourism is a very important engine for the construction industry, for hotels, residences, shopping centres, etc. | Increases taxes, which can affect small business |
| | Promotes the development of crafts and cottage industries | The installation of supermarkets may overwhelm small business that cannot compete, and then disappear |
| | The influx of supermarkets gives people access to cheaper products | |
| | Can dramatically increase the regional domestic product and tax revenues | |
| | Banks are willing to lend money for mortgages to people with a reasonable fixed income | |
| | Tourism will induce local manufacturers to produce more goods such as food, furniture, clothing, beverages, etc. | |

### *8.2.1 The system and the process*

To reach their goals, the community needs a system and a process.

*Do they have a system? Yes, they do.*
> Their system is made up by the people, the town, their assets (lake, river, mountains, tourism infrastructure, etc.), and the local authorities.
> It is a system because the **whole community** is working to make it a better place for them and for their descendants. The system's other components are tourism assets and infrastructure, so they want to rely in this activity to promote well-being.

*Do they have a process? Yes, they do.*
> The process will lead them from **where they are now** to **where the want to be.**
> They have a **plan** and goals. The **goal for the present** is to improve their living conditions such that everyone in the community will gain access to essential services (which is still not the case), to employment and equal employment opportunities, as well as to education and health care. They want to enjoy their beautiful scenario, breathe clean air (which at present

is not very clean), and the other opportunities that the area offers. Their **goal for the future** is to maintain and improve their living conditions for generations to come.

After this introduction, let us start with this case study.

## 8.3 The process

Even if it is difficult to think along a straightforward path from the present to the future to secure a sustainable community, it will be convenient to establish some sort of sequence of activities that need to be performed. This way, it will be easier to organize the work, assign responsibilities, etc. The suggested sequence is as follows:

### A. *Create an agency to be in charge of this whole project*

To manage the entire process, a steering committee has been created called a Sustainability Committee, consisting of volunteer work provided by citizens, stakeholders, industry and commerce representatives, health and education officials, and City Hall bureaucrats. This Committee is an autonomous agency, not reporting to City Hall, and is financed (offices, equipment, etc.), by established contributions from all interested parties. However, it has been agreed that City Hall will defer to their opinions and advice when taking measures that affect the city. Besides, the city is part of a region that has some villages and hamlets within a 35 kilometre radius; consequently, there is also a representative from the region outside the city limits.

### B. *Make an inventory of assets and problems, and determine general orientation and sources of information*

They started devising a plan, for which the community needed to know where precisely they stood. This involved conducting a 'self examination' to determine their strong and weak points, with questions (Q) and answers (A), as follows:

- Q: *What do we have in quantity and in quality?*
  A: *To determine this, we need to undertake an* **inventory of community assets**

- Q: *What do we have that is undesirable?*
  A: *To know this, we need an* **inventory of problems**

- Q: *Where do we want to be at the present and in the future?*
  A: **Goals** *must be clearly articulated*

- Q: *What kind of neighbourhood support, commitment, and skills does the community have?*
  A: *We have commitment from the people, but we also need to gain their* **willingness to dedicate time and effort to this end**

- Q:
  - *What information needs to be collected?*
  - *Where do we find it, and who will analyse it?*
  - *Is it possible to get information from another project like this one?*
  
  A: *Once we have inventories of our assets and problems, we* **need information** *supporting them and reflecting the actual situation. When we know* **what data is needed** *then we will find out where it is and/or how to generate it.*
  *It is indispensable to* **collect information** *from other urban settings. This can be done through Internet with further verbal communication*

- Q:
  - *How can we be sure that we are achieving some progress?*
  - *How often do we monitor progress?*
  - *How can we be sure that our efforts will be continued by the next generation?*
  
  A: **Indicators** *have to be selected.*
  *A* **plan to monitor progress** *towards the objective has to be established, and a frequency must be defined*
  *We cannot be sure that the next generation will follow our efforts; however, we* **must prepare everything** *for them to improve these efforts, considering that there will be new developments, needs, and tools that we do not know about*

- Q: *We reckon that there is a need for indicators to measure our progress towards our goal, but where do we get these indicators from, and how do we select them?*
  
  A: *We can get them from* **international organizations**, *like the United Nations, but also we must also* **develop our own indicators** *based on our unique local conditions. The selection of indicators can be made by expert opinion, or by using some mathematical tool as described in section 7.8.9.*

### C. Determine a general goal and establish a time limit

The citizens want a sustainable community and a sustainable neighbourhood according to the meaning of the term as explained. They need to achieve this goal within the city's present resources, since it is impossible to foresee what the city's main activity will be in the future.

They realize that, although tourism is called the 'industry without chimneys' meaning that it does not pollute, this concept is actually wrong, since they do realize that tourism creates a lot of contamination of the air, water, and soil, and sometimes with a deterioration of living conditions in the site, especially from the social point of view.

The community knows this, so they are thinking about **economic growth and** about **social progress,** yet also about the **environmental effects** that this activity brings. Because of this last, people realize that in order to make educated decisions they need hard figures based on facts to determine the extent of the benefits and drawbacks produced by tourism, and to quantify its importance for employment and commercial activity. For this reason, the community has agreed that the following aspects of the project are paramount, and they have established limits as well as targets for their compliance. These targets are:

*Economic growth*
This concept expresses their desire to **maintain the city's economic structure, but with increasing productivity, greater efficiency, and a little more diversification of its industrial and commercial base.**

Tourism is and will be by far the most important source of revenue, but it can be enhanced (by fighting the competition in nearby areas for tourism dollars), by the development of cottage industries, such as the manufacture of chocolate, goat cheeses, folk art, ceramics, and edibles (viz., home made pastry, jams and marmalade). Based on some modest beginnings, the community is also thinking of developing an informatics hub for the production of specialized software aimed at the hospitality industry. There is a keen interest in this last activity and in others related to the computer industry, such as installing a chip manufacturing plant, because the community knows the hardships they have suffered in the past when in certain years tourism did not meet their expectations. With these measures, they want to try to improve the **economic growth** and enhance the resilience (section 1.9) of the social fabric through a diversification of their industrial base.

They are also studying the possibility of getting natural gas supplied from a nearby pipeline to reduce power and heating costs, mainly in the hospitality industry and homes. Another project calls for the installation of a stationary fuel cell generating plant (section 5.3.5) to replace the actual powerhouse powered by diesel engines, and complementing this with energy generated by a battery of mini-hydro turbines using the abundant creeks in the area. The idea is to become self-sufficient in energy, as far as possible. This way they will reduce their footprint (section 1.6) and have an economic advantage by using free, clean and renewable energy.

*Social progress*
At present, a social stratification exists in the population seen through an economic gap seen in three very distinct sectors of the population:
- Wealthy people who own cottages and do not live in the area.
- Residents, shop and hotel owners.
- People in the service sector.

No-one can claim they live in a situation of complete social equity (although this is nevertheless widely held as a desideratum), but it is recognized that people in the service sector, constituted by low-income families originally from outside the region, are far below others in the social scale. These people provide services, such as help in hotels and restaurants and in some shops, and live in an area that is not yet integrated to town. Therefore, they are looking to correct this situation by promoting **social equity**.

*Environmental protection*
One area of their concern involves looking for solutions that aim to correct:

- Lack of enough capacity of the wastewater treatment plant (WTP);
- Degradation of the air quality, especially during the peak season, due to the high volume of cars from visitors;
- Emissions and noise generated by the old diesel power plant;
- Deterioration of trails made for walkers and horse riding, as well as for the use of off-road recreational vehicles, such as mountain bugs and buggies.

Therefore, they are after **environmental protection**.

### D. Establish definite objectives
After many weeks of debate in workshops with all interested parties, 14 objectives were selected (the order does not imply any rank), and the year 2003 was taken as a baseline or reference.

### Objectives:
1. Integrating City Hall decisions with citizens' input.
2. Decreasing air pollution by 25 percent in 3 years, and by 75 percent in 7 years.
3. Decreasing water pollution by 30 percent within 4 years, and by 95 percent within 8 years.

4. Achieving social justice regarding equity in housing, employment, health care, and education.
5. Increasing social- and industrial-base resilience (section 1.9).
6. Protecting natural assets.
7. Increasing efficiency in the use of energy.
8. Updating the educational system.
9. Updating the health care system.
10. Reducing travel time in the city.
11. Reducing dependency on other places (see 'ecological footprint' in section 1.6).
12. Integrating the city with the region.
13. Extending and enhancing green spaces.
14. Finding a solution for the disposal of solid waste.

### E. Create work commissions and establish responsibilities

Considering the selected objectives, 14 committees were formed, that is, one for each objective. Three to five people form each committee, and reporting is to the Sustainability Committee. The responsibility of each committee includes determining measures to adopt for achieving its own area's objective, and implementing and monitoring those measures after their approval by City Council.

### F. Set up measures, actions, plans and projects to be executed to accomplish the objectives

To reach these objectives, after 7 months of brainstorming meetings the committees have reached a consensus about **measures to take and plans to develop.** These measures strongly suggest considering trade-offs among them, which in fact was one of the hardest issues. These are the measures that apply to each of the objectives listed in D, above:

#### 1. Integrating City Hall with citizens' input

This involves developing a tight relationship between people and public service, with the creation of decentralized municipal offices to facilitate contacts with people, to listen to their complaints and suggestions, and encourage their participation in the town's government. This area has already shown tangible benefits. City Hall has established two branches in distant areas of the city to enhance citizens' more direct access to the municipal government and to express their opinions and voice their suggestions. At present, meetings between citizens and City Hall representatives are scheduled twice a month to discuss solutions to common problems affecting the city.

This initiative that started two years ago is proof that the system is functioning, since people work side by side with local authorities. Important issues are discussed here are, for instance, determining what part of the city budget will be managed by citizens for executing activities that solve some problems detected and proposed by them; for example, the need for better lighting of certain streets, installing traffic lights at an intersection, etc.

## 2. Decreasing air pollution by 25 percent in 3 years, and by 75 percent in 7 years

The solution involves limiting the area's number of vehicles. This is something tricky, since restricting the number of vehicles in circulation might have an effect on the area's economy as it could decrease the number of visitors; or perhaps not, on the other hand, since many people want tranquillity during their vacations. This measure would produce cleaner air but this limitation must be compensated for with good local transportation, and fair fares.

## 3. Decreasing water pollution by 30 percent within 4 years, and by a 95 percent within 8 years

It involves:
- Making it obligatory for hotels to collect and purify their sewage and use the treated water for irrigation.
- Restricting water consumption, and installing water metres for large consumers. There is not much that the hotels and larger-scale consumers can do to restrict water consumption by their guests, although a lot can be accomplished in their kitchens and laundry facilities. It will thereby be necessary to include the hotels managers in this decision, to learn about the feasibility of such a regimen, perhaps including an increase in the price of water.

## 4. Achieving social justice regarding equity in housing, employment, health care and education

- By developing a housing plan for low-income people, or by integrating their neighbourhood in the city (an agreement about the best option has not been reached yet). This is an extensive undertaking that involves not only infrastructure studies for services to be provided (water, sanitation, paving, health care, schools, etc.), but also surveys to be conducted to determine ability and capacity of people to pay for future services rendered as well as to repay loans obtained from local, regional or international institutions. Strong discussions were held on this, because if even everybody agrees that there should be social equity as a condition for sustainability, the question is, up to what point should the town help these people,

considering that other citizens also have rights. The answer in not simple, but one measure that can be effectively used is the Gini Index (see Internet references for Chapter 8), which is an indicator of income inequality.
- By heavily taxing gambling activities and using those revenues to improve the social condition of the city's inhabitants. This is another area that merits careful reflection, considering that some people are in favour of the undertaking, yet other people are also against it. It should contemplate the casino's location, the percentage of proceeds to be collected by local municipality, destination of those revenues, etc.

5. *Increasing social- and industrial-base resilience (section 1.9)*
Efforts are aimed at:
- Making attempts to blend the dedicated service sector by creating more commercial and industrial activities, thereby encouraging a diversified economy. In this respect, some advances have already taken place. The beauty of the place and its quietness have led a small group in the software industry to undertake such work. City Hall has already held informal talks with a multinational in the software business that has showed evidence of interest in installing a facility for producing and exporting specialized software. This early decision has been fuelled by a City Hall offer of a substantive tax reduction for 10 years, and for the construction of a building on a municipal lot for this foreign company.
- Commissioning a study to determine the regional economic multiplier (section 6.9) of the area generated by tourism and by the other potential undertakings. The region needs to know not only the economic benefits brought by tourism but also its social and environmental consequences. Besides, the region wants to know how vulnerable the local economy is, and also its social fabric, prior to undertaking a decrease in its main activity.

6. *Protecting natural assets*
Since the city already has heavily patronized recreational activities, such as scuba diving and snorkelling, trekking in the mountains, fishing, etc., there is a proposal (not approved yet), the decision was to exact a modest fee for users to pay for extra control and conservation of these areas. This is another issue everyone agrees to, since the fees exacted would be moderate and incorporated into the entrance fees to forest trails, scuba diving areas, mountain paths, etc.

7. *Increasing efficiency in the use of energy*
Adopted measures include:

- Commissioning the execution of a study on different energy options, but precluding the installation of thermal plants, except for fuel cells. It is believed that wind energy is not an option.
- Commissioning a study for the inventory of creeks with enough flow to provide hydroelectric generation.
- Analysing the costs, advantages and disadvantages of the gas pipeline.
- Committing City Hall to present a plan to reduce energy consumption in public buildings.

## 8. Updating the educational system
By:
- Preparing material for a survey, and hiring a company to perform it, to determine conditions of schools, projecting demand over the next 15 years for children, teachers, and school personnel. Once the survey is done, obtaining estimates for any repairs and/or construction, including furniture.
- Proceed with an estimate of salaries to be paid to all school personnel, and for maintenance.

## 9. Updating the health care system
Consideration was given to using part of the revenues generated by tourism to build new schools, a hospital and three health care centres. No disagreement about this arose; the only problem lies in determining what amount of revenues should be earmarked for this, and what type of installations are most needed. Information is currently being collected about the number of children projected for 15 years hence, and in the quality of services that will be provided by healthcare centres in the new hospital.

## 10. Reducing travel time in the city
This objective will be achieved by establishing a good public transit system to minimize pollution by private cars. Everyone agrees on this issue, although the economic feasibility of the project is under consideration since there could be insufficient patrons to justify an improved transit system. Regardless, this is being discussed, along with the mode of transportation, i.e. a tram, a fuel-cell bus, or a monorail. This last alternative is appealing since, besides its cleanliness, it offers another tourism incentive, although some stakeholders object to the undertaking on the grounds of its visual contamination.

## 11. Reducing dependency on other places (see ecological footprint in section 1.6)

A study will be commissioned to determine the city's footprint and the main areas where it can be reduced. Preferences should be given to energy by developing local energy sources.

## 12. *Integrating the city with the region*
- This issue is germane here because the current landfill is located near the village of Lyford.
- Analyse, with local authorities of Lyford and the hamlets of Deep Creek, Armstrong, Belleview, and Mallory, the feasibility of a joint effort to upgrade the roads and bridges connecting them, and with Meadow Forest as well.

## 13. *Extending and enhancing green spaces*
This will be done by:
- Establishing new bylaws regarding construction. This is another area under discussion since, from the economic point of view, it may be convenient for the city to make more lodging available, that is encouraging the construction of more hotels and camping facilities. However, pressure from hotel owners already operating within the area raises strong reactions. Besides, and although some vacant lots in prime locations are still available, it is felt by many that these lots would be better used for parks, recreation, green spaces, walking areas, etc.
- Establishing bylaws making it mandatory for hotels to have a certain percentage of green space in their facilities. This has already been approved, although it may be difficult for some hotels to comply with it, especially the older ones.
- Limiting the building of hotels and making the construction of underground parking lots for guest vehicles mandatory. This calls for undertaking the feasibility studies, as well as for preparation of new bylaws.

## 14. *Finding a solution for the disposal of solid waste*
- This is a difficult issue calling for a commission of a study for the final disposal of solid waste. Because the city is located in a valley it does not have strong winds, so an incinerator is not a preferred option, although if it were located in Narrow Fields it might be considered.

- A commission will also direct a study on the feasibility of using methane from the city landfill, and to make an estimate of the remaining available space for this purpose.

All of these measures translate into projects or alternatives to be pursued in a certain period, which will probably involve several years.

### G. Determine type of indicators needed

The community does not yet know how to establish measures of progress for their pursuits, but they do know that this is at the core of sustainable development. Steve Percy, vice-president of BP America, put the matter very well in a speech to the Society of Environmental Journalists in 1998: *"What gets measured gets managed. What gets communicated gets understood"* (quoted in Michael Keating: see Internet references for Chapter 8).

This points to their need for sustainable indicators for measuring certain conditions and relating them to others, most specifically to be able to tell technical and non-technical people **where they are, what the mistakes in the past have been, what things have not gone as expected, and, finally, how they can reach their ultimate goal.**

These indicators, which should be understandable to everyone, can tell the community:

1. Where they stand at present, in comparison with their starting point or baseline.
2. How much they have advanced on some issues, or how much ground they have lost on others.

An example of the first could be, for instance, how the hospitality and catering industry is performing regarding visitors, and the length of stay in town of these last, compared with the situation 10 or 15 years ago. An example of the second might be the degradation of the air over the same period. With these indicators they can ascertain **how far** or **how close** they are to their established goals, and which is very important. They can ascertain if they are within sustainability resilience, that is, near or far from the established thresholds.

In other words, with these indicators of sustainability they will be able to manage their resources and get information about the necessary steps to continue.

### H. Choose indicators to measure progress of actions in (F) and of targets and goals established

At this stage, all goals have presumably been established, plans and undertakings are defined, and costs are agreed upon. The next step is to make a schedule, that is, to define when each task will be completed, to find out the interrelations between the different undertakings, to set up milestones for certain activities, to relate execution with available funds, to define responsibilities for control, and to determine indicators to measure progress.

All of these activities are, of course, important, and require no comment. However, the last one deserves a separate study. As mentioned in section 6.1, indicators are used to let us know about advances or progress in a given activity. For instance, an indicator could be the ratio between household income and mortgage or rent paid. One of the tasks of the Sustainable Committee involves the definition of which indicators will be used.

This is not an easy task, since experience shows that the total number of indicators could be between 100 and 200. If in this case they have, say, an initial list of 128, this is still too many, so a way has to be found to reduce their number to a final list amounting perhaps to no more than 15 or 20. The reason for this abrupt reduction is that nobody can handle, understand, or let alone interrelate, such a large number of indicators, and if the whole set of 128 indicators is used, it will very likely produce more confusion than results.

The initial set of 128 indicators presumably includes those pertaining to all areas of interest — that is, involving social, economic, environmental and sustainable indicators. So the problem becomes complicated because the community needs a final list of, say, 20 indicators but with a balanced representation of the above-mentioned sectors.

This is not a minor problem since, as anyone can guess, thousands of possible combinations of indicators exist that satisfy the restrictions imposed. Fortunately, there is a mathematical tool that can be utilized to address this problem, and its use has been explained and illustrated in section 7.8.9.

Fine, but are they now out of the forest? Not really, because indicators measure a variation or a trend of something regarding some value that is called a threshold (section 7.8.5) and, in the case of the above mentioned example of the ratio between household income and mortgage or rent, the community has to establish a value, and then measure progress (or its lack) with this indicator regarding this threshold.

About selecting indicators, many organizations such as the UN Department of Economics and Social Affairs (see Internet references for Chapter 8), have been working on preparing lists of them. However, it is necessary to remember that indicators are not universal and that each city should make its own listing. For instance, the City of London has prepared a very comprehensive listing that can be consulted (see Internet references for Chapter 8). According to this publication the objectives are:

Taking responsibility.
Developing respect.
Managing resources.
Getting results.
Making up what is known as 'the 4Rs' (Reduction, Reuse, Recycling and Recovery)

## Indicators for this project
Regarding indicators, Meadow Forest has prepared a list of 60 indicators as follows. See Table 8.2.

*Table 8.2*  **Indicators prepared by the city of Meadow Forest**

| # | Goals | Base line conditions | Target results | | |
|---|---|---|---|---|---|
| | | 2003 | 2005 | 2007 | 2009 |
| 1 | *Integrating City Hall with citizens' input* | | | | |
| | Number of joint meetings per year | | | | |
| | Percentage of proposals from citizens accepted and implemented | | | | |
| | Number of City Hall employees per 1,000 inhabitants | | | | |
| | Percentage of city budget allocated to payroll of municipal employees | | | | |
| | Public decentralization, measured by the number of people that visit the three municipal branches around the city, including downtown offices | | | | |
| 2 | *Decreasing air pollution by 25% in 3 years, and by 75% in 7 years* | | | | |
| | Total number of city buses | | | | |
| | Total number of municipal trucks | | | | |
| | Percentage of city buses converted to biofuel | | | | |
| | Percentage of NOx in selected parts of the city | | | | |
| | Percentage of particulate in selected parts of the city | | | | |
| | Number of days per year when there is a clear view of Sunrise Mountain peak from downtown | | | | |
| | Number of accidental releases of noxious gases into the atmosphere | | | | |
| 3 | *Decreasing water pollution by 30% within 4 years, and 95% within 8 years* | | | | |
| | Percentage of $BOD_5$ in the lake | | | | |
| | Visibility (turbidity) of water in the lake | | | | |
| | Average number of trout in seven stations along the river | | | | |
| | Biological content analysis of chemicals in fish | | | | |
| | Chlorine in water (potential source: the pulp mill plant located 75 km upstream, in Salmon River, and discharging into the lake) | | | | |
| | Number of clandestine dumpings of waste or wastewater into the lake or rivers | | | | |
| 4 | *Achieving social justice regarding equity in housing, employment, health care, and education* | | | | |
| | Percentage of home owners | | | | |

|   |   |   |   |   |   |
|---|---|---|---|---|---|
|   | Ratio homeless people/population |   |   |   |   |
|   | Ratio car ownership/population |   |   |   |   |
|   | Ratio household income/payment of rent or mortgage |   |   |   |   |
|   | Value of Gini index |   |   |   |   |
|   | Unemployment percentage |   |   |   |   |
| 5 | *Increasing social- and industrial-base resilience* |   |   |   |   |
|   | Percentage of the top five industries responsible for the 80% of income (measured in amount of wages paid per annum) |   |   |   |   |
|   | Percentage of High School and Technical Institute graduates who stay in town |   |   |   |   |
|   | Percentage of out of town people who work during the tourist season |   |   |   |   |
|   | Economic multiplier for total output, income and employment in the region |   |   |   |   |
| 6 | *Protecting natural assets* |   |   |   |   |
|   | Maintenance of urban forest measured in hectares |   |   |   |   |
|   | Number of wild and fowl life in the city and surrounding area |   |   |   |   |
|   | Number of fines for polluting the forest trails, walking off trails, feeding animals, etc. |   |   |   |   |
|   | Conservation fees collected |   |   |   |   |
| 7 | *Increasing efficiency in the use of energy* |   |   |   |   |
|   | Energy consumption in public buildings |   |   |   |   |
|   | Energy consumption in the streets |   |   |   |   |
| 8 | *Updating the educational system* |   |   |   |   |
|   | Number of children per school room |   |   |   |   |
|   | Number of children per teacher |   |   |   |   |
|   | Number of school buses |   |   |   |   |
| 9 | *Updating the health care system* |   |   |   |   |
|   | Ratio of total number of hospital beds and population |   |   |   |   |
|   | Average waiting time for a surgical operation |   |   |   |   |
|   | Average waiting time for emergency service |   |   |   |   |
|   | Number of ambulances |   |   |   |   |
|   | Number of patients with respiratory diseases |   |   |   |   |
|   | Number of underweight newborns |   |   |   |   |
|   | Number of doctors per 1000 inhabitants |   |   |   |   |
| 10 | *Reducing travel time in the city* |   |   |   |   |
|   | Average travel time in North - Southeast corridor |   |   |   |   |
|   | Average travel time between the northern suburbs and downtown |   |   |   |   |
|   | Average travel time in the cornice road |   |   |   |   |
| 11 | *Reducing dependency on other places* |   |   |   |   |
|   | City, including region footprint |   |   |   |   |
|   | Percentage of food brought from outside the region |   |   |   |   |
| 12 | *Integrating the city with the region* |   |   |   |   |

|    |                                                                                   |  |  |  |  |
|----|-----------------------------------------------------------------------------------|--|--|--|--|
|    | Km of paved roads in the region                                                   |  |  |  |  |
|    | Km of paved road connecting with Lyford village and the hamlets                   |  |  |  |  |
| 13 | **Extending and enhancing green spaces**                                          |  |  |  |  |
|    | Km of bike ways                                                                   |  |  |  |  |
|    | Ratio of $m^2$ of green space per inhabitant                                      |  |  |  |  |
|    | $Km^2$ of parks in the city                                                       |  |  |  |  |
|    | Number of people using the lineal park along the Salmon river                     |  |  |  |  |
|    | $M^2$ of protection work along the river banks                                    |  |  |  |  |
| 14 | **Finding a solution for the disposal of solid waste**                            |  |  |  |  |
|    | Percentage of domestic waste recycled at origin                                   |  |  |  |  |
|    | Percentage of domestic waste recycled at the landfill                             |  |  |  |  |
|    | Number of people who use the pay-per-weight system                                |  |  |  |  |
|    | Opinion of people about the construction of an incinerator                        |  |  |  |  |

Many other effects are not considered in this example, for instance:

- Number of tourists shopping for crafts, postcards, fishing gear, rental bikes, scuba diving equipment, etc.
- Entertainment: Manifested by the number of tourists to the two existing movie theatres, a chairlift, arcade, theatres, etc.
- Increase in the local GDP.
- Increase in taxes collected by City Hall.
- From the social point of view, the school is getting crowded since more resident's children are attending.
- Increase in waiting time for hospitalization and surgery.
- Percentage of people annoyed with the increase of poor neighbourhoods where the seasonal workers live.
- Relationship between existing sewage treatment capacity and increasing wastewater production because of tourism.

In each of these cases, it is possible to derive indicators to take into account the effect of tourism, such as the percentage increase in revenue from taxes.

### I. Develop a schedule detailing on a bar chart each action, from start and finish, listed in (F), including their interrelationships and sequence

The construction of such a bar chart schedule — that is, horizontal bars representing duration of actions in a time scale — is fundamental for keeping track of the whole project, to know who is responsible for each action, the times of launch and completion, interrelations, and, especially, progress. A reporting tool is essential for project control. This was already analysed in section 7.5.

**J. Determine the economic impacts that tourism and the other undertakings will have on the economy, the environment and society**

Section 6.9 illustrates the computation of a multiplier. Generally, this multiplier is calculated for output, income, and employment. In the case of tourism, there is ample information on Internet about the value of this indicator for tourist destinations. However, it must be taken only as a very broad reference, and to provide an idea of orders of magnitude, since each location has its particular characteristics, even within a same country, such that information from one place usually cannot be considered for another.

## 8.4 Impacts created by tourism

Since tourism is the main activity of the city, the Chamber of Commerce has also determined, as a support document, the impacts produced by tourism as seen in Table 8.3. This provides valuable information regarding taxes and demand for services. Besides, the creation of employment in the service sector gives a good advantage to the hospitality and catering industry for their discussions with City Hall.

*Table 8.3*          **Direct impacts produced by tourism**

|   | Areas affected | Indicators | Units | Comments, based on statistical figures in the last 12 years |
|---|---|---|---|---|
| Economic activities |   |   |   |   |
|   | Accommodation | Number of rooms in hotels, bungalows and B&B | Number | The comparison of number of visitors with number of rooms used in the three month period of seasonal activity, shows that one visitor needs 1/3 of a room in hotels and bungalows and 1/2 in B&Bs. There are no trailer facilities, yet City Hall promotes the idea of camping and trailer facilities — which is strongly opposed by the hospitality and catering industry |
|   | Catering | Number of | Number | Statistics show that 42% |

|  |  | tables in restaurants |  | of visitors take their meals in restaurants. The rest shop for food at the supermarket and eat at home in the bungalows, and/or eat on the beach by the lake or rivers |
|---|---|---|---|---|
|  | **Effects** | **Indicators** | **Units** |  |
| **Environment** |  |  |  |  |
|  | Degradation in air quality | Production of NOx | $grNOx/m^3$ | Statistics show that there is a relationship of one car for 3.96 tourists. Each car is assumed to spew about 0.0025 mg of $NOx/m^3$-km |
|  | Increase of solid waste on streets and sidewalks | Production of waste | gr/person-day | There is an average of 100 gr. of solid waste (paper, cigarettes butts, food, etc) in the streets and sidewalks per person and per day. City Hall aims to lower this to 75 gr./person-day through the placing of garbage bins, and an environmental campaign |
|  | Deterioration of soil because activities involving horse riding and walking, and the utilization of dirt bikes, snowmobiles, buggies | Loss of vegetation and erosion |  | This is being appraised at the time of this study. The appraisal wants to determine the carrying capacity (section 6.12) of the affected areas. |
|  | **Effects** | **Indicators** | **Units** |  |
| **Infrastructure** |  |  |  |  |
|  | Increase in water consumption | Average water consumption | Litres/person l day | A new plant has enough capacity to treat raw water and render it potable. Average consumption is about 425 litres/person-day |
|  | Increase in wastewater production | Average wastewater production | Litres/person-day | The old WTP does not have the capacity to process all wastewater during the peak period. Wastewater production is about 202 litres/person-day |
|  | Increase in use of public | Ridership | Number | A limited transport system can transport, as |

|  | | Effects | Indicators | Units | |
|---|---|---|---|---|---|
|  | | transport | | | a maximum over the three month period, 26,000 people |
| Social | | | | | |
|  | | Creation of seasonal job opportunities in the service sector | Number of jobs created | Number of visitors | Statistics show a steady relationship between visitors and number of people employed in service sector (accommodation, restaurant, supermarket, shops, guides, etc.). This ratio amounts to 0.4 persons/visitor. |
|  | | Increase in population in the service sector | Increase in population | % / year | The resort community is rather isolated, and people providing these services come from other places, often distant, so they live in town. The service sector population is increasing at a rate of 2.6%, considering the employee and his/her family, and this population explosion has caused social problems in town. At present, this population is increasing at a greater rate than the demand for services |

Figure 8.1 depicts the chain effect of a driving movement as tourism, for the hospitality industry, and considering car activities.

Row A: Points out that **TOURISM** has a direct impact on the following activities, and each one is assumed to have the same importance or weight, and assigned (1), (An oval above each subject indicates the area it belongs to):

- *Hospitality industry* (hotels and motels, lodging visitors).
- *Catering industry* (restaurants, bars, cafeterias, etc., serving visitors).
- *Land* (visitors purchase land and build houses; a developer builds a go-karting circuit venue; a public swimming pool; etc.).
- *Number of cars* (because of tourists coming to the area). Since this subject of study involves a particular area, the activity of buses carrying passengers to and from the site is not considered, nor is

Nolberto Munier 383

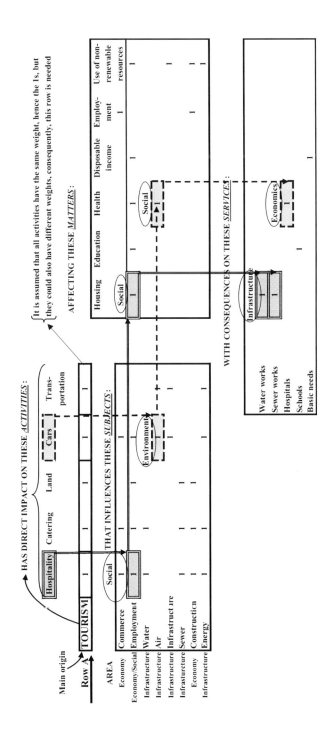

Figure 8.1  Tourism matrix and paths

- account taken of the influence on commerce that these trips produce off the area.
- *Transportation* (local buses).

To select just one activity that is affected, say, the **hospitality** industry (shown as a shadowed dotted rectangle), its column shows that it exerts an influence on the following areas (denoted by the numeral 1):

- *Commerce* (economy area; tourists staying in hotels buy souvenirs, clothing, prescriptions, shoes, etc.).
- *Employment* (economy/social area; there is direct employment for hotels personnel, for cleaning personnel, receptionists, cooks, etc.).
- *Water consumption* (infrastructure area; water consumed by hotel guests and used for cooking, laundry, etc.).
- *Sewer* (infrastructure area; sewage originated by guests, kitchens, laundry, etc.).
- *Construction* (economy area; there could be an increase in the number of rooms, construction of new hotels, construction of swimming pools, tennis courts, etc.).
- Use of *energy* (infrastructure area; consumption by hotel guests, elevators, lighting, kitchens, etc.).

Let us choose now just one subject, such as **employment**, which affects the following matters:

- *Housing* (because people can rent, buy houses, and improve them).
- *Education* (since there could be some money to be spent in learning a new trade, getting a degree, etc.).
- *Health* (for people will have access to prescriptions, better food, and to a more healthy life style)
- *Disposable income* (there will be more money to buy a car, a TV, clothing, or many other things).
- *Employment* (there could be more employment opportunities when people can afford to contract some services, as a gardener, a house helper, etc.)
- *Use of non-renewable resources* (because people will use more water and electricity).

Let us choose now any of the above-mentioned matters, such as **housing:**

Life in a household will have consequence on the following services:

- Water
- Sewage

Dotted arrows follow this circuit from 'Hospitality' to 'Sewage', and arrows in dashes from 'Cars' to 'Economics'.

If another tourism activity is analysed, such as the number of cars in town, arrows in dashes show that they influence subjects like air — in turn affecting a matter such as human health, which has consequences on the health service (hospitals, etc.).

## 8.5 How to measure impacts

As seen in Table 8.3, the community has identified the impacts and their effects on the economy, the environment, and society. Now, the need arises to evaluate them, a task that, as may be expected, is not an easy one. As mentioned before, the Sustainable Committee has requested the calculation of the **economic regional multiplier,** with tourism as a driving force, and has determined areas of their interest. These areas are shown below, and are quantified in Table 8.4.

Hospitality (hotels, motels and cabins)
Catering (restaurants, bars and fast food)
Transportation (urban)
Retail sales
Wholesale sales
Construction
Agriculture
Communications
Real estate
Utilities (electricity, gas, water and sewage)
Health services
Households
Miscellaneous

Four different variables are considered: Output, Income, Employment, and value added. For example, only the multiplier for output has been computed using I/O analysis (see Glossary), as follows:

> Driving force: Tourism, measured in percentage of increase of visitors per year. In this example, it is 1.4%/year

*Output:* Total value is 2.41.

> This means that for every $1 entering the region due to tourism, an additional $1.41 is spent in local business. A look at the values in the output column highlights the importance of activities such as construction and hospitality (shown in bold). The first one has to do with the construction of new hotels, the expansion of old ones, and the construction of private residences for tourists and permanent residents. The second one is, of course, about tourists staying in the area's hotels.

There is also a need to measure the impacts that tourism has on other areas, as follows:

*Environment*
Indicators are built with information from:
Quality of air is automatically measured at certain spots in town, and the information is sent to a control centre.
Quality of water is measured through samples taken periodically
Solid wastes on the street is calculated by weighing the tons of garbage collected over a certain period.
Soil deterioration is done through visual inspection, as well as through the state of vegetable species.

*Infrastructure*
Indicators are constructed with information provided by different sources. For instance, if we want to determine water consumption per capita, information is needed from the Water Department about the volume of water used during a certain period, and also data from the City Hall regarding the number of visitors during such a period. With these two data it is possible to create the indicator.
Another indicator such as the increment in electric utilization generated by each visitor can be developed with said information from City Hall, together with data from the Power Department regarding total consumption during the same period. This indicator could help in planning the city energy supply, when the future number of visitors is estimated.

*Social*
Some information is reflected in the multiplier; there are, however, many other indicators, such as construction permits in certain areas, disposable income, household ratios of income/rent, etc.

*Table 8.4*     Regional economic multipliers generated by tourism

| Industry | Output (Total change in local sales resulting from $1 in sales outside the region) | Income (Total increase in income resulting from a $1 increase in income) | Employment (Total change in employment resulting from a change of basic employment) | Value added (The difference between the value of the output and the value of the inputs purchased) |
|---|---|---|---|---|
| Hospitality | 0.34 | | | |
| Catering | 0.22 | | | |
| Transportation | 0.05 | | | |
| Retail | 0.12 | | | |
| Wholesale | 0.11 | | | |
| Construction | 0.46 | | | |
| Agriculture | 0.02 | | | |
| Communications | 0.01 | | | |
| Real estate | 0.17 | | | |
| Health services | 0.08 | | | |
| Electricity, gas, water and sewage | 0.25 | | | |
| Households | 0.4 | | | |
| Miscellaneous | 0.18 | | | |
| Total | 2.41 | | | |

## 8.6  Conclusions from studies

After a time, measures were implemented and the first report (Table 8.5) was published. An analysis conducted to appraise the state of the basic infrastructure also shows that if gambling were introduced in town, the number of visitors would increase from 1.4 percent to 5 percent — that is, from 89,000 over a three-month period to 93,500. The consequences are detailed in the same Table.

## Social sector

In the service sector, an excess of available personnel was discerned in relation to the actual needs of the hotels and restaurants. Besides, since these workers are from outside the region, they come with their families, increasing unemployment rates due to the presence of the family's non-working members, and creating overcrowding problems at the school and hospital.

## Environment sector

The most important items here are the capacity of the sewer and of the wastewater treatment plant. This last is very unsuited to handling the load since it effectively processes only 68 percent of the wastewater. This fact effectively reduces the total treatment time, so the quality of the treated sewage that is discharged is deteriorated — defeating the main reason for establishing an objective regarding water contamination.

Regarding the atmosphere, studies also show that the pollution caused by visitors' cars exceeds by 28 percent the allowed threshold. This, in turn, might make it mandatory to limit traffic downtown, and perhaps prompt the development of a pedestrian area. Regarding transportation, it turns out that local transport is a little over extended, since it works with an idle capacity of 6 percent.

*Table 8.5*                  **Impacts report**

|  | Number of visitors: 89,000 | Number of visitors: 93,500 | Comments |
|---|---|---|---|
| **Social sector** | | | |
| Employment | 2.7% unemployment | 2.5% unemployment | These figures show improvements regarding other years |
| Demographics | | | A comparison was made between the number of people in the service sector and the number of visitors. That ratio has already been established at 0.4. This relationship will hold if service sector growth keeps pace with the increase in tourism, whatever the number of visitors |
| **Commercial sector** | | | |
| Rooms in hotels | 15% deficit | 19% deficit | This has already produced problems with visitors |
| Rooms in bungalows | 13% deficit | 17% deficit | |
| Rooms in B&B | 38% deficit | 41% deficit | This activity has not been regulated yet, so the quality of service is unknown |
| Activity in restaurants | 9% surplus capacity | 4% surplus capacity | There are no problems with this activity |

| Environmental sector | | | |
|---|---|---|---|
| Waste on streets | 25% more waste than targeted | 28% more waste than targeted | The town does not have separate systems for sewer and stormwater. Due to the increase of waste on the streets, this waste is carried out by rains into the sewer, blocking smaller ducts |
| Sewage | 38% deficit | This is a very serious problem. There will be a 41% deficit | Trunks were not designed to handle the extra flow, so in heavy rains they overflow through the manholes |
| Water treatment plant | Can only process 68% of total wastewater | If no action is taken, the system will be able to process only 59% of the load | Insufficiently treated water discharges into the river. Even if the rapid and steep current produces good aeration, this is not enough, and the water will continue to discharge into the lake with a higher than normal $BOD_5$ |
| Potable water | 22% surplus capacity | There will still be a surplus capacity of 16% | The system will be working at full capacity for 110,000 visitors, and will have a deficit of 3% for 112,000 visitors |
| Air contamination | 28% more that threshold | 31% more than threshold | Residents and visitors alike are complaining about air pollution |
| Infrastructure sector | | | |
| Transportation | 6% idle capacity | 1% idle capacity | |
| Parking | No more capacity on streets | No more capacity on streets | This is a significant problem. Because of this lack of space, people are starting to park on sidewalks |

### K. Establish a reporting mechanism to communicate results to people and for feedback

Reporting was discussed in section 6.14 and Table 6.10, and can be applied to this case in the same format. Naturally, reporting is more than to inform people, since their feedback is necessary to make readjustments of the schedule or targets when necessary.

The agreement was reached to publish the reports on the Internet.

**Internet references for Chapter 8**

**Source:** Minister of Environment, Ontario (2001)
**Author:** Michael Keating
Title: *Review and analysis of best practices in public reporting on environmental performance. A report to Executive Resource Group*
Comment: Of particular importance is the emerging trend in environmental performance reporting and indicators to measure change and progress.
Address:
http://www.ene.gov.on.ca/envision/ergreport/downloads/report_paper9.htm)

**Source:** Urban Ecology Coalition (Minneapolis) (1999)
Title: *Neighbourhood sustainability indicators guidebook*
Address:
http://www.crcworks.org/crossroads/guide.pdf

**Source:** English Tourism Council (2002)
Title: *National sustainable tourism indicators*
Address:
http://www.crcworks.org/crossroads/guide.pdf

**Source**: New Economics Foundation (1998)
Title: *Communities count! A step-by-step guide to local community indicators*
Address:
http://www.neweconomics.org/gen/

**Gini Index**
**Source**: Development Data Group, The World Bank (2002), World Development Indicators 2002 online
(see http://www.worldbank.org/data/)Washington, D.C.: The World Bank.
Comment: Explanation of how the Gini index is computed.
Address:
http://www.earthtrends.wri.org/text/ECN/variables/353notes.htm

**Source**: UN Department of Economic and Social Affairs – Division for Sustainable Development (2003)
Title: *Chapter 7: Promoting Sustainable Human Settlement Development*
Comment: Explains the rate of growth of urban population indicators, its relevance to sustainable/unsustainable development, and linkages to other indicators.
Address:
http://www.un.org/esa/sustdev/natlinfo/indicators/indisd/english/chapt7e.htm

**Source**: London Sustainable Development Commission (2003)
Title: *Consultation on sustainable development framework indicators – February 2003*
Address:
http://www.london.gov.uk/mayor/sustainabledevelopment/docs/lsdc_consultation.pdf

**Source**: OECD (2001)
Title: *Using the pressure-state-response model to develop indicators of sustainability – OECD framework for environmental indicators*
Address:
http://destinet.ewindows.eu.org/aManagement/2/OECD_P-S-R_indicator_model.pdf

**Author**: G. Reza Arabsheibani and Alvaro Delgado – Aparicio Labarthe (2002)
Title: *Tourism multiplier effects on Peru*
Comment: Very good paper detailing de use of I/O tables to calculate forward and backward linkages, output, income, and employment multipliers.
Address:
http://www.ucb.br/prg/economia/Revista/TOURISM%20MULTIPLIER%20EFFECTS%20ON%20PERU.PDF

**Source**: Vermont Department of Tourism and Marketing
Title: *IMPLAN methodology for the study of the impact of tourism on the Vermont economy*
Comment: Valuable and concise analysis of the use of I/O models to compute multipliers. Mathematically-inclined readers will find a clear example of the Input-Output transaction table.
Address:
http://www.uvm.edu/~snrvtdc/publications/implan_method.pdf

**Author**: Vivian C. Choi
Title: *On the multiplier effect*
Comment: This paper provides a table with the different tourism income multipliers for different countries and region. It usefully provides an idea of order of magnitude and range of variation.
Address:
http://www.admin.gov.gu/commerce/multiplier.htm

**Authors**: J. Fletcher & S. Wanhill (2003)
Title: *Input-Output study of Gibraltar*

Comment: One particular interest of this study lies in the population size it affects (about 28,000 inhabitants in 2002), which could be very valuable for tourist areas of a similar size and resources.
Address:
http://www.gfsb.gi/archive/input.htm

**Source**: Macaulay Land Use Research Institute, with Geoff Broom Associates (2000)
Titled: *Forest's role in tourism - Phase I – Final report - Research contract for the Forestry Commission*
Comment: This interesting paper deals with a subject rarely found in the literature: how a region's forest contributes to the economic output that is generated by tourism.
http://www.forestry.gov.uk/website/pdf.nsf/pdf/tourism.pdf/$FILE/tourism.pdf

**Source**: Australian Bureau of Statistics (2002)
Title: *Indirect economic contribution of tourism, 1997-98 to 2000-01*
Comment: As the title implies, this paper deals with the indirect effects of tourism, and the most important methods for estimating it.
Address:
http://www.abs.gov.au/Ausstats/ABS@.nsf/94713ad445ff1425ca25682000192af2/bdc075a8b9b27d7dca256c10000157d9!OpenDocument

**Author**: Murray Peterson – Massey University (2001)
Title: *Environmental performance and eco-efficiency of the New Zealand tourism sector*
Comment: Excellent paper regarding the effects of tourism on the environment using Life Cycle Assessment.
Address:
http://eerg.massey.ac.nz/adobefiles/wordAM2.1_Patterson.pdf

# APPENDIX

## A.1 The Zeleny method for determining weights

This method, developed by Milan Zeleny (Zeleny, 1982), can compute weights for criteria based on the data provided by the performance matrix. Zeleny's method considers both the weight derived from the values in the matrix, and those weights assigned by expert opinion, but the latter modified with the weights obtained from the matrix. To do this, Zeleny's method is based on Information Theory (see Glossary) to finds out which criterion has the minimum **entropy** (see Glossary), that in this context has the meaning of 'discrimination'. The lower the entropy for a criterion, the better, because the greater the discrimination obtained between projects. According to Zeleny, entropy of information reflects, *"...the average amount of intrinsic information conveyed by a given information source"*

To understand this better, assume for instance the following example with five alternatives and only one criterion, such as aesthetics.

|           | Alt. 1 | Alt. 2 | Alt. 3 | Alt. 4 | Alt. 5 |
|-----------|--------|--------|--------|--------|--------|
| Criterion |        |        |        |        |        |
| Aesthetics | 7 | 7 | 7 | 7 | 7 |

That is, all alternatives are evaluated as the same (on a 1-10 scale, and the higher the better) regarding the criterion 'aesthetics'.
Is it possible to select an alternative <u>based on this criterion</u>? Obviously not, since the value or appraisal that each alternative has regarding this criterion is the same for all alternatives. In other words, the amount of information obtained from this **criterion is zero,** that is a **minimum,** and consequently there is **maximum lack of knowledge** for the decision-maker. It also corresponds to the **maximum entropy**, and with a value of **1.61**.

Suppose now that there is another criterion such as durability, also measured on the 1-to-10 scale, but with these values:

|           | Alt. 1 | Alt. 2 | Alt. 3 | Alt. 4 | Alt. 5 |
|-----------|--------|--------|--------|--------|--------|
| Criterion |        |        |        |        |        |
| Durability | 5 | 2 | 7 | 3 | 6 |

It is possible then to gauge again the alternatives but using this criterion now. Consequently, alternative 3 should be chosen, because it has the highest mark. Therefore, this criterion **does provide** some information that can be measured using its entropy, which value is **1.52**, which allows one to say that this criterion has a higher weighting that the aesthetic criterion.

Assume now that another criterion calls for expressing how people measure satisfaction, considering each alternative.

|  | Alt. 1 | Alt. 2 | Alt. 3 | Alt. 4 | Alt. 5 |
|---|---|---|---|---|---|
| **Criterion** |  |  |  |  |  |
| People's satisfaction | 3 | 3 | 3 | 3 | 10 |

Obviously, there is a marked preference for alternative number 5, and then the computed entropy of the criterion is **1.45**.

Finally, suppose that there is another criterion, such as the opinion from other communities with experience of the same five alternatives.

|  | Alt. 1 | Alt. 2 | Alt. 3 | Alt. 4 | Alt. 5 |
|---|---|---|---|---|---|
| **Criterion** |  |  |  |  |  |
| People satisfaction | 1 | 10 | 1 | 1 | 1 |

In this case, there is a unanimous rejection for all of them, but for number 2, so the entropy for this criterion is **0.99**. This is the **minimum entropy,** which gives us the **maximum of information**.

It can be seen how the entropy has been decreasing from the case where all the values were identical to the last one, where there is a large discrepancy, and this is the concept used to find weights for criteria.

Entropy is a notion that comes from Thermodynamics, and is loosely understood as 'disorder'. For instance, water with molecules at low speed, as in ice, has low entropy. When ice heats up, its water molecules increase their velocity and, as one bumps or collides into another, their disorder increases, liquid water is produced, and its entropy is augmented. If the heating continues, its water molecules will step up their velocity and more violent collisions will occur — leading to more disorder — the entropy is augmented, and steam is produced.

Information Theory (IT), developed by Claude Shannon (Shannon, 1948), relates to the amount of information in a message, which is measured in 'bits'. To quantify this amount of information he used the entropy concept and the following formula, where 'S' denotes entropy:

$$S = - \sum_{i=1}^{n} p_i \ln p_i$$

Where:

$p_i$ = Probability of occurrence
$\ln$ = Natural logarithm of the probability of occurrence

Zeleny's method is applied here to the example in section 7.8, and starts with data contained in Table 7.4, which is reproduced in Table A.1.1. (there is a slight difference in zero values with Table 7.4, in order to avoid the division by zero).

Table A.1.1            Performance matrix (Table 7.4)

|  |  |  | Solar based | Small hydro | Wind energy | Bio-mass |
|---|---|---|---|---|---|---|
|  | Cost | $/kWh | 0.048 | 0.037 | 0.044 | 0.045 |
| **Attributes** | **Area** | **Units** |  |  |  |  |
| Generation potential | Economics | MW | 1.6 | 1.82 | 3.1 | 2.1 |
| Land use | Environmental | ha | 40 | 3 | 29 | 45 |
| Dependability | Economics | hours/day | 11 | 24 | 9 | 24 |
| Environmental impact | Environmental | % | 3 | 5 | 10 | 20 |
| Human impact | Social | % | 0.01 | 0.01 | 5 | 12 |
| Social impact | Social | % | 2 | 0.01 | 5 | 12 |
| Maintenance cost | Economics | $/kW | 0.00043 | 0.00017 | 0.00072 | 0.00068 |
| Return on investment (ROI) | Economics | % | 4.8 | 6.2 | 5.8 | 5.1 |
| Sharing land with other uses | Economics | ha | 0.01 | 41.6 | 28 | 0.01 |
| People's opinion | Social | No. | 4 | 9 | 7 | 3 |

It is necessary first to normalize the coefficients. This is done by dividing each value on each row of Table A.1.1 by the sum of the values on each row. These values are shown in Table A.1.2.

Table A.1.2  Sum of each row

| Attributes | Sum for each row |
|---|---|
| Generation potential | 8.62 |
| Land use | 117 |
| Dependability | 68 |
| Environmental impact | 38 |
| Human impact | 17.02 |

| | |
|---|---|
| Social impact | 19.01 |
| Maintenance cost | 0.002 |
| Return on investment (ROI) | 21.9 |
| Sharing land with other uses | 69.62 |
| People's opinion | 23 |

Table A.1.3 is the result of this division.

*Table A.1.3*    Criteria weights determination

| | Solar based | Small hydro | Wind energy | Bio-mass | Entropy | Diversification | Weight |
|---|---|---|---|---|---|---|---|
| | 0.048 | 0.037 | 0.044 | 0.045 | $S_j$ | $d_j$ | $w_j$ |
| **Attributes** | | | | | | | |
| Generation potential | 0.19 | 0.21 | 0.36 | 0.24 | 1.58 | -0.58 | 0.03 |
| Land use | 0.34 | 0.03 | 0.25 | 0.38 | 2.78 | -1.78 | 0.10 |
| Dependability | 0.16 | 0.35 | 0.13 | 0.35 | 1.17 | -0.17 | 0.01 |
| Environmental impact | 0.08 | 0.13 | 0.26 | 0.53 | 2.07 | -1.07 | 0.06 |
| Human impact | 0.00 | 0.00 | 0.29 | 0.71 | 7.44 | -6.44 | 0.37 |
| Social impact | 0.11 | 0.00 | 0.26 | 0.63 | 6.99 | -5.99 | 0.35 |
| Maintenance cost | 0.22 | 0.09 | 0.36 | 0.34 | 2.27 | -1.27 | 0.07 |
| Return on investment (ROI) | 0.22 | 0.28 | 0.26 | 0.23 | 1.32 | -0.32 | 0.02 |
| Sharing land with other uses | 0.00 | 0.60 | 0.40 | 0.00 | 0.52 | 0.48 | -0.03 |
| People's opinion | 0.17 | 0.39 | 0.30 | 0.13 | -1.08 | -0.08 | 0.00 |
| | | | | | Sum diversification | 17.20 | |

The values for entropy are found in the column "Entropy".
As an example of calculation, entropy is computed for the attribute 'Generation potential':

Entropy:
$$S = -(0.19 \times \ln(0.19) + 0.21 \times \ln(0.21) + 0.36 \times \ln(0.36) + 0.24 \times \ln(0.24)) =$$
$$= 1.58$$

A diversification is computed as follows:
  D = 1 − s = − 0.58

Then a weight is derived by dividing each value into the total value, i.e.
-17.20
Thus, the weight is:

  W = -0.58 /-17.20 = 0.03

If expert opinion establishes cardinal preferences of alternatives in percentages, these subjective weights can then be multiplied by the objective weights found, and then one divides each product by the sum of the products for all criteria. In other words, it finds an average between the product of objective and subjective weights divided by the sum of these products.

### A.2 Determination of Return on Investment and Net Present Value

The reader is probably acquainted with the techniques of Net Present Values and Return on Investment. Nevertheless, to refresh these concepts, this section deals with the calculation of both concepts for the example depicted in Table 7.8, that is, the selection of renewable alternatives. Table A.2.1 shows the information gathered about the four alternatives. Consider that:

- Different numbers of units of each type have been selected, in accordance with the geological, geographical, and population distribution of the region.

- In the case of Biomass, 10 different landfills have been considered, and assuming that methane in each site will be used to power diesel engines.

- Notice that solar and wind sources are considered not to work 24 hours a day, but as an average only 11 and 9 hours, respectively, according to 50-year statistics.
  The same source indicates that there is a yearly average of 310 days of clear skies (85 percent), and 259 windy days (71 percent), which is strong enough to drive wind turbines. These values have been considered in the calculations.

- The calculations assume that the other plants work 365 days a year.

- Maintenance and operative costs are computed as fixed costs plus variable costs, the latter depending on the amount of energy generated.

- Total production is calculated as production hours per year, times output.

- A residual value has been calculated for each type of equipment. This value corresponds to the proceeds from the sale of equipment at the end of its life minus the costs for their removal.

- Table A.2.2 computes the Net Present Values for each alternative, at different rates of interest. This Table is used to build Figure A2.1.

Table A.2.2 shows the necessary investments (with a negative sign), as well as the net income for each year and for each alternative. Net benefits have been computed considering total value for generated energy minus costs. For the last year, the residual value has been added to each benefit.

Table A.2.3 shows the corresponding Return on Investment for each alternative. This calculation has been made using Excel® financial functions, and these results are applied to the ROI row in Table 7.4, in Chapter 7, section 7.8.9. Table A.2.4 shows the corresponding values for constructing the graphic depicted in Figure A.2.1, which shows four different curves, one per option. These curves indicate the value of the Net Present Value (NPV) for each alternative at a specific interest rate.

Each curve starts with a maximum NPV corresponding to the lowest interest rate, and decreases with increasing rates, intercepting the x-axis at a certain point. This point is the ROI for that particular alternative. That is, if the return on funds invested is higher that the interest obtained when placing these funds in the financial market to earn interest, then the option is economically feasible. The highest ROI corresponds to hydro, with 6.2 percent, which is then, from the commercial point of view, the most favourable.
The second most favourable alternative from this point of view is wind energy.
It follows that for higher rates of interest the NPV is negative

*Table A.2.1*  Computation of main parameters

| Source | Output (kW) | Number of units (#) | Production hours per day (hours) | Production hours per year (hours) | Maintenance and operation costs ($/kWh) | Total maintenance and operation costs ($) | Total production (kW) | Residual value ($) |
|---|---|---|---|---|---|---|---|---|
| Solar | 100 | 45 | 11 | 153,574 | 0.00043 | 295,165 | 15,357,375 | 90,000 |
| Hydro | 400 | 4 | 24 | 35,040 | 0.00017 | 9,531 | 14,016,000 | 110,000 |
| Wind | 370 | 9 | 9 | 20,991 | 0.00072 | 49,979 | 7,766,726 | 456,000 |
| Biomass | 120 | 10 | 24 | 87,600 | 0.00068 | 70,956 | 10,512,000 | 567,000 |

*Table A.2.2*  Computation of Net Present Values

| Source | Price (kW/h) | 2005 | 2006 | 2007 | 2008 | 2009 | 2010 | 2023 | 2024 | 2025 |
|---|---|---|---|---|---|---|---|---|---|---|
| | | Investments | | Net income | | | | | Net income | |
| Solar | 0.037 | -1,834,829 | -1,443,156 | 271,058 | 271,058 | 271,058 | 271,058 | 271,058 | 271,058 | 361,058 |
| Hydro | 0.041 | -4,050,790 | -1,942,076 | 565,125 | 565,125 | 565,125 | 565,125 | 565,125 | 565,125 | 675,125 |
| Wind | 0.035 | -1,307,410 | -1,285,810 | 221,857 | 221,857 | 221,857 | 221,857 | 221,857 | 221,857 | 677,857 |
| Biomass | 0.045 | -1,984,034 | -2,956,034 | 402,084 | 402,084 | 402,084 | 402,084 | 402,084 | 402,084 | 969,084 |

*Table A.2.3*  Values of Return on Investment (ROI)

| Source | ROI |
|---|---|
| Solar | 4.8 |
| Hydro | 6.2 |
| Wind | 5.8 |
| Biomass | 5.1 |

*Table A.2.4*     **Computation of NPV for different rates of interest**

|  | Rate of interest (%) | | | |
|---|---|---|---|---|
|  | 2 | 4 | 6 | 8 |
| Source | Net Present Values | | | |
| Solar | 958,155 | 232,428 | -297,112 | -686,547 |
| Hydro | 2,750,787 | 1,220,074 | 94,495 | -740,917 |
| Wind | 1,126,502 | 448,198 | -40,450 | -395,688 |
| Biomass | 1,646,816 | 490,447 | -342,991 | -948,333 |

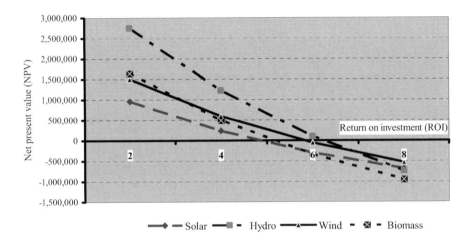

*Figure A.2.1* Determination of the Rate of Return (ROI) for a project or alternative Values from Table A.2.4

## A.3   A guide to strategic planning

When people in a community gather and make designs for the future, they are delineating a plan, a strategic plan, since it comprises the policies and approaches to reach the goal. This strategic planning includes goals, projects, plans, and programs, criteria to gauge them, as well as indicators to measure progress against certain standards, norms, or thresholds. Since everything is interlocked, it can be rather confusing to understand the whole picture; for this reason, figure A.3.1 shows how the intervening parts interact and interlock. Of course, one starts with the objective defined and the projects, plans or programs to achieve it.

*Projects, plans, and programs*
These are the product of the interaction between the grassroots and the policy-makers, as mentioned in section 6.1. They may be called 'Municipal input', just to give them a name, since in reality they should be a consequence of the joint action between policy makers and citizens. Citizens can be a factor, with their knowledge of the problems and consequences, since they also suffer them, while the municipality contributes with information that is usually not available to the public, such as about financing, which usually comes from several sources.

Considering the two approaches mentioned in section 6.2, that is, the bottom-up and top-down, indicators are also determined in this part to measure progress on problems detected (bottom-up), and to measure progress in plans established (top-down).

*Impacts*
Expert opinion, citizens, and policy-makers analyse positive or negative impacts from every project, plan or alternative, and consider social, economic, and environmental effects. Most projects involve people participating directly or indirectly, so the social consequences and effects have to be carefully appraised.

Effects are usually temporal, that is, related to some time in the future, such as establishing a new education curriculum, or the development of a new activity, like the creation of an informatics pole. Impacts are also related to space, such as the emissions from an incinerator, particulates that can precipitate far from the source. Figure A.3.1 simulates this geographical space as shadowed.

*Impacts valuation*
Impacts are measured according to their consequences; therefore, we need a cardinal value in different units to keep track of their effects. Thus, we can have impacts measured in $mg/Nm^3$ for air pollution, or in $m^3/sec$ for capacities, in number of units per inhabitants, etc. These values are in reality indicators, which not only inform us about a certain condition, but even more importantly link that value with a norm or threshold. Information for indicators comes from the very first stage commented, what we call municipal input.

*Thresholds*
Thresholds make the assessment normative, in that they establish limits to something, like society's carrying capacity.

*Projects selection*
Once all of this information is available, it constitutes a database from where values are taken, according to the method used for selection, transformed and

combined, after which their results are analysed. This book has proposed the use of the multicriteria analysis that allows the study of the trade-offs between all of the intervening elements. However, it is not expected that the solution given by a certain procedure should contemplate all the subjectivities and problems emerging in a real-life scenario. For this reason, there must be a feedback process where the stakeholders are informed about the results, and these are eventually corrected or amended in accordance with their opinion and wishes.

*Monitoring*
Once a plan has been established, it reaches the implementation stage. The best plan can become useless if it is not controlled, regarding not only its physical progress, but also mainly considering how the information extracted and representing this progress relates to the targets, both in value and in time, and with the thresholds. The latter is perhaps the most important concept, since it reflects how a plan must address a certain issue — such as fossil fuel-generated energy, for instance — reflects the progress in implementing a policy to switch to renewable energy sources.

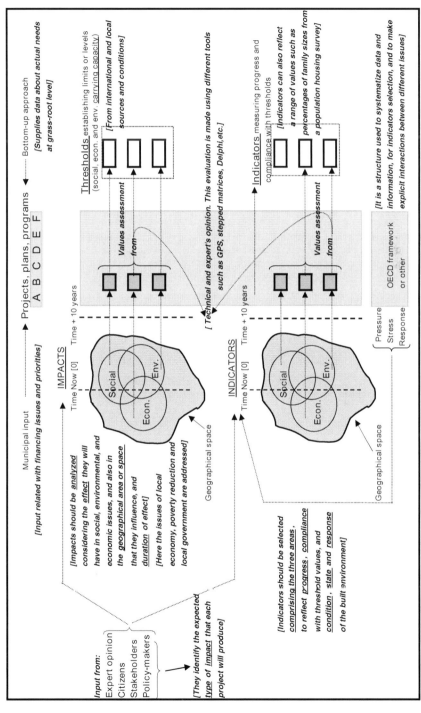

*Figure A.3.1* **Strategic Planning**

## A.4 Visualizing progress towards sustainability goals

It would be ideal to have a single number representing the whole interaction of gains and losses of the complex environment taken as a whole. That number should be obtained as an amalgamation or mixture of indicators, and could give a general idea of progress leading to sustainability. Such a number, composed of many indicators, is called an **index**. Efforts have been made to develop such an index, but without much success. A number such as the ecological footprint (section 1.6) is a close approximation, although it does not reflect the social aspects.

Another approach is called **'the amoeba approach'**, and though it does not give a single number, it can combine many indicators in a single diagram that can be easily read. This approach has been applied by the Dutch to sustainable issues, based on the original diagram that is commonly used in marketing efforts, especially in the car industry. Here, several common important car characteristics are considered, such as price, comfort, manoeuvrability, time to reach a certain speed, fuel consumption, etc., and it can be used to see how well a car matches standard values.

An example of this tool is proposed now to monitor progress towards sustainability, for the reader to appreciate its advantages. Assume that a final set of indicators to reach sustainability has been agreed upon, using the analysis detailed in section 7.8.

Table A.4.1 shows the list of 11 indicators selected. Assume that the time now is the year 2008, four years after the initiation of the sustainability process.

*Table A.4.1*  **Comparison between target and actual values**

| Indicators | Units | Baseline data for year 2004 | Target standards for year 2008 | TIME NOW Actual values for year 2008 | Percentage of achievement (+) or underachievement (-) |
|---|---|---|---|---|---|
| Annual growth of regional output (regional GDP) | % | 1.67 | 2 | 1.56 | - 0.44 |
| Strengthening small enterprises | No. | 36 | 44 | 51 | + 7 |
| Ratio of home rental to income | ratio | 0.43 | 0.39 | 0.42 | - 0.03 |
| Under-five mortality | % | 0.8 | 0.6 | 0.6 | 0 |
| Hospital beds/1,000 inhabitants | No. | 98 | 143 | 78 | - 65 |

| | | | | |
|---|---|---|---|---|
| Domestic recycling | % | 22.5 | 26 | 23 | - 3 |
| Average BOD$_5$ in water | mg/Nm$^3$ | 239 | 180 | 180 | 0 |
| Average NOx | ppm | 67 | 54 | 43 | - 11 |
| Crime rate | % /1,000 inhab. | 0.112 | 0.09 | 0.08 | +0.01 |
| Houses connected to sewer | % | 78 | 84 | 88 | + 4 |
| Flooding after heavy rains | No. | 12 | 10 | 9 | - 1 |

If the actual value is computed as a function of the target standard, then it is possible to construct Table A.4.2. Notice that target standards are depicted as '1'; consequently, actual values are related with these unitary figures. For instance, for the first indicator it will be 1.56/2 = 0.780. For the second indicator it is 51/44 = 1.159 and so on.

Table A.4.2      Actual values as a percentage of target standards

| Indicators | Target standards for year | TIME NOW Actual values for year | Target standards for year |
|---|---|---|---|
| | 2008 | 2008 | 2012 |
| Annual growth of regional output (regional GDP) | 1 | 0.780 | 1.05 |
| Strengthening small enterprises | 1 | 1.159 | 1.182 |
| Ratio of home rental to income | 1 | 0.880 | 0.769 |
| Under-five mortality | 1 | 1.000 | 0.667 |
| Hospital beds/1,000 inhabitants | 1 | 0.545 | 1.084 |
| Domestic recycling | 1 | 0.885 | 1.308 |
| Average BOD$_5$ in water | 1 | 1.000 | 0.444 |
| Average NOx | 1 | 0.790 | 0.741 |
| Crime rate | 1 | 0.889 | 0.667 |
| Houses connected to sewer | 1 | 1.048 | 1.190 |
| Flooding after heavy rains | 1 | 0.900 | 0.2000 |

Using graphics from Excel® and selecting the radial diagram, it is very easy to draw the diagram shown in Figure A.4.1.

The larger circle represents the target standards for year 2008, that is, 1, while the irregular, thicker line (amoeba) shows actual values. The difference between the large circle and the amoeba indicates the percentages of achievement and underachievement, just as depicted in the last column of Table A.4.1.

Using this diagram, it is possible to have an instant overview of progress towards sustainability. An indicator could be at its best at its highest value, for instance 'domestic recycling', while another could be at its best at its lowest value, for instance 'crime rate'. For this reason each indicator has a (+) or (-) sign to indicate these two different meanings.

In this example, it can be seen at a glance that when an indicator shows the (+) sign, its best is when it is equal or greater than the value corresponding to the circle. Conversely, if an indicator shows the (-) sign, its best is when it is equal or smaller that the value corresponding to the circle.

It is evident, on inspecting the amoeba diagram, that in this whole scenario there are two areas where the process is not going well: the annual growth of the region is less than expected, and the ratio of hospital beds per 1,000 persons is way smaller that the target. Of course, there could be very rational explanations for these, which have nothing to do with the sustainability process. For instance, the smaller-than-expected growth could be blamed on the closing of two important manufacturing plants, for whatever reasons, while the non-compliance with the hospital bed ratio could result from a fire in a wing of the hospital, which destroyed the installations, including beds.

If there is, as usual, another target standard for a further date, say for 2012, as shown in Table A.4.2 it is also possible to represent these target standards, but in this case the figure will not be a circle but a polygon, as depicted in Figure A.4.2.

As a few examples, it can be seen that 'strengthening small enterprises' have achieved in 2008 almost the values established for 2012.

'Under-five mortality' is, at 2008, right on schedule, but still has a long way to go to reach the 2012 standard.

It is already known that for 'hospital beds' actual values fall short for the 2008 standard, and are far indeed from the 2012 standards, although since the construction of a new hospital is planned for 2009 and 2010, it is thought that this target will be reached.

Regarding 'average NOx in 2008, this is above standard, and it appears that it will meet the 2012 standard.

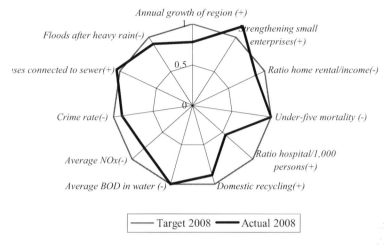

*Figure A.4.1* **Amoeba diagram with targets and actual values for 2008**

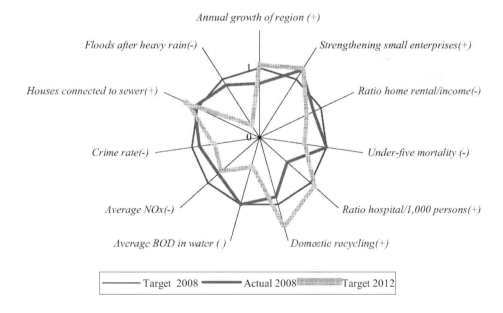

*Figure A.4.2* **Amoeba diagram with targets for 2008 and 2012 and actual values for 2008**

## A.5 Life Cycle Assessment (LCA)

This is a systematic and quantitative analysis of the whole life of a product, tracing it backwards to the extraction of the raw materials necessary for its manufacture and forward to its final disposal. Figure A.5.1 shows this process for a product (a fridge), following the classic cradle-to-grave approach. As can be seen, a main component is transportation between industries but also within an industrial sector.

| Process | Direction from product fabrication | Example | Transportation |
|---|---|---|---|
| Extraction of raw material | ↑ Backwards | - Aluminium<br>- Hydrocarbons (for plastics)<br>- Copper (for electric motor). Etc. | ⇔ ⇓ |
| Processing raw material | | Smelting and refining | ⇔ ⇓ |
| Manufacturing products | | Aluminium sheets, profiles, copper wire, plastic sheets, moulded plastic, insulation, etc. | ⇔ ⇓ |
| **Fabrication of product under analysis** | | **Fridge** | ⇔ ⇓ |
| Packaging and Marketing | ↓ Forwards | Cardboard box, plastic, wood | ⇔ ⇓ |
| Product utilization | | Electric energy and maintenance | ⇓ |
| Recycling | | Unit back to manufacturer to recover elements | ⇓ |
| Final disposal | | Refrigerant extraction | ⇓ |

⇔ Intra-industry transportation     ⇓ Inter-industry transportation

*Figure A.5.1*     **Backwards and forward analysis for a product**

The main purpose of Life Cycle Assessment is to assess quantitatively the impact that a product has on the environment. More often than not, this analysis is also a very good tool to determine how a product can be improved, from not only the environmental point of view but also considering costs, weight, reliability, durability, etc. These environmental impacts are measured in kg of carbon dioxide ($CO_2$), ammonia ($NH_3$), nitrogen oxides (NOx), particulates, and other gases spewed into the atmosphere by the different processes, referred to the output dollar value of the good produced.

In the US, this calculation is updated periodically for 485 commodities, with the intervention of different federal agencies such as the Environmental Protection Agency (EPA), the Department of Energy (DoE), and the US Geological Survey.

Internationally, this activity is regulated by the
ISO 14040 Norm (see Glossary). http://www.iso-14001.org.uk/iso-14040.htm
This Norm establishes a framework involving:

- *Goal and scope definition*
  Establishes goals for the study and defines boundaries for the product.
- *Inventory analysis*
  Analyses material and energy flows, prepares flow sheets and collect data, and determines emissions.
  *Impact assessment*
- Analyses the different impacts, spatially and temporally, and how they affect the environment.
- Interpretation.

### A.5.1  Example of application in industrial complex

This methodology is not only useful for a new product but also finds application in other undertakings, such as processes, as shown in the following example.

Assume that an industrial complex, with three industrial plants in a 50 km radius, produces a large quantity of spent chemical solvents, as well as paint sludge, non-recyclable plastic packing, and some chemical and flammable by-products with no commercial value. The plants have been storing the chemical solvents and paint sludge in large tanks and burning the flammable gases, however, new, more stringent regulations prohibit this burning. Besides, there are two additional problems: the plants are running out of storage space for the spent chemical solvents, and insurance costs have skyrocked.

After studying different alternatives, it appears that the installation of an industrial high temperature incinerator can solve these problems, since all the material will be incinerated. There is, besides, the advantage of utilization of the hot exhaust gases to produce steam for driving a turbine generator, and, at the same time, to produce enough hot water to be used in part of the industrial processes.

With this aim, an LCA study was performed to utilize an industrial incinerator that is available on the market. The flow sheet that was prepared with all the information assumes that the incinerator manufacturer provided the data about contamination produced by the supply chain for its incinerator and ancillary equipment. The manufacturer assures that because of the high temperature (1,000°C), there will be no production of dioxins, which is the main concern from the municipality and neighbours, although it is known that during cooling, the exhaust gases' dioxins can eventually form again; for this reason, a dioxin filter has been added. Nevertheless, a **maximum threshold** of about 0.11 nanograms per cubic meter of dioxins in the exhaust gas has been established and accepted by the manufacturer.

The waste produced by the plant does not contain mercury; therefore, there is no danger of spewing this chemical into the air as could happen with municipal incinerators.

The material fed to the incinerator will be an appropriate and balanced mix of the mentioned waste, as well as some initial amount of fossil fuel to start combustion. The height of the chimney and the direction of prevailing winds is another factor that has been carefully considered. Figure A.5.2 depicts the flowsheet.

It can be seen that the supply chains of the following components are to be considered as the contamination to the environment:

- Incinerator construction (not included in Figure A.5.2).
- Fresh water treatment.
- Extraction, process and transportation for the fossil fuel used in the incinerator.
- Manufacturing, transportation, handling, and sorting equipment for the three plants.
- Burning fossil fuel for trucks and equipment.
- Manufacturing of the dioxin filter, the heat exchanger and the metal recovery plant.
- Manufacturing of the gas turbine and electric generator.
- Using landfill space to deposit ashes.

Incineration, while avoiding a dangerous problem (viz., the storage of chemicals in tanks that can corrode and leak to the ground), also has these beneficial effects for the environment:

- Hot gases from the incinerator are used to drive a turbine generator that then generates electricity. Consequently, there are benefits to the environment when no supply chain is used to generate electricity for the complex, or at least for part of it.
- The same applies to the heating of water to be utilized in the processes; otherwise, fossil fuels would be used to heat this water.
- Scrubber sludge as well as ashes are a source of valuable metals. This, of course, continues to benefit the environment since it does not use the supply chain to obtain metals.
- Chemicals and flammable products are no longer burnt, therefore their contribution to global warming and climate change is eliminated.

This very basic example nonetheless shows the principles on which LCA is based; however, although this information is useful to make decisions, it has to be complemented with quantitative values.

To compute the contribution to environmental contamination using the supply chains, it is necessary to use dedicated computer software. There are various on the market, and one of the pioneers in this kind of analysis is Carnegie Mellon University's Green Design Initiative. It is possible to access their software via the Internet and then calculate, based on dollar values, the amounts of contamination produced by the manufacture of some product. (See Internet references for this Appendix.)

The US Environmental Protection Agency (EPA) has also developed a software called TRACI (Tool for the Reduction and Assessment of Chemical Impacts). As it name indicates, this allows characterising potential effects from chemical products. (See Internet references for this Appendix.)

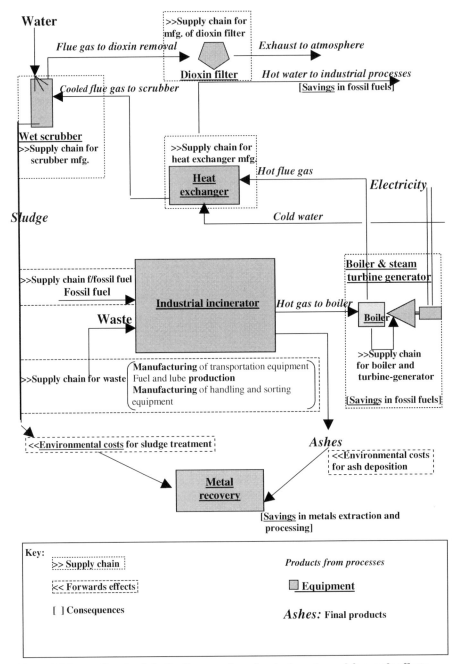

*Figure A.5.2*    Supply chain for the operation of an incinerator and forwards effects

## A.6 Regression analysis for weights determination

More often than not, the single value given by an indicator is not enough, so **indexes** are developed that are a **composite of indicators.** One of the main problems is to ascertain the weight attributable to each component. This author proposes the use of **regression analysis** (see Glossary) to determine not only the composite value of an index but also to find out the weights corresponding to each component.

For example, if a community wants to gauge the sustainability of their transportation, that is, to have some measure of the quality of their service, they can naturally turn to many indicators to find such a value, although none will reflect the contribution made in other areas. For instance, even it is recognized that they have good frequency of service, what about the travel time, or the comfort given by the units, or the fare?

Therefore, the community has prepared a list of all the factors that they consider affect the service, in the three dimensions. However, one question arises. Do all the factors have the same importance, or weight? For instance, it is assumed a priori that fare is a very important factor, but how does it compare with travel time, or with accessibility for old and disabled people? In other words, which is the relative importance of each one and how do they combine to have a final number?

They have already prepared a short list of indicators, and now it is time to find out the weight of each one. To keep this demonstration from getting too complicated, assume that five indicators (which are now called **variables**) are chosen to measure sustainable quality of transportation, as follows:

*Dependent variable:*
**Ridership:** This is the value one wants to determine, and it is called the 'dependent variable' because its importance depends on the values taken by other variables, which are called 'independent variables'.

It is assumed that people express their satisfaction by the number of patrons that ride the system; if not, they would look for other options, such as using their own car, carpooling, etc. If the service is good, many people driving their cars to work may opt to use the park-and-ride system and travel instead by bus, increasing the present number of regular passengers.

For this example, values of ridership haven been considered from 1996 to 2003 included.

*Independent variables:*
These are the variables whose change will affect the dependent variable. They are:

**Frequency:** This reflects the average regularity of the bus service, and, of course, a system with small intervals between successive buses at a certain bus stop are a sign of its quality. This can be measured in number of buses per hour, as in this example, or its inverse, that is, the time (in minutes) a patron has to wait at a bus stop between two consecutive buses.

**Accessibility:** There is an increasing need to provide better boarding and alighting for the infirm, old and disabled people, parents with prams, etc. The solution could be lowering the footpath's height, or run buses with flat floors, or providing access for wheelchairs and safety restraints for prams and wheelchairs, etc.

**Modernization:** Buses deteriorate with age, and have a certain life limit, after which they must be replaced. It is therefore important to know the rate at which the transportation company is renewing its units. This is also a measure of comfort, since most new units will probably have air conditioning and better quality seats. This could be a measure of environmental concern, if there is a commitment to replace old buses with units with more environmentally-friendly engines, such as electric buses, or that use the new fuel-cell technology (section 5.3.5).

**Fare**: In this example, a passenger pays a single fare that permitting them transfer rights during a certain period.

**Analysis for two variables**

To understand this problem better, let us begin with the analysis of the relationship between only two variables: **Ridership** (y) and **frequency** ($x_1$). Values for this as well as other Tables have been obtained from City Hall logs.
With this information, regression analysis is easily performed using one of the many software packages that are available. In this case, a Multiple Linear Regression Package that is included in a Corel Quatro Pro® spreadsheet program was used.

The relationship between these two variables is expressed through a straight line, called the **regression line,** which has a certain **slope.** This slope indicates how much the **dependent variable changes** with a **unitary change of the independent y variable.** Figure A.6.1 represents this situation in two dimensional, coordinate axis. The line intercepts one of the axes in a certain point, and its value is called a **constant.**
Table A.6.1 shows the information gathered for the two variables considered.

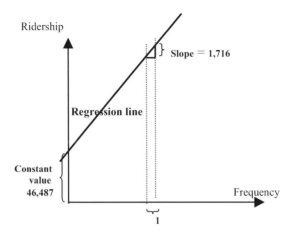

*Figure A.6.1*   Linear relationship between ridership and frequency (not in scale)

*Table A.6. 1*   Values for two variables

| Year | Dependent variable (y) Ridership (Number of passengers x 000s) | Independent variable ($x_1$) Frequency (Number of buses per hour) |
|---|---|---|
|  | (y) | ($x_1$) |
| 1996 | 67,765 | 10 |
| 1997 | 65,375 | 14 |
| 1998 | 68,865 | 15 |
| 1999 | 76,448 | 18 |
| 2000 | 78,504 | 18 |
| 2001 | 79,298 | 18 |
| 2002 | 80,900 | 19 |
| 2003 | 79,565 | 19 |

What is the meaning of the computed slope? The slope indicates how much the ridership will change for a change of one unit in frequency; that is, if the frequency goes from 19 to 20 buses per hour, the number of passengers will increase by 1,716 patrons. The regression line will intercept the ridership axis at a constant value equal to 46,487.

**Analysis for five variables**

The same reasoning can be applied to a larger number of variables. Let us now consider the five variables mentioned before. Table A.6.2 show the information gathered on these five variables.

*Table A.6.2*             Variables data

|  | Dependent variable | Independent variables | | | |
|---|---|---|---|---|---|
| Year | Ridership (Number of passengers x 000's) | Frequency (Number of buses per hour) | Accessibility (Number of buses with easy access) | Modernization (Percentage of buses renewed per year) | Fare ($/trip) |
|  | (y) | ($x_1$) | ($x_2$) | ($x_3$) | ($x_4$) |
| 1996 | 67,765 | 10 | 56 | 3 | 1.70 |
| 1997 | 65,375 | 14 | 59 | 5 | 1.90 |
| 1998 | 68,865 | 15 | 61 | 5 | 1.90 |
| 1999 | 76,448 | 18 | 65 | 5 | 1.95 |
| 2000 | 78,504 | 18 | 65 | 5.2 | 1.95 |
| 2001 | 79,298 | 18 | 65 | 7 | 1.95 |
| 2002 | 80,900 | 19 | 70 | 7 | 2.20 |
| 2003 | 79,565 | 19 | 120 | 7 | 2.20 |

Again, the purpose of this analysis is to find the coefficient — i.e. the slope — of each variable. Each one indicates how much the dependent passenger variable (y) changes with a change by one unit in any of the independent variables ($x_1$, $x_2$, $x_3$, $x_4$). Assuming again that one wants to know the effect on ridership if the frequency of the service is increased by one unit, that is, to 20 buses per hour instead of 19 buses. Solving the system using the mentioned software it is found that the ridership increases in 1,774 passengers for each additional bus that is added on per hour. Table A.6.3 gives the resulting values for the other variables, as well as the constant value.

*Table A.6.3*             Results from regression analysis

|  |  | Frequency (Number of buses per hour) | Accessibility (Number of buses with easy access) | Modernization (Percentage of buses renewed per year) | Fare ($/trip) |
|---|---|---|---|---|---|
|  | Constant | ($x_1$) | ($x_2$) | ($x_3$) | ($x_4$) |
| Coefficient value | 53,778 | **1,774** | 35 | 254 | (6,129) |
| Standardized values (weight) |  | 1.231 | 0.157 | 0.078 | -0.222 |
| Ranking |  | First | Fourth | Third | Second |

The reader will notice that both the slope for variable frequency ($x_1$) and the constant have changed when compared with the two variables case-analysed above.

Why?

Because the model is now considering the joint action of five variables, and not only two of them. Where is the graphic representation of this more complete case? There is none, because it is impossible to represent a space with five dimensions. How do we know that slopes really represent an increase in the dependent variable? How can this be proven?
Considering all coefficients depicted in Table A.6.3, the formula that expresses the relationship of the dependent and independent variables can be written as:

$$y = \text{constant} + (\text{coeff. for frequency})x_1 + (\text{coeff. for accessibility})x_2 + \\ + (\text{coeff. for modernization})x_3 + (\text{coeff. for fare})x_4 \quad [1]$$

Replacing values (The sign (*) utilized in computing to indicate multiplication is also used here with the same meaning to avoid confusion with variables that are represented as x's):

$$y = 53{,}778 + 1{,}774 * x_1 + 35 * x_2 + 254 * x_3 - 6{,}129 * x_4 \quad [2]$$

Just to verify the reliability of this formula, let us replace de x's with the values depicted in Table A.6.2, for any year, say 2003:

$$y = 53{,}778 + 1{,}774 * \mathbf{19} + 35 * 120 + 254 * 7 - 6{,}129 * 2.20 = 79{,}934 \quad [3]$$

This value for ridership, analytically computed through formula [3], is then compared with the actual value for the same year (79,565 passengers) (second column of Table A.6.2).

The difference is therefore:

$$79{,}934 - 79{,}565 = 369 \text{ passengers}$$

The erroneous estimate regarding the actual figure for the year 2003, is then negligible, since it is:

$$369/\,79{,}565 - 0,4\%$$

Therefore, the formula works well. The reason for this closeness is that the correlation coefficient (see Glossary), that is, the degree of association between the number of passengers — or dependent variable — and the independent variables, is quite high (77 percent).

Now that the accuracy of the formula has been shown, let us return to the original problem, and replace in formula [3] the coefficient 19 (buses per hour) by 20. The result is indicated in formula [4]:

$$y = 53{,}778 + 1{,}774 * \mathbf{20} + 35 * 120 + 254 * 7 - 6{,}129 * 2.20 = 81{,}707 \quad [4]$$

Comparing this value with the number of passengers, when the frequency was of 19 buses, formula [2] (that is 79,934 passengers) leads to this result:

$$81{,}707 - 79{,}934 = 1{,}774 \text{ passengers (rounded)}$$

This number is the same as the coefficient for variable $x_1$ or 'Number of buses per hour'. See Table A.6.3.

Naturally, any variable can be changed. Assume that in the original problem one wants to assess the variation of passengers when the fare (variable $x_4$) is increased by $0.10: that is, from $2.20 to $2.30. We already know the answer, the number of passengers will decrease (note that this coefficient has a negative sign), so the decrease will be $6{,}120/10 = 613$ passengers.

To double-check this result, a replacement is made in formula [3], placing $2.30 instead of $2.20:

$$y = 53{,}778 + 1{,}774 * 19 + 35 * 120 + 254 * 7.00 - 6{,}129 * \mathbf{2.30} = 79{,}321 \quad [5]$$

Therefore:

$$79{,}321 - 79{,}934 = -613 \text{ passengers}$$

…which is a 1/10 of the slope value for variable $x_4$. See Table A.6.3.

**Variable weights**

The question that now arises, which constitutes the main reason for the development of this section, is this: Is it thereby possible to assume that the coefficients of the variables are weights? The answer is: no.

Why not?

Because each of them has different units of measure (number of buses/hr, number of buses with easy access, % of buses replaced, $/trip).

The way to solve this problem is simple. It is necessary to standardize these different values, and the way to do that can be found in any book on Statistics, and an easy calculation from data provided by computer software, such as Excel®, Quatro Pro® or Lotus 123®. For this example the following standardized values have been obtained and depicted in Table A.6.3.

According to this data, the most important factor is 'frequency', which has the largest absolute value, followed by 'fare'.

Observe that all the coefficients but 'fare' have positive values, while 'fare' has a negative coefficient. This is easily explained considering that:

- The greater the frequency, the more passengers in the transit system;
- The better the accessibility, the more seniors will be able to ride the system;
- The more modern the buses, the greater comfort for passengers, who are getting better service for their money;
- The higher the fare, the greater the number of potential passengers that think twice before making a trip, decreasing ridership.

**Forecasting**

Regression analysis is a very good tool for forecasting. For instance, it could be necessary to predict the ridership in 2007. To compute this, value formula [3] is used and the x variables are assigned estimated values for that year; therefore, one could end up with something like this:

Frequency: Will be maintained if there are 19 buses per hour.
Accessibility: a total of 360 buses with have this convenience.
Modernization: percentages will be held at 7 percent per year.
Fare: It is assumed that this will increase to $3.20.

The calculation is then:

$$y = 53,778 + 1,774 * 19 + 35 * 360 + 254 * 7.00 - 6,129 * 3.2 = 82,249 \qquad [6]$$

### A.7 Discharges and their effect on the environment

It would be quite impossible to list all household, municipal and industrial discharges into the environment, and to then evaluate their effects. This section will show in a very condensed manner the main discharges and effects on the air, soil and water.

Approximate limits allowed for discharges into water courses are shown in Table A.7.1.

*Table A.7.1*        **Approximate discharges in water**

| Discharge of | Unit of measure | Values |
|---|---|---|
| Acidity or alkalinity (measured in pH) (see Glossary) | Number | 6 to 9 |
| Antimony | ppb | 6 |
| Arsenide | mg/l | 0.5 |

| | | |
|---|---|---|
| Cadmium | ppb | 5 |
| Chromium | mg/l | 10 |
| Copper | ppm | 1.3 |
| Cyanides | ppm | 0.2 |
| Hydrocarbons | mg/l | 20 |
| Lead | mg/l | 1 |
| Mercury | ppb | 2 |
| Nickel | mg/l | 4 |
| Nitrates | ppm | 10 |
| Nitrites | ppm | 1 |
| Nitrogen | mg/l | 80 |
| Oils and fats | mg/l | 150 |
| Phosphorous | mg/l | 10 |
| Sulphates | mg/l | 1,000 |
| Sulphides | mg/l | 5 |
| Suspended solids | mg/l | 300 |
| Temperature | Degrees Centigrade (°C) | 35 |
| Zinc | mg/l | 5 |

Table A.7.2 shows the origin, composition and effects of some contaminants. The Internet has many sites where these values can be found, especially form the EPA (Environmental Protection Agency).

*Table A.7.2*       **Origin, composition and effects of contaminants**

| Origin | Composition | Effect |
|---|---|---|
| **Emissions in the atmosphere** | | |
| Cooling towers used to cool or condense fluids | - Water vapour ($H_2O$) | Has no environmental effects, although when combined with sulphur from flue gases from powerhouses and industrial stacks, can produce acid rain, which attacks people, buildings, crops, etc. |
| Flue gases from powerhouses and industrial stacks | - Carbon dioxide ($CO_2$)<br>- Sulphur dioxide ($SO_2$)<br>- Particulate | Contributes to global warming.<br>When combined with water vapour, produces acid rain<br>Dangerous for human health inducing pulmonary diseases |
| From landfills, cattle and agricultural operations | - Methane ($CH_4$)<br><br>- Greenhouse gases ($CO_2$, $CH_4$) | Produces fires but most important is its contribution to depleting the ozone layer<br>Climate change |
| Gases and leaks from | - Volatile organic compounds (VOC)<br>- Nitrogen oxides (NOx) | The Bophal disaster in India (see Appendix) is a grim |

| | | |
|---|---|---|
| petrochemical plants and industries | - Sulphur oxides (SOx)<br>- Halocarbons | reminder of this emission<br><br>Ozone layer depletion |
| Cars | - Carbon monoxide (CO) | |
| Air conditioning | - Fluorocarbons (HFCs) | Destroys the ozone layer |
| Industrial and domestic incinerators | - Dioxins and Furans | Dangerous for human health |
| Cement plants | - Particulates, dust | Affects human health and can deposit in water sources, reducing photosynthesis processes |
| **Discharges to soil** | | |
| Cars, machinery, leaks | - Oils and fuels | Contaminates the soil and percolates to the groundwater |
| Herbicides and pesticides | - Inject Nitrogen (N), Phosphorous (P) and Potasium (K) into the soil | Contaminates water and favours eutrophication |
| Landfills Industries | Heavy metals into the soil | Contaminates sources of water |
| Agriculture | - Salinisation | Decreases crops yield and produces erosion |
| Industry | - Generation of hazardous wastes | When stored in drums they can leak and contaminate soil and groundwater |
| **Discharges to water** | | |
| Industries | - Chemicals<br>- Organic matter<br><br>- Acid/Alkaline discharges<br><br>- Suspended solids<br><br>- Sand<br><br>- High temperature<br><br>- Colored water<br><br>- Salinity | Kills aquatic life<br>Consumes oxygen in the water that fish need<br>Affects the fish environment, which needs a neutral pH (see Appendix) for health<br>Affects photosynthesis processes in water, and then the generation of oxygen<br>Produces turbidity and affects the photosynthesis process<br>Since oxygen solubility in water decreases with a higher temperature, less oxygen is available for aquatic life<br>Affect the photosynthesis process in water, and then the generation of oxygen<br>Affects aquatic life because it alters the osmotic pressure |
| Domestic waste | - Wastewater | Even treated with a secondary treatment, it deposits into water sources nitrogen, |

| | | |
|---|---|---|
| | - Detergents | potassium and phosphorous, which produce eutrophication (see Glossary) in rivers and lakes. The eutrophication results from too many nutrients going to phytoplankton, which when dead becomes organic matter that consumes oxygen |
| | | Produces foam and then alters the photosynthesis process in water courses |

For our purposes, it is important to keep track of how the values of contaminants vary with time; perhaps even more important, there is a need to monitor by what percentages the contamination exceeds established maximum limits. For example, Figure A.7.1 shows this evolution over a 10-year period of exceedances of air contaminants.

It appears that substantial progress has been made to decrease the levels of CO and particles, with drastic reductions in both their excess levels, although success in reducing CO is greater than for particle reductions. It seems that $SO_2$ production has not followed suit with this reduction, and has even increased — indicating that more severe measures have to be taken to reduce these emissions. This is an excellent and dynamic tool to show progress corresponding to measures taken, as a function of time.

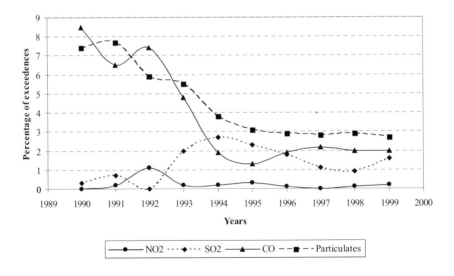

*Figure A.7.1*     **Temporal evolution of exceedances air contaminants**

## Internet references for Appendix

**Source:** Carnegie Mellon Green Design Initiative (1997)
Title: *Economic input-output life cycle assessment*
Comment: An extremely useful and highly recommended document. It covers 27 categories from Government to Banking, and main industrial activities. In turn, each category for instance, "Transportation and Utilities", is broken down into several others, such as "Railroads and Related Services". For each one of these, results can be obtained regarding:

- Economics.
- Conventional pollutants.
- OSHA Safety in employment.
- Water use.
- Global warming gases.
- Fertilizers.
- Energy.
- Ores.
- External costs.
- Toxic releases by sector.
- Weighted toxics, by sector.

For example, assuming that it is of interest to know the amount of conventional pollutants released by a project related to railroads, introducing as our data the cost of our project in US$. The result will indicate the released contaminants and their weights for this activity and for that amount of investment, as follows:

$SO_2$ = Sulphur Dioxide
CO = Carbon Monoxide
$NO_2$ = Nitrogen Dioxide
VOC = Volatile Organic Compounds
Lead = Lead particulate emissions (air)
PM10 = Particulate Matter (less than 10 microns in diameter)

If, for instance, there is a need to know the economic effect that the investment will have on all sectors of the economy, choose 'Economics' and a Table will display this information. In other words, one gets the multiplier effect of our project.
Highly recommended document.

Address:
http://www.eiolca.net/

**Source:** U.S. Environmental Pretection Agency (2004)
Title: *Life-Cycle assessment – Tool for the reduction and assessment of chemical impacts (TRACI)*
Comment: This very important document contains a wealth of information for potential effects such as warming, acidification, eutrophication, tropospheric ozone, smog formation, human particulate effects, human carcinogenic effects, human non-carcinogenic effects, fossil fuel depletion, and land use effects. TRACI is also available for downloading from an address provided in this paper. The whole document is also valuable because it provides information on another LCA applications, as well as case studies.
Visiting this Website is highly recommended.
Address:
http://www.epa.gov/ORD/NRMRL/lcaccess/ongoing.htm

# GLOSSARY

**Adsorption:** Natural phenomenon by which molecules of liquids and gases adhere to the surface of solid or liquid bodies.
**Algorithm:** A method, formula or a sequence of steps for doing something; mainly used in Mathematics.
**Aerobic:** Usually used in connection with 'aerobic bacteria', which produce decay in the presence of oxygen.
**Anaerobic:** Usually used in connection with 'anaerobic bacteria', which live without oxygen.
**Aquifer:** A water-bearing layer of rocks or gravel beneath the earth's surface.
**Bessemer converters:** Invented by Henry Bessemer, these are used to obtain steel from pig iron. It consists in an ovoid-shaped vessel lined with clay or dolomite. The bottom of the device is perforated in order to allow air to pass through the perforations. The load of pig iron is placed on the converter and air is blown from below. The oxygen in the air oxidizes the impurities in the pig iron, and also raises the temperature of the load, which is then kept molten.
**Bhopal:** City in India where a human disaster occurred due to the release of noxious chemicals on December $3^{rd}$, 1984.
**Biodegradable:** A substance than can be broken down into relatively harmless products by the action of micro-organisms. For instance, some organic material in the soil can be broken down into methane.
**Biota:** Refers to all plants and animal life in a certain area.
**$BOD_5$:** When organic matter is in water it requires oxygen to break it down (biochemical oxidation). The amount of oxygen that organic matters needs to biodegrade within a five-day period at 20° C is called Biochemical Oxygen Demand, and it is measured in milligrams per litre (mg/l).
**BTU:** British Thermal Unit. A measure of heat whose quantity is approximately equivalent to 252 calories.
**Carbon cycle:** A complex process, fundamental for life. It begins with photosynthesis, during which process atmospheric $CO_2$ is converted into organic matter. Other organisms feed with this organic matter and release $CO_2$ waste gas into the atmosphere, reinitiating the cycle.
**Chernobyl:** Site in the Ukraine of an explosion of nuclear reactor No. 4, on April 26, 1984.
**CHP:** Abbreviation of Combined Heat and Power. Very often used in the technical literature especially referring to the joint production of heat and electricity. For instance, after powering a steam turbine, heat can be utilized in many applications, such as for water heating, heating houses, etc.
**Clinker:** Main component in the manufacture of Portland cement.

**Chlorophyll**: Large molecule found in plants, composed of oxygen, carbon, hydrogen, magnesium, and nitrogen. Its main function is to initiate photosynthesis, a complex process that uses solar energy to convert water and carbon dioxide into glucose and other components.

**Coke:** High quality fuel and metal reducer for blast furnaces obtained by burning coal at high temperatures and with little air, in order to get concentrated carbon.

**$CO_2$ sinks:** Method to compensate for the emission of $CO_2$, premised on the development of gas-absorbing trees.

**Comparative advantage:** Very important theory in Economics related to the production cost of the same product in two different regions.

**Contingent valuation:** There are circumstances in which the values of some environmental assets such as clean water, land, clean air, etc, as well as benefits received from ecosystems such as watching wildlife in tropical forests, ice fields, etc., are difficult to evaluate in economic terms. One way to solve this difficult issue is to assign a dollar amount to these benefits, **not in the sense of allocating to them a market value,** but instead assessing how much they mean to people. This is called contingent valuation.

**Correlation:** The degree of association between two or more variables.

**Cradle to grave:** Concept that makes a particular industry environmentally responsible for the products it manufactures, from its suppliers up to the disposal of the product at the end of its life.

**Critical Path Schedule:** A planning tool used by engineers to establish the correct sequence of activities as a function of time and their inter-relationships. It produces a graph, and its computation determines a path or paths between the project's start and its termination, involving the most important activities (thereby called 'critical activities') pertaining to the completion time. This path is called the 'critical path', and it identifies not only all the critical activities but indicates the shortest time needed to execute a project considering all the existent restrictions, including manpower and equipment availability, and funding.

**dB:** Decibel, a measure used in Acoustics for sound power.

**Digesters:** Sludge taken from settling ponds in wastewater treatment facilities is thickened and then pumped into vessels where anaerobic digestion takes place. This process involves the action of micro-organisms that, aided by heat, break down the solid matter into bio-solids, water, carbon dioxide and methane gas. The methane gas can then be used to generate electricity through a dual-fuel engine, and to provide the heat necessary for the digestion process.

**Earth Summit**: Refers to the Convention on Global Diversity, held in Rio de Janeiro in 1992. This Convention was attended by 152 world leaders, and led to the signing of agreements on biological diversity and many other sustainable issues including *Agenda 21*.

**Ecology:** The science that investigates the relationship between organisms and their environments.

**Ecosystem:** An ecosystem consists of a dynamic set of living organisms (plants, animals and micro-organisms) all interacting among themselves and with the environment in which they live (soil, climate, water and light). (Natural Resources Canada).

**Economy of scale:** There are economies of scale when the production cost per unit diminishes as the output increases.

**EEC:** International organization formed by European countries.

**Elasticity:** Deformation in a solid that does not change its physical characteristics.

**Electrolysis:** Electrochemical process that provokes chemical changes in an electrolyte by the passage of an electric current. When using water, it splits it into hydrogen and oxygen.

**Electrolyte**: A non-metallic electric conductor where current is produced by ion movements.

**Entropy:** This is a fundamental concept of heat dynamics (thermodynamics) and in Information Theory. In thermodynamics, entropy measures the quantity of concentrated energy that is spread out as a function of time. It means a dispersion of energy. In Information Theory it is used to measure the quantity of information in a message.

**Eutrophication:** Condition that develops in an aquatic ecosystem when high nutrient concentrations stimulate algae growth. This growth consumes dissolved oxygen in the water.

**Expert opinion or expert judgment:** Opinion or judgment by people who are knowledgeable about some subject.

**Floatation:** Mechanical process used to separate minerals, employing pulverized ore, water and chemicals.

**Fission** (nuclear): The split of an atom, such as uranium-235, producing lighter elements and energy.

**Fission reactor**: A device where fission is produced. The generated energy is used to heat a fluid that in turns boils water, generating steam that is used to drive a steam turbine and thereby generate electricity.

**Freon:** Commercial name of dichlorodifluoromethane, which is utilized as a refrigerant for refrigerators and air conditioning units forcommercial uses and houses.

**Fusion** (nuclear): The combination of light nuclei to make a heavier element and produce energy. This process is used by stars, such as the sun.

**GDP:** Gross Domestic Product, System to measure economic development.

**GW:** Giga Watts, a measure of electrical output. 1 GW equals 1,000 MW.

**GIS:** Geographic Information System, involving geo-referenced data.

**Global Urban Observatory:** System developed by the United Nations to keep track of advances in the use of urban indicators.

**Global warming potential (GWP):** The potential of a component to increase global warming. It is the capacity of each type of emission to store heat. It is computed by comparing a gas to the same mass of carbon dioxide which is assigned a GWP value of one.

**Greenhouse effect:** The increase in temperature within a greenhouse. Sunrays enter the greenhouse as short radiation waves, and because the sunlight crosses the glass, it changes its frequency to a longer wavelength. This last cannot go back through the glass, so it stays inside the greenhouse, heating it up. The same happens with the earth when some gases, mainly $CO_2$, allow sunrays in but not out, producing global warming. This is why $CO_2$ is considered a contaminant.

**Groundwater:** Water penetrating the earth's surface and found accumulated in layers.

**Gyproc:** Lightweight building material for wall linings (plasterboards), made of gypsum (hydrate calcium sulphate).

**Impact:** Can be defined as the change of some conditions in human health and in the ecosystem caused by the development and implementation of a project.

**Indicators:** Values or metrics established to measure some issues.

**Information Theory (IT):** Branch of mathematics using statistics, founded by Claude Shannon (see Bibliography: Shannon, C.E.). It deals with aspects of communications related with the measurement of the quantity of information that can be transmitted through channels and the efficiency of the transmission process.

**Ion:** An atom with a negative charge due to its gain of electrons (anion), or with a positive charge due to a loss of electrons (cation**)**. A free electron is also an ion.

**Input-Output analysis:** Refers to the use of a tool developed in 1936 by Wassily Leontieff, in a work entitled *"The structure of American economy, 1919-1939"*, $2^{nd}$ edition, Oxford University Press, New York, 1951. In essence the model uses a matrix or table format, with an equal number of rows and columns, each representing one sector of the economy of a country. Also known as the industrial interrelationship matrix because it shows goods transferred to one industrial sector from another (Munier, 2004).

**Hydrologic cycle:** One of the three most important cycles in nature. The hydrologic cycle, the carbon cycle, and the nitrogen cycle are together so important that life would not be possible without them. The hydrologic cycle stars with the sun evaporating water from the oceans and forming clouds. Winds transport those clouds, which at a certain point cannot hold the water and pour it out as rain. Some rainwater falls on the ground and finds its way down to recharge aquifers, which eventually void into the ocean. Most rainfall feeds the rivers that end up in the ocean, restarting the cycle.

**ISO 1040:** Norm from the International Standard for Organization.

**Kilocalorie (Kcal):** The amount of heat needed to raise the temperature of 1 kg of water by 1° Centigrade.

**Kyoto Protocol:** International Conference where 160 countries signed a binding agreement to reduce greenhouse gases in accordance to specified targets.
**Landfill:** A piece of land devoted to the dumping of waste.
**Leaching:** The action of dissolving through a percolating fluid. For instance, in a landfill, rainwater finds its way through the waste and can dissolve minerals, producing a leaching.
**Life Cycle Assessment:** Technique that analyses the whole life of a product, from the 'cradle to the grave'. In so doing, it determines all the inputs for manufacturing the product, starting with raw materials, energy, water, etc., continues with parts made from these raw materials, and goes on with the components made from parts (sub-assemblies and assemblies), until the final product stage is reached.
**Local Urban Observatory**: A Global Urban Observatory at local level.
**Methane:** Flammable gas produced by the decomposition of organic matter.
**Montreal Protocol:** An international agreement signed in 1987 designed to protect the ozone layer by banning the production and consumption of such chemical compounds as chlorofluorocarbons, halons (fire extinguishing agents), carbon tetrachloride, and methyl chloroform.
**Multiplier effect:** A certain project can generate an increase in the spending habits of the population it affects. This spending, in turn, produces more spending, which also creates further expenditures. The relationship between the total spending and the original one is called 'multiplier effect'. Different multiplier effects can be calculated for such activities as tourism, construction of large industrial complexes, etc. See Vivian C. Choi's brief article, with a very clear explanation of this metric, at:
http://www.admin.gov.gu/commerce/multiplier.htm
**Nitrogen cycle:** Probably the most important nutrient cycle in nature. Living organisms (man, animals and plants) use nitrogen to produce amino acids, proteins, and nucleic acids. The waste product or organic matter, such as dung or manure left by animals on the ground, contains a large proportion of ammonia ($NH_4$), which contains nitrogen — a basic nutrient for plants. However, plants cannot use ammonia, but only inorganic versions of it, such as *nitrates*, which are produced when aerobic organisms break down the ammonia contained in the waste. Plants further use nitrates, and then nitrogen finds its way into living organisms when man and animals eat them. Another product of the break-down process is *nitrites,* which are very soluble in water and then leach to groundwater and rivers.
**NOx:** General formulation to indicate oxides of nitrogen.
**OECD:** Organization for Economic Co-Operation and Development.
**Oxymoron:** A combination of contradictory words.
**PCB:** Polychlorinated Biphenyl — a dangerous chemical that was once used in transformers.
**PDAs:** Sort of hand-held computer used for specific tasks.

**Pernada (right of) (also droit du seigneur):** The right attributed to noblemen to impose the sexual act with vassal women during the XIV century.
**pH:** Hydrogen potential. It is a measure of hydrogen concentration. A pH = 7 means that a solution is neutral. A value below 7 indicates an acidic solution, while any value above 7 is an alkaline solution. This concept is also used for soils, with the same meaning.
**Pyrolisis:** The process of heating a substance in the absence of oxygen.
**Ramie fibre:** A natural plant fibre, also known as China grass. Found in China, Japan, Korea, Brazil, and some parts of Europe, it is a very strong white, lustrous, stain-resistant fibre with many different uses.
**Reformer:** Device used to extract hydrogen from hydrocarbons or alcohol fuels. Naturally, the use of a reformer decreases the efficiency of a fuel cell, yet, on the other hand, it allows the use of hydrogen carriers instead of expensive pure hydrogen.
**Regression analysis:** Statistical technique utilized to find relationships between independent and dependent variables.
**Refrigeration ton:** Unit of measure for the refrigeration capacity of a plant. One ton of refrigeration is the cooling effect of 1 ton of ice, at 0 °C, melting in 24 hours.
**Responsible care:** Approach adopted by the chemical industry in 1988 concerning the manufacture and use of chemical products. With this approach, member companies are committed to responsible management of chemicals.
**Reverse osmosis:** Water purification technique that uses a pump to create pressure for untreated water to pass through a membrane. The membrane allows the passage of water, yet not of its impurities.
**Scrubber:** Device to remove impurities in a flue gas.
**Sludge**: Solid matter and water in a sewage treatment plant, subsequent to the action of aerobic bacteria on organic matter in raw sewage. The solid matter is about 2 percent dry weight.
**SOx:** General formulation to indicate chemical compounds formed with sulphur and oxygen, usually originating in the combustion of fuels containing sulphur. For instance, $SO_2$, sulphur dioxide, produces sulphuric acid when mixed with the air's moisture.
**Stakeholders:** People who may be affected by the impact of a project, that is, government, decision-makers, community associations, industry, and the public in general.
**Standard deviation:** In a probability distribution, the square root of the variance, in other words it shows how tightly the values around the mean value are.
**Social rate of return:** Reflects all the benefits accrued when spin-offs are considered and that benefit society.

**Supply chain:** The network of goods and services, including the raw materials' extraction and all suppliers, manufacturing, transportation, and distribution activities that are involved in making a certain product.

**Thermal convection:** When air is heated, there is a transfer of heat within the fluid due to a movement of molecules from one region to another. This is thermal convection.

**Transpiration:** Water that evaporates from plants into the atmosphere.

**United Nations Habitat:** United Nations Agency dealing with housing.

**Water table:** Boundary between two layers of ground water, the upper or aeration zone, and the lower or saturation zone.

**Weighting**: Assigning a degree of importance or establishing a hierarchy to projects or alternatives and criteria.

**Wells turbine:** A turbine, invented by Alan Wells, that works with air for use in wave generating plants.

# BIBLIOGRAPHY

Aall, C. (1998) Directional analysis for sustainable development in municipal planning and politics. A summary of a research and development project. Songdal, VF-F report 2.

Alberti, M., *et al.* (1996). Measuring urban sustainability. Environmental Impact Assessment Review, **16**, 381-424.

Ayres, R., and Ayres, L. (2001) A handbook of industrial ecology. Publisher Edward Elgar, Cheltenham, UK.

Bartelmus, P., and Vesper A. (2000) Green accounting and material flow analysis. Alternatives or complements? Wuppertal Institute, Wuppertal, Germany.

Bossel, H. (1998) Indicators for sustainable development – Theory, method, applications, International Institute for Sustainable Development (iiSD), Winnipeg, Canada.

Cellamare, C. (2001) Contexts of project interaction and local development: an experience in the province of Rieti (Italy). The International Symposium 'The region: Approaches for a Sustainable Development', ENSURE The European Network for Sustainable and Regional Development. Temi Editrice, Trento, Italy. Edited by C. Diamantini.

Chambers, N., and Lewis K., (2001) Ecological Footprinting Analysis: Towards a Sustainability Indicator for Business – ACCA Research Report No. 65, Glasgow, U.K.

Coenen, F. (2001) Participation and effectiveness at the juncture of local and regional sustainable governance. The International Symposium 'The region: Approaches for a Sustainable Development', ENSURE The European Network for Sustainable and Regional Development. Temi Editrice, Trento, Italy. Edited by C. Diamantini.

DEFRA (2002) Regional quality of life counts – Regional versions of the national Headline Indicators of sustainable development - Department of Environment, Food and Foreign Affairs. London, UK.

Dilks, D. (1995) Measuring urban sustainability: Canadian indicators workshop. Workshop Proceedings, State of the Environment Directorate, Environment Canada and Centre for Future Studies in Housing and Living Environments, Canada Mortgage and Housing Corporation, Ottawa, Canada.

Gamlin, L. (1988) Sweden's factory forests, *New Scientist* (January 28, 1988), pp. 41-47, cited in Tibbs (1991).

Frosch, R., and Gallopoulos N. (1989) Strategies for manufacturing, *Scientific American* (Special Edition, September 1989), pp. 144-152.

Georgescu-Roegen (1971) The entropy law and the economic process, Harvard University Press, Cambridge, Mass, USA.

Hardin, G. (1968) The tragedy of the commons. Science (162) 1243-1248.

Hawken P. (1993) The ecology of commerce – A declaration of sustainability, Harper Business – A division of Harper Collins Publishers.

Leontieff, W: (1951) *The structure of American economy, 1919-1939* - $2^{nd}$ edition, Oxford University Press, New York.

Lewanski. R. (2001) Environmental sustainability policies. Are they politically sustainable? The International Symposium "The region: Approaches for a Sustainable Development", ENSURE The European Network for Sustainable and Regional Development. Temi Editrice, Trento, Italy. Edited by C. Diamantini.

Lindseth, G. (2001) Participation, discourse and consensus: Local Agenda 21 in a deliberative democracy perspective. The International Symposium 'The region: Approaches for a Sustainable Development', ENSURE The European Network for Sustainable and Regional Development. Temi Editrice, Trento, Italy. Edited by C. Diamantini.

Meadows, D. H. (1993) Conference at Bowdoin College. Toward a sustainable Maine: The politics, economics, and ethics of sustainability.

Meadows, D.H., Meadows, and D., Randfers, J. (1972) The limits to growth, Universe Press, New York.

Munier, N. (2004) Multicriteria environmental assessment: A practical guide. Kluwer Academic Publishers, Dordrecht, The Netherlands.

Odum, H. (1996) Environmental Accounting: Emergy and environmental decision making. Published by John Wiley & Sons, New York.

Pareglio, S. (2001) Sustainable development and urban governance. The International Symposium 'The region: Approaches for a Sustainable Development', ENSURE The European Network for Sustainable and Regional Development. Temi Editrice, Trento, Italy. Edited by C. Diamantini.

Rao, P. (2000) Sustainable development: economics and policy. Blackwell Publishers, Malden, Mass., USA.

Rees, W. (1996) Revisiting carrying capacity. Area based indicators of sustainability. The University of British Columbia, Vancouver, BC. (http://www.dieoff.org/page110.htm).

Robinson, J., *et al.*, 1996: Life in 2030: Exploring a sustainable future for Canada. UBC Press, Vancouver. The International Symposium 'The region: Approaches for a Sustainable Development', ENSURE The European Network for Sustainable and Regional Development. Temi Editrice, Trento, Italy. Edited by C. Diamantini.

Rogers, J., and Feiss G. (1998). People and the Earth: Basic issues in the sustainability of resources and environment, Cambridge University Press, Cambridge, Mass., USA.

Sachs, W. Loske, R, and Linz, M. (1998). Greening the North: A Post-Industrial Blueprint for Ecology and Equity. Zed Books, London.

Serageldin, I. Grootaert C. (2000). Defining social capital: an integrating view. Pages 40-58 in *Social Capital: A Multifaceted Perspective*. Partha Dasgupta and Ismaïl Serageldin, eds. World Bank, Washington, DC. Cited in http://www.eeexchange.org/sustainability/content/f4.html

Shannon, C.E. (1948) A mathematical theory of communication. *The Bell System Technical Journal,* **27**, 379-423.

Smith, W. (1991). Elephant song. McClelland & Stewart Inc., Toronto, Canada.

Tibbs H. (1991). Industrial Ecology: An Environmental Agenda for Industry, published by Arthur D. Little, Inc., Boston, Mass., USA.

The World Bank (1997). Expanding the measure of wealth – Indicators of environmentally sustainable development – Environmentally Sustainable Development – Studies and Monographs Series No. 17, Washington, D.C.

Vaughan R., P. Bearse (1981) 'Federal Economic Development Programs: A Framework for Design and Evaluation,' in Robert Friedman and William Schweke (editors), *Expanding the Opportunity to Produce: Revitalizing The American Economy Through New Enterprise Development* (Washington: D.C., 1981), 309.

Wackernagel, M., *et al.* (1996). An ecological footprint: Reducing human impact on Earth. New Society Publishers, Gabriola Island, B.C., Canada.

Wald M. (2004). Questions about a hydrogen economy, *Scientific American*, **(290) (5)** (68-73).

World Commission on Environment and Development (1987) *Our Common Future*. Oxford University Press. The Brundtland Report, Oxford, UK

Zeleny, M (1982). Multicriteria decision making, McGraw-Hill Series in Quantitative Methods for Management.

# INDEX

## A

Air
   Contamination by dioxins and furans .................................. 60
   Fluidization to eliminate contaminants ..................... 122
   In a sustainable development. vi, 120

Assessment
   Appraisal of progress towards sustainable development .... 39
   Assets in a community ..... ix, 12, 35, 298, 313, 316, 323, 325, 326, 365, 366, 367, 370, 372, 378, 428
   Backwards and forwards linkages for a product ...... 274
   Input-Output analysis to quantify impacts .............. 203
   Problems in a community .... 359
   Respending effect in the local economy because a new undertaking .............. 282, 283
   Return on investment for local projects ............ 240, 335, 340
   The Life Cycle Model ......... 162
   The use of thresholds ........... 403

## B

Biomass
   Example of using pig manure .......................................... 260
   Largest plant in Europe ....... 247

## C

Capital (kinds of) in a country .... v, 29, 35, 190, 343, 437

Carrying capacity case examples
   Carrying capacity in the social fabric .............................. 290
   The carrying capacity of the River Rhine basin ............ 289
   The Ogallala aquifer in the USA ........................................ 201

Case studies
   Albertslund, Denmark -People participation ................... v, 22
   Amersfoort, the Netherlands - Energy from photovoltaics .......................................... 241
   Belize, Belize - Energy from industrial waste .......... vii, 174
   Bogotá, Colombia – Urban transportation ............. vii, 198
   Clearlake, USA – Energy from wastewater .................... vi, 80
   Curitiba, Brazil – Urban transportation .............. vi, 114
   Esquel, Argentina – People's against corporations ........... 23
   Georgia, USA – Carpets recycling .......................... 143
   Göteborg, Sweden – Heat recovery from incinerators . v, 63, 99
   Groton, USA –Light from garbage .......................... v, 69
   Guadalajara, Mexico – Urban indicators selection .... ix, 302, 303, 312

Kalundborg, Denmark-
  Industrial integration .vii, 172
South America – Recovery of
  components in flue gas.... 173
St. John, Virgin Islands, USA –
  Best use of resources .vii, 209
Commerce
  Merchandises without packing
    ......................................... 134
  Paper bags.................... 164, 165
  Plastic bags... 90, 106, 135, 165, 182
  Unsustainable commerce.... 134, 135
Construction
  Multi-family buildings... vii, 205
  Single dwellings .. 207, 208, 209
Criteria
  Selecting criteria for gauging
    projects or plans................ 334
  The need of attributes to gauge
    projects ....................... ix, 328

# D

Decision making
  Data
    Coefficients for the database
      ........................ ix, 340, 350
    Selection of alternatives .ix, 341
  Dematerialization .............. vii, 175

# E

Economy
  Economic growth.. 6, 16, 17, 18, 19, 20, 36, 70, 209, 290, 293, 303, 321, 322, 368
  Economic impact
    Measurement ................... 281
Energy
  Consumption reduction . 76, 141
  Emergy accounting as a value
    of quantity of energy used. 30
  Heat pumps as energy savers 63, 138, 208, 227, 228, 229, 230, 231, 261
  Renewable source(RES)
    Energy from Photovoltaics
      ............................. viii, 242
  Renewable sources (RES)
    Energy from biomass....... viii, 176, 225, 243, 264, 333, 334, 343, 344, 345, 347, 350, 351, 399, 401, 402
    Energy from fuel cells ..... viii, 102, 139, 247, 252, 260
    Energy from photovoltaics
      ............................. viii, 238
    Energy from the wind...... viii, 234, 235, 246, 260, 332, 342, 343, 344, 351, 397, 398
    Tidal and wave sources for
      energy ................... viii, 255
  Sustainable use of energy ..... vii, 217
Environmental sustainability .. 312, 357
Expert opinion
  How to select criteria........... 334
  How to select weights.......... 337
  Mechanism to evaluate impacts
    ......................................... 403

# F

Fuel cells as clean energy - types
  AFC – Alkaline ...viii, 139, 249, 254
  DMFC – Direct methanol.... viii, 139, 249, 254
  MCFC – Molten carbonate.. viii, 139, 249, 253

PAFC – Phosphoric acid ..... viii, 139, 249, 252, 253
PEM – Proton exchange membrane viii, 139, 245, 249, 250, 253, 260
Regenerative type . viii, 249, 255
SOFC – Solid oxide ..... viii, 139, 253, 254, 263

## G

Goal
  Agreeing on the goal for sustainability ............... ix, 320

## H

Health in a sustainable society ... 3, 61, 62, 120, 191, 196, 198, 299, 305, 309, 314, 357, 384, 385, 387

## I

Impacts
  Created by an activity (Tourism) ..................... x, 380
  Externalities of ................. v, 34
  Importance of thresholds ..... 286
  Measurements of ............. x, 385
Incinerators, advantages and disadvantages .. 8, 55, 61, 62, 91, 99, 100, 101, 152, 331, 412, 423
Indicators
  Choosing indicators ..... viii, 267, 278
  Ecological footprint .. 27, 28, 41, 213, 406
  Integration of sustainable indicators ........................ 275
  Linkages of .................. viii, 274
  Multipliers .... 10, 281, 283, 387, 391
  Selection of a set of final indicators ................... ix, 293
  Types of .......................... x, 375
  Weight of ..................... viii, 276
Industry
  Chemical industry software to reduce chemical impacts . 413
  Cleaner production vii, 151, 155, 179, 218
  Cradle-to-grave policy for waste reduction ........................ 428
  Design for the environment .. vii, 177, 180
  Eco-efficiency ..... 166, 216, 392
  Industrial complex .. x, 171, 411, 431
  Industrial integration 8, 171, 172
  Industrial metabolism ........ 8, 98
  Materials flow analysis .. vii, 168
  Responsible Care Program for the chemical industry ....... 151
  Sequential use of inputs ......... 98
  Styria case example of industrial integration 171, 217
  Supply chain in manufacturing ..... vi, 54, 74, 86, 88, 89, 159, 161, 162, 163, 179, 412, 413
Internet references for Appendix ............................................ 425
Internet references for chapter 1 39
Internet references for chapter 2 ........................................ 53, 98
Internet references for chapter 3 ............................................ 142
Internet references for chapter 4 ............................................ 212
Internet references for chapter 5 ............................................ 258
Internet references for chapter 6 ............................................ 309
Internet references for chapter 7 ............................................ 354

Internet references for chapter 8 ............................................. 390

## L

Land use
    Reducing land usage ............ 142
    Related to housing .......... vi, 112
    Related to transportation ...... 113
Life cycle analysis and access to software to compute contamination ....... 162, 163, 413
Local Agenda 21- Principles .. v, 7, 22, 36, 38, 436

## M

Mathematical Programming
    Applied to projects selection ................................. 341, 352
Metabolism
    Definition .............................. 68
    In society ........................ v, 68
Monitoring
    Controlling compliance of standards .......................... 404
    Using a bar chart to plan and control activities .......... x, 379

## N

Natural capital
    Decrease in ........................... 36
    Related with flow of resources ........................................... 36
Nature's cycles
    The carbon cycle .... 64, 123, 430
    The hydrologic cycle ........... 430
    The nitrogen cycle ... 64, 67, 430
    The photosynthesis process .. 65, 123, 226, 243, 423, 427, 428

Nutrients. Their fundamental role in the life cycle ... 24, 25, 26, 54, 107, 127, 136, 167, 222, 225, 246, 423

## O

Objectives to reach a goal x, 9, 12, 21, 39, 213, 215, 312, 318, 319, 323, 334, 349, 350, 353, 359, 360, 362, 369, 370, 376

## P

Photovoltaics
    The largest commercial application in the world ... 239
Population
    Participation in the decision making process ................ 129
    Relocation of people from slums ............................... 118
    Social ratios ........................ 296
    Sustainable issues ................ 123
    Use of resources by ............. 287

## Q

Quality of life
    Definitions and attributes .... 311
    Indicators for ................... 14, 22
    Report for ............................. 41

## R

Raw materials
    Actions to reduce consumption of .................................. vi, 90
Recycling
    Current waste treatment for cars ................... 52, 73, 74, 91, 97

Current waste treatment for glass ........... 52, 73, 74, 91, 97
Current waste treatment for metals ......... 52, 73, 74, 91, 97
Current waste treatment for paper .......... 52, 73, 74, 91, 97
Current waste treatment for plastic ......... 52, 73, 74, 91, 97
Tires recycling into energy .... vi, 85

Recycling: ....................... 182, 183
Reducing consumption
  In energy ................... vi, 75, 141
  In land use ....................... vi, 112
  In raw materials ..................... 78
  In water ............................ vi, 77
Reengineering a process ..... 78, 93, 154, 156, 157, 158, 164, 186
relationships ............................ 428
Reporting
  Example from a city ............ 148
  Using amoeba diagrams for reporting .......................... 408
  Using indicators of sustainable development ..................... 309
  Visualizing progress in a sustainability process... x, 406
Resilience
  Social, economical and political ........................................ v, 32
Resources
  Inventory of ........................ 322
Risk
  To human health .................. 100

---

**S**

Scheduling activities for local objectives x, 267, 306, 326, 362, 375, 379, 389, 408
Sludge composition and treatment ............................................. 54
Slums

Relocation ........................... 118
Society
  Social equity ..... 5, 10, 122, 132, 136, 192, 198, 312, 318, 320, 360, 369, 371
  Social justice .. 14, 35, 370, 371, 377
  Social progress ..................... 10
Soil
  In a sustainable environment. vi, 126
Soil contamination
  Pesticides used in crops .... 3, 50, 85, 93, 124, 150, 423
Sustainability
  At individual level .......... vi, 105
  Definition ..................... v, 6, 10
  In agriculture ...................... 200
  In education .................. vi, 132
  In forestry .................... vii, 202
  In public administration.. vi, 128
  In public health ............... vi, 131
  In the household ............. vi, 106
  In transportation ........... vii, 187
  Measures of sustainability .. 6, 9, 11, 13, 16, 23, 35, 38, 39, 43, 45, 80, 83, 112, 127, 133, 154, 155, 187, 203, 205, 213, 221, 251, 272, 275, 280, 292, 293, 303, 310, 312, 315, 316, 321, 335, 336, 363, 375, 391, 436, 437
  Non-market values assets ...... 35
  The Bellagio Principles v, 7, 37, 39, 315
  The Brundtland report .. 10, 312, 315
  The system and the process necessaries to reach it . ix, 365
  Urban and regional sustainability ............... ix, 315
Sustainable development ....... v, 16
  Strengthens and weakness in a community ..................... 317

Sustainable development XE "Sustainable development"
   vs. economic growth ...... v, 16
Sustainable development:v, 10, 16, 17, 313, 356, 357, 437
Sustainable fuels
   Biodiesel ....................... viii, 245
   Biogas ................................. 333
   Ethanol ......................... viii, 245
   Methane ........................ viii, 245
   Methanol ....................... viii, 244
Sustainable vs. common indicators
   ....................................... viii, 268

## T

Team effort for a sustainable environment .................... vi, 127
Technical ............................... 437
The baseline concept used for comparisons .................... ix, 316
The tragedy of the commons
   Resources belong to everybody
   .................................. 358, 436
Thresholds
   Selection ............................. 338
Transportation
   Between intra and inter industries ......................... 410
   Factors to be considered for sustainable transportation 415

## U

Urban space
   Indicators for ... 9, 285, 302, 429
   Strategic planning in the city .. x, 402
   Sustainability of .................. 315

Transportation for187, 193, 195, 250

## W

Wastes
   Composting in the household52, 73, 74, 91, 97
   Distribution of domestic waste
   ........................................... 47
   From construction sites .. 52, 73, 74, 91, 97
   From households 52, 73, 74, 91, 97
   Generation of wastes43, 64, 280
   Hazardous wastes .................. 59
   How to decrease waste generation ..................... v, 64
   Human and animal ................ 53
   Industrial wastes . 52, 73, 74, 91, 97
   Institutional wastes ... 52, 73, 74, 91, 97
   Municipal wastes 52, 73, 74, 91, 97
   Originated in farms ... 52, 73, 74, 91, 97
   Production ...................... 6, 154
   Radioactive wastes 45, 221, 222
   Solid waste in the household . vi, 107
   Vitrification of wastes . v, 56, 59
   Waste exchanges ................... 96
   Waste incinerators .. 61, 99, 100, 152
   WTP residues 52, 73, 74, 91, 97
Water
   Use in the household ...... vi, 107
   Use of water in industry ... vi, 78
   Water overuse ..................... 201
   Water reuse ...................... vi, 79

Made in the USA
San Bernardino, CA
05 January 2015